Operator Theory
Advances and Applications
Vol. 80

Editor
I. Gohberg

Operator Theory and Boundary Eigenvalue Problems

International Workshop in Vienna, July 27–30, 1993

Edited by

I. Gohberg
H. Langer

Springer Basel AG

I. Gohberg
School of Mathematical Sciences
Tel Aviv University
69978 Tel Aviv
Israel

H. Langer
Institut für Analysis, technische Mathematik
und Versicherungsmathematik
TU Wien
1040 Wien
Austria

A CIP catalogue record for this book is available from the Library of Congress, Washington D.C., USA

Deutsche Bibliothek Cataloging-in-Publication Data
Operator theory and boundary eigenvalue problems : international workshop in vienna, July 27 – 30, 1993 / ed. by I. Gohberg ; H. Langer. - Basel ; Boston ; Berlin : Birkhäuser, 1995
(Operator theory ; Vol. 80)
ISBN 978-3-0348-9909-3 ISBN 978-3-0348-9106-6 (eBook)
DOI 10.1007/ 978-3-0348-9106-6

NE: Gochberg, Izrail' C. [Hrsg.]; GT

Printed on acid-free paper produced from chlorine-free pulp ∞
Cover design: Heinz Hiltbrunner, Basel

9 8 7 6 5 4 3 2 1

Table of Contents

EDITORIAL INTRODUCTION

This volume contains proceedings of the Workshop on Operator Theory and Boundary Eigenvalue Problems which was held at the Technical University of Vienna, July 27 to 30, 1993. The workshop preceeded the International Symposium on the Mathematical Theory of Networks and Systems in Regensburg, Germany, August 2 to 6, 1993. It was the seventh workshop of this kind. Following is a list of the six preceeding workshops with reference to their Proceedings:

1981 Operator Theory (Santa Monica, California, USA)
1983 Applications of Linear Operator Theory to Systems and Networks (Rehovot, Israel), OT 12
1985 Operator Theory and its Applications (Amsterdam, the Netherlands), OT 19
1987 Operator Theory and Functional Analysis (Mesa, Arizona, USA), OT 35
1989 Matrix and Operator Theory (Rotterdam, the Netherlands), OT 50
1991 Operator Theory and Complex Analysis, (Sapporo, Japan) OT 59

The next workshop in this series will be held at the University of Regensburg, Germany, July 31 to August 4, 1995 (IWOTA 1995).

At the workshop in Vienna there were presented 76 lectures, covering a wide range of topics on operator theory. The main topics were interpolation problems and analytic matrix functions, operator theory in spaces with an indefinite scalar product, boundary value problems for differential and functional-differential equations and systems theory and control. The workshop covered different aspects, starting with abstract operator theory up to concrete applications. The papers in these proceedings present an accurate cross section of the lectures presented at the workshop.

Due to the recent political changes and the central position of Vienna in Europe there was a comparatively large number of participants from Eastern Europe.

The financial support of the following institutions and companies is highly appreciated.

Austrian Ministry of Science and Research,
Austrian Academy of Sciences,
City of Vienna,
Austrian Mathematical Society,
Jewish Community of Vienna,
Creditanstalt Bankverein,
Austrian National Bank,
First Austrian Savings Bank,
Minerva, Scientific Bookshop, Vienna,
Technical University of Vienna,
Department of Analysis, Technical Mathematics and Actuarial Theory,
of the Technical University of Vienna,
International Science Foundation, U.S.A.,
Society of Applied Mathematics and Mechanics (GAMM),
Birkhäuser Verlag, Basel.

I. Gohberg

H. Langer

Operator Theory:
Advances and Applications, Vol. 80

COISOMETRICALLY VALUED RATIONAL MATRIX FUNCTIONS

D. ALPAY and M. RAKOWSKI

Given signature matrices J_1 and J_2 of possibly different sizes, and a rational matrix function W without a pole or zero at infinity, we indicate a necessary and sufficient condition for the equation $J_1 = W(z)J_2W(z)^*$ to hold on the imaginary axis or the unit circle. The condition makes it possible to characterize inner rational matrix functions without a zero at the origin. We also discuss an application to factorization.

1 Introduction

Consider a linear, time invariant system of the form

$$\begin{cases} x'(t) = Ax(t) + Bu(t), \\ y(t) = Cx(t) + Du(t), \end{cases} \quad \text{or} \quad \begin{cases} x_{k+1}(t) = Ax_k(t) + Bu_k(t), \\ y_k(t) = Cx_k(t) + Du_k(t). \end{cases} \tag{1.1}$$

After taking Laplace transform of the signals in (1.1) in the continuous time case, or z-transform in the discrete time case, we can find the transfer function $u \to y$,

$$W(z) = D + C(z - A)^{-1}B. \tag{1.2}$$

The basic problem considered below is to determine when W takes coisometric values on the imaginary axis $i\mathbf{R}$ or the unit circle $\mathbf{T}$, that is, when

$$W(z)W(z)^* = I, \qquad \Re e\, z = 0 \tag{1.3}$$

or

$$W(z)W(z)^* = I, \qquad |z| = 1. \tag{1.4}$$

In fact, we consider a slightly more general problem. Suppose we have two signature matrices J_1 and J_2, i.e., matrices such that $J_i^* = J_i^{-1} = J_i$ ($i = 1, 2$), and a rational matrix function W of the form (1.2). Find the necessary and sufficient condition for the equation

$$W(z)J_2W(z)^* = J_1 \tag{1.5}$$

to hold on the imaginary axis or the unit circle. For practical reasons, we want to obtain a condition which can be computationally verified.

This research was partially supported by the National Science Foundation Grant DMS-9302706.

The problem can be rephrased as follows. If $H \in \mathbf{C}^{n\times n}$ is a hermitian matrix, endow $\mathbf{C}^n$ with the indefinite inner product

$$[x,y]_H = y^*Hx. \tag{1.6}$$

Let $J_1 \in \mathbf{C}^{m\times m}$ and $J_2 \in \mathbf{C}^{n\times n}$ be signature matrices and let $W: \mathbf{C}^n \to \mathbf{C}^m$ be a linear transformation. The transformation W is a (J_2, J_1)-coisometry if W^* is an isometry from $(\mathbf{C}^m, [\cdot,\cdot]_{J_1})$ to $(\mathbf{C}^n, [\cdot,\cdot]_{J_2})$, i.e., if

$$[x,y]_{J_1} = [W^*x, W^*y]_{J_2} \tag{1.7}$$

for all $x, y \in \mathbf{C}^m$. The last condition is equivalent to

$$J_1 = WJ_2W^*. \tag{1.8}$$

Thus, (1.3) (resp. (1.4)) holds if and only if the rational matrix function W takes (J_2, J_1)-coisometric values on the imaginary line (resp. unit circle). Note that if $J_1 = J_2 = J$, (1.8) holds if and only if JW^*J is the inverse of W, which holds if and only if $J = W^*JW$. In this case the linear transformation W is said to be J-unitary.

We will use the following notation. A quadruple of matrices (A, B, C, D) such that (1.2) holds is called a *realization* of W. Each proper (i.e., analytic at infinity) rational matrix function has a realization. A realization (A, B, C, D) is said to be *observable* if the pair (C, A) is observable, i.e., if

$$\ker \begin{bmatrix} C \\ CA \\ \vdots \\ CA^k \end{bmatrix} = (0) \tag{1.9}$$

for sufficiently large integers k. The realization is *controllable* if the pair (A, B) is controllable, i.e., if the image of the matrix $[\,A^kB \quad A^{k-1}B \quad \dots \quad B\,]$ fills out the whole space for sufficiently large k. An observable and controllable realization is said to be *minimal*. The domain of A in a minimal realization has the smallest possible dimension. This dimension is called the *McMillan degree* of the function. The McMillan degree of (a not necessarily proper) W can be also defined as the sum of multiplicities of all the poles of W. Minimal realizations of W are unique up to *similarity*, that is, up to a nonsingular matrix S such that

$$W(z) = D + CS(z - S^{-1}AS)^{-1}S^{-1}B. \tag{1.10}$$

Various special cases of the basic problem have been considered previously in the literature. The case where $J_1 = J_2 = I$ has been treated in [Gl] in the context of model reduction. It has been shown in [Gl] that if $(A, B, C, 0)$ is a minimal realization, TFAE

a) there exists matrix D such that (A, B, C, D) is a realization of a function with unitary values on the imaginary axis,
b) there exists a nonsingular hermitian matrix P such that

$$AP + PA^* + BB^* = 0 \qquad \text{and} \qquad A^*P^{-1} + P^{-1}A + C^*C = 0. \tag{1.11}$$

Moreover, the matrix D in (b) satisfies equations

$$D^*D = I, \qquad D^*C + B^*P^{-1} = 0, \qquad DB^* + CP = 0, \tag{1.12}$$

and, for any matrix D which satisfies equations (1.12), (A, B, C, D) is a realization of a function with unitary values on the imaginary axis.

The case $J_1 = J_2 = J$ has been treated in [AG]. Let (A, B, C, D) be a minimal realization of a function W. It has been shown in [AG] that TFAE

a) W takes J-unitary values on the imaginary axis,

b) D is J-unitary and there exists an invertible hermitian solution H of the Lyapunov equation

$$A^*H + HA = -C^*JC \quad \text{such that} \quad B = -H^{-1}C^*JD, \tag{1.13}$$

c) D is J-unitary and there exists an invertible hermitian solution H of the Lyapunov equation

$$GA^* + AG = -BJB^* \quad \text{such that} \quad C = -DJB^*G^{-1}. \tag{1.14}$$

This result has a circle analogue. If W does not have a zero at infinity, TFAE

a) W takes J-unitary values on the unit circle,

b) W is analytic and has a nonsingular value at the origin, and there exists an invertible hermitian matrix H such that

$$\begin{bmatrix} A & B \\ C & D \end{bmatrix}^* \begin{bmatrix} H & 0 \\ 0 & -J \end{bmatrix} \begin{bmatrix} A & B \\ C & D \end{bmatrix} = \begin{bmatrix} H & 0 \\ 0 & -J \end{bmatrix}, \tag{1.15}$$

c) W is analytic and has a nonsingular value at the origin, and there exists an invertible hermitian matrix H such that

$$\begin{bmatrix} A & B \\ C & D \end{bmatrix} \begin{bmatrix} H^{-1} & 0 \\ 0 & -J \end{bmatrix} \begin{bmatrix} A & B \\ C & D \end{bmatrix}^* = \begin{bmatrix} H^{-1} & 0 \\ 0 & -J \end{bmatrix}. \tag{1.16}$$

Below, we extend the results in [AG] to the case $J_1 \neq J_2$. The important part of the assumption $J_1 \neq J_2$ is that J_1 and J_2 may have different sizes. Indeed, the proofs in [Gl] and [AG] rely on the formula for multiplicative inverse of a rational matrix function with nonsingular value at infinity. This tool is not available when W is rectangular. Based on results in [R], one can follow the square case argument using one-sided inverses. This leads to a generalization of condition (1.13) to rectangular functions with zero defect (Theorem 2.1 below), where the *defect* is the difference between the sum of multiplicities of all the poles and the sum of multiplicities of all the zeros of a function [K]. Condition (1.14) turns out to be more general and its rectangular analogue can be used to characterize all proper rational matrix functions with (J_2, J_1)-coisometric values on the imaginary axis (Theorem 2.2 below). The generalization of condition (1.16) to rectangular case is given in Theorem 3.1 below. We note that condition (1.15) cannot be extended to rectangular matrix functions in the same way because the rank of the matrix on the right hand side would have to be greater than the rank of the middle factor.

The paper is organized as follows. In Section 2, we consider functions which take (J_2, J_1)-coisometric values on the imaginary axis. The proof of the main theorem can be found in [AR] and we do not include it here. We also recall two results on factorization of functions with coisometric values on the imaginary axis. Section 3 contains results on functions with (J_2, J_1)-coisometric values on the unit circle. In Section 4, we discuss rational matrix functions inner in the unit disk.

2 Functions with coisometric values on the line

Denote by $\mathcal{R}$ the field of scalar rational functions, and by $\mathcal{R}^{m\times n}$ the space of $m \times n$ matrices over $\mathcal{R}$. It follows immediately from the definition that a function $W \in \mathcal{R}^{m\times n}$ takes (J_2, J_1)-coisometric values on the imaginary line if and only if

$$J_1 = W(z)J_2W(-\overline{z})^*, \qquad z \in i\mathbf{R}. \tag{2.1}$$

Since $W(-\overline{z})^* \in \mathcal{R}^{n\times m}$, by analytic continuation condition (2.1) is equivalent to

$$J_1 = W(z)J_2W(-\overline{z})^* \tag{2.2}$$

for all points z in the complex plane at which W is analytic and surjective. Hence, if J_1 and J_2 have the same size, (2.1) holds whenever the function

$$J_2W(-\overline{z})^*J_1 \tag{2.3}$$

is a multiplicative inverse of W. If W takes nonsingular value at infinity and (A, B, CD) is any realization of W, the multiplicative inverse of W is given by the formula

$$W(z)^{-1} = D^{-1} - D^{-1}C(z - A + BD^{-1}C)^{-1}BD^{-1}. \tag{2.4}$$

The necessary and sufficient condition for identity (2.2) to hold in this case has been derived [AG, Gl] by comparing minimal realizations of the function $W(z)^{-1}$ given by (2.3) and (2.4).

If J_1 and J_2 have different sizes, (2.1) holds whenever the function (2.3) is a right inverse of W. If D^R is a right inverse of D, a straightforward computation shows that

$$W(z)^R = D^R - D^RC(z - A + BD^RC)^{-1}BD^R \tag{2.5}$$

is a multiplicative right inverse of W. The natural approach to the rectangular case is to adapt the argument used in the square case, i.e., to compare minimal realizations of functions (2.3) and (2.5). This leads to a solution in a special case where the defect of W equals 0. Recall that the defect of a rational matrix function W is the difference between its McMillan degree and the sum of multiplicities of all its zeros. Let δ_π denote the McMillan degree of a function $W \in \mathcal{R}^{m\times n}$, let δ_ζ be the sum of multiplicities of all its zeros, and let α (resp. β) be the sum of degrees of vector polynomials in a minimal polynomial basis [F] for the left (resp. right) kernel of W. Then (see Theorem 3 in [VDK], cf. Theorem 5.1 in [WSCP])

$$\delta_\pi - \delta_\zeta = \alpha + \beta.$$

Hence W has zero defect if and only if $\alpha = \beta = 0$, i.e., if and only if the row span and column span of $W(z)$ do not depend on z.

Theorem 2.1 *Let J_1 and J_2 be signature matrices and let (A, B, C, D) be a minimal realization of a rational matrix function W. Then TFAE*

(i) *D is a (J_2, J_1)-coisometry and there exists a nonsingular hermitian solution H of the Lyapunov equation*

$$A^*X + XA = -C^*J_1C \tag{2.6}$$

such that

$$B = -H^{-1}C^*J_1D, \tag{2.7}$$

(ii) W *takes* (J_2, J_1)*-coisometric values on the imaginary axis and its defect* 0.

Proof Suppose (i) holds. Then

$$DJ_2D^* = J_1, \qquad BJ_2D^* = -H^{-1}C^*,$$

and

$$BJ_2B^* = H^{-1}C^*J_1CH^{-1} = -H^{-1}A^* - AH^{-1}.$$

Hence

$$\begin{aligned} W(z)J_2W(z)^* &= DJ_2D^* + DJ_2B^*(\overline{z} - A^*)^{-1}C^* + C(z-A)^{-1}BJ_2D^* + \\ &\quad + C(z-A)^{-1}BJ_2B^*(\overline{z} - A^*)^{-1}C^* \\ &= J_1 - CH^{-1}(\overline{z} - A^*)^{-1}C^* - C(z-A)^{-1}H^{-1}C^* + \\ &\quad - C(z-A)^{-1}(H^{-1}A^* + AH^{-1})(\overline{z} - A^*)^{-1}C^* \\ &= J_1 - C(z-A)^{-1}((z-A)H^{-1} + H^{-1}(\overline{z} - A^*) + H^{-1}A^* + AH^{-1}) \\ &\quad \cdot (\overline{z} - A^*)^{-1}C^* \\ &= J_1 - (z + \overline{z})C(z-A)^{-1}H^{-1}(\overline{z} - A^*)^{-1}C^* \end{aligned}$$

and $W(z)J_2W(z)^* = J_1$ for z on the imaginary axis. Also, by (2.7), the row span of B is contained in the row span of D. Hence the row span of $W(z)$ does not depend on z and the defect of W equals 0. Thus, (i) implies (ii).

Conversely, suppose (ii) holds. Then the function

$$W^R(z) = J_2W(-\overline{z})^*J_1 \tag{2.8}$$

is a right multiplicative inverse of W. Since W has zero defect, the column span of $W(-\overline{z})^*$ does not depend on z. Hence the column span of W^R is independent of z and W^R has zero defect as well. Consequently, (cf. Theorem 4.3 in [R]) there exists a right inverse D^R of D such that $W^R(z)$ has the form (2.5). Since zeros of W are necessarily the poles of W^R, of at least the same multiplicity, and the defect of W equals 0, the McMillan degree of W^R is at least equal to the McMillan degree of W. Hence formulas (2.8) and (2.5) provide two minimal realizations of W^R,

$$(-A^*, C^*J_1, -J_2B^*, J_2D^*J_1) \tag{2.10}$$

and

$$(A - BD^RC, BD^R, -D^RC, D^R). \tag{2.11}$$

Consequently, $D^R = J_2D^*J_1$ and there exists a unique nonsingular matrix S such that

$$-S^{-1}A^*S = A - BD^RC, \qquad S^{-1}C^*J_1 = BD^R, \qquad J_2B^*S = D^RC. \tag{2.12}$$

The first two equations in (2.12) imply

$$A^*S + SA = C^*J_1C \tag{2.13}$$

and

$$BD^RD = S^{-1}C^*J_1D. \tag{2.14}$$

Since the defect of W equals 0, the row span of B is contained in the row span of D and $BD^RD = D$. Thus, (2.6) and (2.7) hold with $H = -S$. It remains to show that S is hermitian.

¿From the last two equations in (2.12) we obtain

$$J_1C = D^{R*}B^*S^* \qquad \text{and} \qquad BJ_2 = S^{*-1}C^*D^{R*}. \tag{2.15}$$

Since $D^R = J_2D^*J_1$, equations (2.15) imply

$$C = DJ_2B^*S^* \qquad \text{and} \qquad B = S^{*-1}C^*J_1D, \tag{2.16}$$

or

$$D^RC = D^RDJ_2B^*S^* \qquad \text{and} \qquad BD^R = S^{*-1}C^*J_1. \tag{2.17}$$

Since the image of J_2B^* is contained in the image of D^R (by the defect assumption), $D^RDJ_2B^* = J_2B^*$ and it follows that the last two equations in (2.12) hold with S^* instead of S. Now multiplying the first equation in (2.12) on the left side by S and on the right side by S^{-1}, and taking conjugates, we obtain

$$-S^{*-1}A^*S = A - S^{*-1}C^*J_1DJ_2B^*S^* = A - BD^RDD^RC = A - BD^RC.$$

Thus, all three equations in (2.12) hold with S^* instead of S. By the uniqueness of the similarity transformation, $S = S^*$. □

We note that condition (i) in Theorem 2.1 is useless for functions with nonzero defect. For example, if J_1 and J_2 are identity matrices and

$$W(z) = \frac{1}{\sqrt{2}}\begin{bmatrix} \frac{z-1}{z+1} & 1 \end{bmatrix},$$

each minimal realization of W gives rise to a unique hermitian solution of equation (2.6) which does not satisfy equation (2.7) (cf. Example 2.4 in [AR]).

In [AR], the following necessary and sufficient condition for identity (2.2) in the general case has been obtained.

Theorem 2.2 *Let (A, B, C, D) be an observable realization of a function $W \in \mathcal{R}^{m\times n}$, and let J_1 and J_2 be signature matrices. Then W takes (J_2, J_1)-coisometric values on the imaginary axis if and only if D is a (J_2, J_1)-coisometry and there exists a hermitian solution G of the Lyapunov equation*

$$XA^* + AX = -BJ_2B^* \tag{2.18}$$

such that

$$CG = -DJ_2B^*. \tag{2.19}$$

Moreover, if the hermitian matrix G satisfying equations (2.18) *and* (2.19) *exists, it is unique.*

The reader may wish to compare conditions in Theorems 2.1 and 2.2. Suppose realization (A, B, C, D) in Theorem 2.2 is minimal. If a nonsingular hermitian matrix H satisfies equations (2.6) and (2.7), H^{-1} solves equations (2.18) and (2.19). On the other hand, existence of the hermitian solution of equations (2.18) and (2.19) does not imply that equations (2.6) and (2.7) have a solution. There are two difficulties involved. First, the solution G of (2.18) and (2.19) does not have to be invertible. The second difficulty is related to the fact that D need only be right invertible. Also note that if J_1 and J_2 are identity matrices and A is stable, equations (2.6) and (2.18) have unique solutions, namely, the observability gramian

$$P = \int_0^\infty e^{A^*t} C^* C e^{At} dt$$

and the controllability gramian

$$Q = \int_0^\infty e^{At} B B^* e^{A^*t} dt$$

(see e.g. [BGR]). If, in addition, W is square, $PQ = I$.

The matrix G in Theorem 2.2 is called the *hermitian matrix associated with a realization* (A, B, C, D), or, if it is clear which realization we have in mind, the *associated hermitian matrix.* If the matrix J_2 is definite, or when W has zero defect (in particular, when W is square), G is nonsingular. In the general case, G may be singular. The number of positive (resp. negative) eigenvalues of G is equal to the number of positive (resp. negative) squares of the function

$$K_W(z, \lambda) = \frac{J_1 - W(z) J_2 W(\lambda)^*}{z + \overline{\lambda}},$$

i.e., the maximum number of positive (negative) eigenvalues of a matrix

$$\left[c_i^* K(\lambda_i, \lambda_j) c_j \right]_{i,j=1}^r$$

where $c_1, c_2, \ldots c_r$ are constant vectors and $\lambda_1, \lambda_2, \ldots, \lambda_r$ are points in the domain of W.

The associated hermitian matrix can be used to factor function $W \in \mathcal{R}^{m \times n}$ with (J_2, J_1)-coisometric values on the imaginary axis into a product $W_1 W_2$ where W_i takes (J_i, J_1)-coisometric values on the line $(i = 1, 2)$. We will denote by $G^\dagger$ the Moore-Penrose inverse of a matrix G. If M is a subspace of $\mathbf{C}^N$, $M^{[\perp]}$ will denote the orthogonal companion of M with respect to the inner product $[\cdot, \cdot]_{G^\dagger}$, i.e.,

$$M^{[\perp]} = \{x \in \mathbf{C}^N : \ [x, m]_{G^\dagger} = 0 \ \text{ for all } \ m \in M\}.$$

We recall two results from [AR].

Theorem 2.3 *Let J_1 and J_2 be signature matrices, and let $W \in \mathcal{R}^{m \times n}$ be a function analytic at infinity with (J_2, J_1)-coisometric values on the imaginary line. Pick a minimal* *realization (A,B,C,D) of W, let G be the associated hermitian matrix, and let $A^{\times} = A + GC^*J_1C$. Let M_1 be a subspace non-degenerate with respect to the inner product $[\cdot,\cdot]_{G^\dagger}$ and such that $A(M_1) \subset M_1 \subset \operatorname{im} G$, and suppose $\operatorname{im} G \cap M_1^{[\perp]}$ can be extended to a complementary to M_1 subspace M_2 such that $A^{\times}(M_2) \subset M_2$. Let $S = [\,S_1 \quad S_2\,]$ where S_1 contains a basis for M_1 and S_2 a basis for M_2, and suppose*

$$S^{-1}AS = \begin{bmatrix} A_{11} & A_{12} \\ A_{21} & A_{22} \end{bmatrix}, \qquad S^{-1}B = \begin{bmatrix} B_1 \\ B_2 \end{bmatrix}, \qquad CS = [\,C_1 \quad C_2\,]. \tag{2.20}$$

If $B_1J_2D^*J_1D = B_1$, then W admits a minimal factorization W_1W_2 where W_1 takes J_1-unitary values and W_2 takes (J_2,J_1)-coisometric values on the imaginary line. Moreover, if $D = D_1D_2$ where the matrices D_1 and D_2 are such that $J_1 = D_1J_1D_1^* = D_2J_2D_2^*$, then possible factors are given by the formulas

$$W_1(z) = D_1 + C_1(z - A_{11})^{-1}B_1J_2D_2^*J_1 \tag{2.21}$$

and

$$W_2(z) = D_2 + D_1^{-1}C_2(z - A_{22})^{-1}B_2. \tag{2.22}$$

Conversely, if W_1W_2 is a minimal factorization of W such that W_1 takes J_1-unitary and W_2 takes (J_2,J_1)-coisometric values on the line, then $D_1 = W_1(\infty)$ and $D_2 = W_2(\infty)$ satisfy $J_1 = D_1J_1D_1^* = D_2J_2D_2^*$ and there exist subspaces M_1 and M_2 as above such that W_1 and W_2 have minimal realizations (2.21)-(2.22).

Note that $J_1 = D_1J_1D_1^*$ implies $J_1 = D_1^{-1}J_1D_1^{*-1}$. Hence the factor W_1 in Theorem 2.3 can be normalized so that $D_1 = I$. Then the formulas (2.21)-(2.22) become

$$W_1(z) = I + C_1(z - A_{11})^{-1}B_1J_2D^*J_1 \tag{2.23}$$

and

$$W_2(z) = D + C_2(z - A_{22})^{-1}B_2. \tag{2.24}$$

Corollary 2.4 *Let J_1 and J_2 be signature matrices, and let $W \in \mathcal{R}^{m\times n}$ be a function which is analytic at infinity and takes (J_2,J_1)-coisometric values on the imaginary line. Pick a minimal realization (A,B,C,D) of W, suppose the associated hermitian matrix G is nonsingular, and let $A^{\times} = A + GC^*J_1C$. Let v be an eigenvector of A with the corresponding eigenvalue λ such that $[v,v]_{G^{-1}} \neq 0$, and suppose*

(i) *$[A^{\times}x, v]_{G^{-1}} = 0$ whenever $[x,v]_{G^{-1}} = 0$,*

(ii) *$B_1J_2D^*J_1D = B_1$ where $B_1 = v^*G^{-1}B$.*

Then W admits a minimal factorization W_1W_2 where W_2 takes (J_2,J_1)-coisometric values on the imaginary line and

$$W_1(z) = I - \frac{Cvv^*C^*J_1}{(z-\lambda)[v,v]_{G^{-1}}} \tag{2.25}$$

has J_1-unitary values on the line. Moreover, if $\Re e\,\lambda \neq 0$,

$$W_1(z) = I - P + \frac{z+\overline{\lambda}}{z-\lambda}P \quad \text{where} \quad P = \frac{yy^*J_1}{y^*J_1y} \quad \text{with} \quad y = Cv. \tag{2.26}$$

Note that if $\Re e\,\lambda > 0$, the multiplicative inverse of $W_1(z)$

$$W_1(z)^{-1} = I - P + \frac{z-\lambda}{z+\overline{\lambda}}P \tag{2.27}$$

is the Blaschke-Potapov factor in the half-plane.

3 Functions with coisometric values on the circle

Let J_1 and J_2 be signature matrices. It follows from definition that a rational matrix function W takes (J_2, J_1)-coisometric values on the unit circle if and only if

$$J_1 = W(z)J_2W\left(\frac{1}{\overline{z}}\right)^*, \qquad z \in \mathbf{T}. \tag{3.1}$$

Since $W(\frac{1}{\overline{z}})^*$ is a rational matrix function, by analytic continuation (3.1) holds if and only if

$$J_1 = W(z)J_2W\left(\frac{1}{\overline{z}}\right)^* \tag{3.2}$$

for all z where $W(z)$ and $W(\frac{1}{\overline{z}})^*$ have finite values.

There are many examples of rational matrix functions with (J_2, J_1)-coisometric values on the unit circle. If $J_1 = J_2 = I$ and P is an orthogonal projection, the function

$$W(z) = I - P + \frac{z-\lambda}{1-z\overline{\lambda}}P, \qquad |\lambda| < 1, \tag{3.3}$$

takes (J_2, J_1)-coisometric (in this case, unitary) values on the unit circle. More generally, if $J_1 = J_2 = J$ and P is a projection such that $PJ = (PJ)^*$, then the function (3.3) takes J-unitary values on the unit circle. The function (3.3) is called a *Blaschke-Potapov factor.* Note that condition $|\lambda| < 1$ can be replaced by $|\lambda| \neq 1$.

One can use Blaschke-Potapov products to define rectangular coisometrically valued functions. If $J_1 = 1$ and $J_2 = I_2$, and λ_1, λ_2 are two points in the unit disk, the function

$$W(z) = \tfrac{1}{\sqrt{2}}\begin{bmatrix} \frac{z-\lambda_1}{1-z\overline{\lambda}_1} & \frac{z-\lambda_2}{1-z\overline{\lambda}_2} \end{bmatrix} \tag{3.4}$$

takes coisometric values on the unit circle. As another example, let

$$J_1 = [\,-1\,], \qquad J_2 = \begin{bmatrix} 1 & & & 0 \\ & -1 & & \\ & & -1 & \\ 0 & & & 1 \end{bmatrix}, \tag{3.5}$$

and let $W(z) = \begin{bmatrix} \frac{1}{z^2} & 1 & \frac{\sqrt{2}}{z} & 1 \end{bmatrix}$. Then

$$W(z)J_2W(z)^* = \frac{1}{z^2\overline{z}^2} - \frac{2}{z\overline{z}} = -1, \qquad \text{if } |z| = 1,$$

and W takes (J_2, J_1)-coisometric values on the unit circle.

We derive now the analogue of Theorem 2.2 in the circle case.

Theorem 3.1 *Let $J_1 \in \mathbf{C}^{m\times m}$ and $J_2 \in \mathbf{C}^{n\times n}$ be signature matrices, suppose $W \in \mathcal{R}^{m\times n}$ does not have a pole or zero at infinity, and let (A, B, C, D) be an observable realization of W. Then W takes (J_2, J_1)-coisometric values on the unit circle if and only if there exists a hermitian matrix H such that*

$$\begin{bmatrix} A & B \\ C & D \end{bmatrix} \begin{bmatrix} H & 0 \\ 0 & J_2 \end{bmatrix} \begin{bmatrix} A & B \\ C & D \end{bmatrix}^* = \begin{bmatrix} H & 0 \\ 0 & J_1 \end{bmatrix}. \tag{3.6}$$

Moreover, if W takes (J_2, J_1)-coisometric values on the unit circle, the matrix H is unique.

Proof Equation (3.6) is equivalent to the three equations

$$H - AHA^* = BJ_2B^* \tag{3.7}$$

$$CHA^* = -DJ_2B^* \tag{3.8}$$

$$J_1 - DJ_2D^* = CHC^*. \tag{3.9}$$

Suppose equations (3.7)-(3.9) hold. Then

$$\begin{aligned} W(z)J_2W(z)^* &= DJ_2D^* + DJ_2B^*(\overline{z} - A^*)^{-1}C^* + C(z-A)^{-1}BJ_2D^* + \\ &\quad + C(z-A)^{-1}BJ_2B^*(\overline{z} - A^*)^{-1}C^* \\ &= J_1 - CHC^* - CHA^*(\overline{z} - A^*)^{-1}C^* - C(z-A)^{-1}AHC^* + \\ &\quad + C(z-A)^{-1}(H - AHA^*)(\overline{z} - A^*)^{-1}C^* \\ &= J_1 + C(z-A)^{-1}(-(z-A)H(\overline{z} - A^*) - (z-A)HA^* + \\ &\quad - AH(\overline{z} - A^*) + H - AHA^*)(\overline{z} - A^*)^{-1}C^* \\ &= J_1 + (1 - z\overline{z})C(z-A)^{-1}H(\overline{z} - A^*)^{-1}C^*. \end{aligned} \tag{3.10}$$

Hence $J_1 = W(z)J_2W(z)^*$ for $z \in \mathbf{T}$.

Conversely, suppose W takes (J_2, J_1)-coisometric values on the unit circle. By Theorem 3.6 in [AG] (see also Theorem 1.9 in [BGK]), for any $\alpha \in \mathbf{T}\backslash\sigma(A)$

$$W\left(\alpha\frac{\lambda - 1}{\lambda + 1}\right) = \tilde{D} + \tilde{C}(\lambda - \tilde{A})^{-1}\tilde{B} \tag{3.11}$$

where

$$\tilde{A} = (\alpha - A)^{-1}(\alpha + A), \tag{3.12}$$

$$\tilde{B} = \sqrt{2}(\alpha - A)^{-1}B, \tag{3.13}$$

$$\tilde{C} = \sqrt{2}\alpha C(\alpha - A)^{-1}, \tag{3.14}$$

$$\tilde{D} = D + C(\alpha - A)^{-1}B. \tag{3.15}$$

Let $R(\lambda) = W\left(\alpha\frac{\lambda-1}{\lambda+1}\right)$. Then R takes (J_2, J_1)-coisometric values on the imaginary line and, by Theorem 2.2, there exists a hermitian matrix G such that

$$G\tilde{A}^* + \tilde{A}G = -\tilde{B}J_2\tilde{B}^* \qquad \text{and} \qquad \tilde{C}G = -\tilde{D}J_2\tilde{B}^*. \tag{3.16}$$

The first equation in (3.16) implies

$$(\alpha - A)G(\alpha + A)^* + (\alpha + A)G(\alpha - A)^* = -2BJ_2B^*.$$

Hence

$$G - AGA^* = -BJ_2B^*. \tag{3.17}$$

The second equation in (3.16) gives

$$\alpha C(\alpha - A)^{-1}G(\overline{\alpha} - A^*) = -(D + C(\alpha - A)^{-1}B)J_2B^*.$$

Since, by (3.17),

$$\alpha(\alpha - A)^{-1}G(\overline{\alpha} - A^*) + (\alpha - A)^{-1}BJ_2B^* = -GA^*, \tag{3.18}$$

we obtain

$$CGA^* = DJ_2B^*. \tag{3.19}$$

By (3.13) and (3.15), $D = \tilde{D} - \frac{1}{\sqrt{2}}C\tilde{B}$. Using the second equation in (3.16), we obtain

$$DJ_2D^* = \tilde{D}J_2\tilde{D}^* + \frac{1}{\sqrt{2}}\tilde{C}GC^* + \frac{1}{\sqrt{2}}CG\tilde{C}^* + \frac{1}{2}C\tilde{B}J_2\tilde{B}^*C^*.$$

Since $\tilde{D}$ is a (J_2, J_1)-coisometry, and $\tilde{C}$ and $\tilde{B}$ satisfy (3.13) and (3.14),

$$DJ_2D^* = J_1 + C\big(\alpha(\alpha - A)^{-1}G + G\overline{\alpha}(\alpha - A)^{*-1} + (\alpha - A)^{-1}BJ_2B^*(\alpha - A)^{*-1}\big)C^*.$$

By (3.18),

$$DJ_2D^* = J_1 + C(G\overline{\alpha}(\alpha - A)^{*-1} - GA^*(\alpha - A)^{*-1})C^* = J_1 + CGC^*. \tag{3.20}$$

Equations (3.17), (3.19) and (3.20) imply (3.7)-(3.9) with $H = -G$.

To show uniqueness, suppose (3.7)-(3.9) hold for two hermitian matrices H_1 and H_2, and let $H_\Delta = H_1 - H_2$. By (3.7) and (3.8),

$$H_\Delta = AH_\Delta A^* \qquad \text{and} \qquad CH_\Delta A^* = 0. \tag{3.21}$$

The first equation in (3.21) implies $\ker A^* \subset \ker H_\Delta$, and so we may assume that A^* is nonsingular (this assumption is needed when the realization (A, B, C, D) is merely observable). Then $CA^jH_\Delta = 0$ for all integers j and, by the observability of the pair (C, A), $H_\Delta = 0$. □

It follows from the second part of the proof of Theorem 3.1 that if a function W without a pole or zero at infinity takes (J_2, J_1)-coisometric values on the unit circle and H satisfies (3.6), then $-H$ is the hermitian matrix associated with the realization (3.11) of the function $R(\lambda) = W(\alpha\frac{\lambda-1}{\lambda+1})$. For consistency with the imaginary axis case (and with the case $J_1 = J_2 = J$), we will call the matrix $G = -H$ the *hermitian matrix associated with the realization* (A, B, C, D) of W. Thus, if W takes (J_2, J_1)-coisometric values on the unit circle, then G is the hermitian matrix associated with an observable realization (A, B, C, D) of W if

$$\begin{bmatrix} A & B \\ C & D \end{bmatrix} \begin{bmatrix} G & 0 \\ 0 & -J_2 \end{bmatrix} \begin{bmatrix} A & B \\ C & D \end{bmatrix}^* = \begin{bmatrix} G & 0 \\ 0 & -J_1 \end{bmatrix}. \tag{3.22}$$

Note that controllability is not assumed in Theorem 3.1. To illustrate this point, let $J_1 = 1$, let $J_2 = I_2$, and let $W(z) = \frac{1}{\sqrt{2}}[1 \quad \frac{1}{z}]$. The function W has a minimal realization

$$([0],\ [0 \quad 1],\ [\tfrac{1}{\sqrt{2}}],\ [\tfrac{1}{\sqrt{2}} \quad 0]). \tag{3.23}$$

Since

$$\begin{bmatrix} 0 & 0 & 1 \\ \frac{1}{\sqrt{2}} & \frac{1}{\sqrt{2}} & 0 \end{bmatrix} \begin{bmatrix} -1 & 0 & 0 \\ 0 & -1 & 0 \\ 0 & 0 & -1 \end{bmatrix} \begin{bmatrix} 0 & \frac{1}{\sqrt{2}} \\ 0 & \frac{1}{\sqrt{2}} \\ 1 & 0 \end{bmatrix} = \begin{bmatrix} -1 & 0 \\ 0 & -1 \end{bmatrix}, \tag{3.24}$$

the associated hermitian matrix $G = [-1]$. The function W has an observable nonminimal realization

$$\left(\begin{bmatrix} 0 & 0 \\ 0 & 1 \end{bmatrix}, \begin{bmatrix} 0 & 1 \\ 0 & 0 \end{bmatrix}, [\tfrac{1}{\sqrt{2}} \quad \tfrac{1}{\sqrt{2}}], [\tfrac{1}{\sqrt{2}} \quad 0] \right). \tag{3.25}$$

Since

$$\begin{bmatrix} 0 & 0 & 0 & 1 \\ 0 & 1 & 0 & 0 \\ \frac{1}{\sqrt{2}} & \frac{1}{\sqrt{2}} & \frac{1}{\sqrt{2}} & 0 \end{bmatrix} \begin{bmatrix} -1 & 0 & 0 & 0 \\ 0 & 0 & 0 & 0 \\ 0 & 0 & -1 & 0 \\ 0 & 0 & 0 & -1 \end{bmatrix} \begin{bmatrix} 0 & 0 & \frac{1}{\sqrt{2}} \\ 0 & 1 & \frac{1}{\sqrt{2}} \\ 0 & 0 & \frac{1}{\sqrt{2}} \\ 1 & 0 & 0 \end{bmatrix} = \begin{bmatrix} -1 & 0 & 0 \\ 0 & 0 & 0 \\ 0 & 0 & -1 \end{bmatrix},$$

the hermitian matrix associated with realization (3.25) has the form

$$G = \begin{bmatrix} -1 & 0 \\ 0 & 0 \end{bmatrix}.$$

Suppose $W = W_1 W_2$ and (A_i, B_i, C_i, D_i) is a realization of a function W_i, $i = 1, 2$. Then

$$\left(\begin{bmatrix} A_1 & B_1 C \\ 0 & A_2 \end{bmatrix}, \begin{bmatrix} B_1 D_2 \\ B_2 \end{bmatrix}, [C_1 \quad D_1 C_2], D_1 D_2 \right) \tag{3.26}$$

is a realization of W. Suppose W does not have a zero at infinity, the realizations of W_1, W_2 are minimal, and the McMillan degree of W equals the sum of McMillan degrees of W_i. Then (3.26) is a minimal realization of W. Let J_1 and J_2 be signature matrices, and suppose W_i takes (J_i, J_1)-coisometric values on the unit circle, $i = 1, 2$. If G_1, G_2 are the associated hermitian matrices,

$$\begin{bmatrix} A_i & B_i \\ C_i & D_i \end{bmatrix} \begin{bmatrix} G_i & 0 \\ 0 & -J_i \end{bmatrix} \begin{bmatrix} A_i & B_i \\ C_i & D_i \end{bmatrix}^* = \begin{bmatrix} G_i & 0 \\ 0 & -J_i \end{bmatrix} \tag{3.27}$$

for $i = 1, 2$. Then

$$G_i - A_i G_i A_i^* = -B_i J_i B_i^*, \quad C_i G_i A_i^* = D_i J_i B_i^*, \quad J_1 - D_i J_i D_i^* = -C_i G_i C_i^*, \tag{3.28}$$

$i = 1, 2$, and

$$\begin{bmatrix} A_1 & B_1C_2 & B_1D_2 \\ 0 & A_2 & B_2 \\ C_1 & D_1C_2 & D_1D_2 \end{bmatrix} \begin{bmatrix} G_1 & 0 & 0 \\ 0 & G_2 & 0 \\ 0 & 0 & -J_2 \end{bmatrix} \begin{bmatrix} A_1 & B_1C_2 & B_1D_2 \\ 0 & A_2 & B_2 \\ C_1 & D_1C_2 & D_1D_2 \end{bmatrix}^* = \begin{bmatrix} G_1 & 0 & 0 \\ 0 & G_2 & 0 \\ 0 & 0 & -J_1 \end{bmatrix}.$$

Thus,

$$G = \begin{bmatrix} G_1 & 0 \\ 0 & G_2 \end{bmatrix} \tag{3.30}$$

is the hermitian matrix associated with realization (3.26). That is, the following holds.

Proposition 3.2 *Suppose a function W_i without a pole or zero at infinity takes (J_i, J_1)-coisometric values on the unit circle, $i = 1, 2$. If (A_i, B_i, C_i, D_i) is a minimal realization of W_i with the associated hermitian matrix G_i, and $W = W_1 W_2$ is a minimal factorization, then* (3.30) *is the hermitian matrix associated with the realization* (3.26).

Similarly, if (3.22) holds and S is a nonsingular matrix, then

$$\begin{bmatrix} SAS^{-1} & SB \\ CS^{-1} & D \end{bmatrix} \begin{bmatrix} SGS^* & 0 \\ 0 & -J_2 \end{bmatrix} \begin{bmatrix} SAS^{-1} & SB \\ CS^{-1} & D \end{bmatrix}^* = \begin{bmatrix} SGS^* & 0 \\ 0 & -J_1 \end{bmatrix}.$$

That is, the following holds.

Proposition 3.3 *Suppose a rational matrix function W without a pole or zero at infinity takes (J_2, J_1)-coisometric values on the unit circle. If G is the hermitian matrix associated with an observable realization (A, B, C, D) of W, and S is a nonsingular matrix, then SGS^* is the hermitian matrix associated with the realization $(SAS^{-1}, SB, CS^{-1}, D)$.*

Propositions 3.2 and 3.3 suggest that the hermitian matrix associated with a minimal realization of a (J_2, J_1)-coisometrically valued function W can be used to factor W. We consider minimal factorizations $W = W_1 W_2$ where W_i takes (J_i, J_1)-coisometric values on the unit circle, so that W_1 is square. If $\alpha \in \mathbf{T}$ is not a pole of W, there is no loss of generality in requiring $W_1(\alpha) = I$. Indeed, if $W = W_1 W_2$ is any "coisometric" factorization, then $W_1(\alpha) J_1 W_1(\alpha)^* = J_1$ and $W_1(\alpha)^{-1} J_1 W_1(\alpha)^{*-1} = J_1$, so that the factorization

$$W(z) = (W_1(z) W_1(\alpha)^{-1})(W_1(\alpha) W_2(z)) =: \widehat{W}_1(z) \widehat{W}_2(z)$$

has the properties $\widehat{W}_i(z) J_i \widehat{W}_i(z)^* = J_1$ $(i = 1, 2,\ z \in \mathbf{T})$ and $\widehat{W}_1(\alpha) = I$. Theorem 2.3 has the following circle case analogue. Note that if a function W without a pole or zero at infinity takes (J_2, J_1)-coisometric values on the unit circle then, by (3.2), W has neither a pole nor zero at the origin. Hence, if (A, B, C, D) is a minimal realization of W, the matrix A is necessarily invertible. Identity (3.2) implies also

$$J_1 = DJ_2(D^* + B^*A^{*-1}C^*).$$

Thus, $J_2(D^* + B^*A^{*-1}C^*)J_1$ is a right inverse of D.

Theorem 3.4 *Let $J_1 \in \mathbf{C}^{m\times m}$ and $J_2 \in \mathbf{C}^{n\times n}$ be signature matrices, and suppose a function $W \in \mathcal{R}^{m\times n}$ without a pole or zero at infinity takes (J_2, J_1)-coisometric values on the unit circle. Pick a minimal realization (A, B, C, D) of W, let G be the associated hermitian matrix, and let $A^\times = A - BD^RC$ where D^R is a right inverse of D, e.g., $D^R = J_2(D^* + B^*A^{*-1}C^*)J_1$. Let M_1 be a subspace nondegenerate with respect to the inner product $[\cdot,\cdot]_{G^\dagger}$ and such that $A(M_1) \subset M_1 \subset \operatorname{im} G$, and suppose $\operatorname{im} G \cap M_1^{[\perp]}$ can be extended to a complementary to M_1 subspace M_2 which is invariant under $A^\times$. Let $S = [\,S_1 \quad S_2\,]$ where S_i contains a basis for M_i, $i = 1, 2$, and suppose*

$$S^{-1}AS = \begin{bmatrix} A_{11} & A_{12} \\ A_{21} & A_{22} \end{bmatrix}, \quad S^{-1}B = \begin{bmatrix} B_1 \\ B_2 \end{bmatrix}, \quad CS = [\,C_1 \quad C_2\,] \tag{3.31}$$

with the partitioning of matrices conformal to the partitioning of S. If $B_1D^RD = B_1$, then W admits a factorization W_1W_2 where W_i takes (J_i, J_1)-coisometric values on the unit circle, $i = 1, 2$. Moreover, if $\alpha \in \mathbf{T}\backslash\sigma(A)$ and

$$D_1 = (I + C_1(\alpha - A_{11})^{-1}B_1D^R)^{-1}, \tag{3.32}$$

then possible factors with $W_1(\alpha) = I$ are given by the formulas

$$W_1(z) = (I + C_1(z - A_{11})^{-1}B_1D^R)D_1 \tag{3.33}$$

and

$$W_2(z) = D_1^{-1}D + D_1C_2(z - A_{22})^{-1}B_2. \tag{3.34}$$

Conversely, any minimal factorization $W = W_1W_2$ where W_i takes (J_i, J_1)-coisometric values on the unit circle ($i = 1, 2$), and $W_1(\alpha) = I$, can be obtained in this way.

Proof Suppose first there exists a minimal factorization $W = W_1W_2$ where W_i takes (J_i, J_1)-coisometric values on the unit circle ($i = 1, 2$) and $W_1(\alpha) = I$. Let $(A_{\rm i}, B_{\rm i}, C_{\rm i}, D_{\rm i})$ be a minimal realization of W_1, let $(A_{\rm ii}, B_{\rm ii}, C_{\rm ii}, D_{\rm ii})$ be a minimal realization of W_2, and let G_1 and G_2 be the associated hermitian matrices. Then there exists a nonsingular matrix T such that

$$T^{-1}AT = \begin{bmatrix} A_{\rm i} & B_{\rm i}C_{\rm ii} \\ 0 & A_{\rm ii} \end{bmatrix}, \quad T^{-1}B = \begin{bmatrix} B_{\rm i}D_{\rm ii} \\ B_{\rm ii} \end{bmatrix}, \quad CT = [\,C_{\rm i} \quad D_{\rm i}C_{\rm ii}\,]. \tag{3.35}$$

By Proposition 3.3,

$$G = T\begin{bmatrix} G_1 & 0 \\ 0 & G_2 \end{bmatrix}T^*. \tag{3.36}$$

Let

$$M_1 = T\left(\operatorname{im}\begin{bmatrix} I \\ 0 \end{bmatrix}\right). \tag{3.37}$$

Since W_1 is square, G_1 is nonsingular [AG] and $M_1 \subset \operatorname{im} G$. By the first equation in (3.35), $A(M_1) \subset M_1$. If

$$G = [\,T_1 \quad \hat{T}_2\,]\begin{bmatrix} G_1 & 0 \\ 0 & \hat{G}_2 \end{bmatrix}\begin{bmatrix} T_1^* \\ \hat{T}_2^* \end{bmatrix} =: \hat{T}\hat{G}\hat{T}^* \tag{3.38}$$

is a full-rank factorization, the Moore-Penrose inverse of G is given by the formula (cf. Theorem 5 in [BG], p. 23)

$$G^\dagger = \hat{T}(\hat{T}^*\hat{T})^{-1}\hat{G}^{-1}(\hat{T}\hat{T})^{-1}\hat{T}^*. \tag{3.39}$$

Using (3.39), one can verify that M_1 is nondegenerate with respect to the inner product $[\cdot,\cdot]_{G^\dagger}$ and $M_1^{[\perp]} \cap \operatorname{im} G = \operatorname{im} \hat{T}_2$. Hence $M_1^{[\perp]} \cap \operatorname{im} G$ can be extended to $M_2 = \operatorname{im} T_2$, a subspace complementary to M_1. By Theorem 3.1 in [R], $A^\times(M_2) \subset M_2$. Thus, we may take $S = T$ and formulas (3.31) and (3.35) coincide.

We verify that W_1 and W_2 have representations (3.33) and (3.34). Comparing (3.31) and (3.35) we see that

$$A_{\mathrm{i}} = A_{11}, \qquad B_{\mathrm{i}}D_{\mathrm{ii}} = B_1, \qquad C_{\mathrm{i}} = C_1 \tag{3.40}$$

and

$$A_{\mathrm{ii}} = A_{22}, \qquad B_{\mathrm{ii}} = B_2, \qquad D_{\mathrm{i}}C_{\mathrm{ii}} = C_2. \tag{3.41}$$

Also, D_{i} is nonsingular and $D_{\mathrm{i}}D_{\mathrm{ii}} = D$. Thus,

$$W_1(z) = D_{\mathrm{i}} + C_1(z - A_{11})^{-1}B_1D_{\mathrm{ii}}^R, \tag{3.42}$$

where D_{ii}^R is any right inverse of D_{ii}, and

$$W_2(z) = D_{\mathrm{i}}^{-1}D + D_{\mathrm{i}}C_2(z - A_{22})^{-1}B_2. \tag{3.43}$$

Since $W_1(\alpha) = I$,

$$D_{\mathrm{i}} = I - C_1(\alpha - A_{11})^{-1}B_1D_{\mathrm{ii}}^R. \tag{3.44}$$

Since $D_{\mathrm{i}}D_{\mathrm{ii}} = D$,

$$D_{\mathrm{ii}}^R = D^R D_{\mathrm{i}} \tag{3.46}$$

is a right inverse of D_{ii}. The formulas (3.42), (3.43) and (3.46) will imply (3.33) and (3.34) once we show that D_{i} coincides with D_1 in (3.32). But, since $W_1(\alpha) = I$, this follows from formulas (3.44) and (3.46). To conclude the proof of the converse statement, note that by (3.40) the row span of B_1 is contained in the row span of D, so $B_1D^RD = B_1$ for any right inverse D^R of D.

Suppose now $M_1 \subset \operatorname{im} G$ is nondegenerate with respect to the inner product $[\cdot,\cdot]_{G^\dagger}$, $A(M_1) \subset M_1$, and M_2 is a subspace complementary to M_1 such that $M_1^{[\perp]} \cap \operatorname{im} G \subset M_2$ and $A^\times(M_2) \subset M_2$. If $B_1D^RD = B_1$, by Theorem 3.1 in [R] the factorization

$W = W_1W_2$ with W_1 and W_2 as in (3.33) and (3.34) is minimal. Since $M_1 \subset \operatorname{im} G$ and $M_1^{[\perp]} \cap \operatorname{im} G \subset M_2$, it follows that

$$S^{-1}GS^{*-1} = \begin{bmatrix} G_1 & 0 \\ 0 & G_2 \end{bmatrix}$$

is block-diagonal. By Theorem 3.1,

$$\begin{bmatrix} A_{11} & A_{12} & B_1 \\ 0 & A_{22} & B_2 \\ C_1 & C_2 & D \end{bmatrix} \begin{bmatrix} G_1 & 0 & 0 \\ 0 & G_2 & 0 \\ 0 & 0 & -J_2 \end{bmatrix} \begin{bmatrix} A_{11}^* & 0 & C_1^* \\ A_{12}^* & A_{22}^* & C_2^* \\ B_1^* & B_2^* & D^* \end{bmatrix} = \begin{bmatrix} G_1 & 0 & 0 \\ 0 & G_2 & 0 \\ 0 & 0 & -J_1 \end{bmatrix}.$$

Hence

$$\begin{bmatrix} A_{22} & B_2 \\ C_2 & D \end{bmatrix} \begin{bmatrix} G_2 & 0 \\ 0 & -J_2 \end{bmatrix} \begin{bmatrix} A_{22}^* & C_2^* \\ B_2^* & D^* \end{bmatrix} = \begin{bmatrix} G_2 & 0 \\ 0 & -J_1 - C_1G_1C_1^* \end{bmatrix}.$$

By (3.28), $J_1 + C_1G_1C_1^* = -D_1J_1D_1^*$. Hence

$$\begin{bmatrix} A_{22} & B_2 \\ D_1^{-1}C_2 & D_1^{-1}D \end{bmatrix} \begin{bmatrix} G_2 & 0 \\ 0 & -J_2 \end{bmatrix} \begin{bmatrix} A_{22} & B_2 \\ D_1^{-1}C_2 & D_1^{-1}D \end{bmatrix}^* = \begin{bmatrix} G_2 & 0 \\ 0 & -J_1 \end{bmatrix}$$

and W_2 takes (J_2, J_1)-coisometric values on the unit circle. □

Similarly as in Corollary 2.4, the assumptions in Theorem 3.4 simplify if G is nonsingular and the space M_1 is one-dimensional.

4 Inner functions

In this section, we consider rational matrix functions which are (J_2, J_1)-*inner* in the unit disk, that is functions W such that

$$W(z)^*J_2W(z) \leq J_1, \quad \text{if } |z| < 1, \tag{4.1}$$
$$W(z)^*J_2W(z) = J_1, \quad \text{if } |z| = 1, \tag{4.2}$$
$$W(z)^*J_2W(z) \geq J_1, \quad \text{if } |z| > 1. \tag{4.3}$$

If J_1 and J_2 are identity matrices (of possibly different sizes), W which satisfies conditions (4.1)-(4.3) is said to be *inner*. In the case of an inner function, similarly as in the case $J_1 = J_2$ [GDKDM], conditions (4.1)-(4.3) are redundant: any two of them imply the third one.

If condition (4.2) alone holds, W takes (J_2, J_1)-isometric values on the unit circle, i.e., values of W at those points of $\mathbf{T}$ where W is defined are isometries from $(\mathbf{C}^n, [\cdot,\cdot]_{J_1})$ to $(\mathbf{C}^m, [\cdot,\cdot]_{J_2})$. Plainly, there is a close connection between (J_2, J_1)-isometrically and (J_2, J_1)-coisometrically valued functions, and Theorem 3.1 provides the following characterization of (J_2, J_1)-isometrically valued functions.

Theorem 4.1 *Let $J_1 \in \mathbf{C}^{n\times n}$ and $J_2 \in \mathbf{C}^{m\times m}$ be signature matrices, suppose $W \in \mathcal{R}^{m\times n}$ does not have a pole or zero at infinity, and let (A, B, C, D) be a controllable realization of*

W. Then W takes (J_2, J_1)-isometric values on the unit circle if and only if there exists a hermitian matrix H such that

$$\begin{bmatrix} A & B \\ C & D \end{bmatrix}^* \begin{bmatrix} H & 0 \\ 0 & J_2 \end{bmatrix} \begin{bmatrix} A & B \\ C & D \end{bmatrix} = \begin{bmatrix} H & 0 \\ 0 & J_1 \end{bmatrix}. \tag{4.4}$$

Moreover, if W takes (J_2, J_1)-isometric values on the unit circle, the matrix H is unique.

Proof The function W takes (J_2, J_1)-isometric values on the unit circle if and only if W^T with an observable realization (A^T, C^T, B^T, D^T) takes (J_2^T, J_1^T)-coisometric values on the unit circle. By Theorem 3.1, this happens if and only if

$$\begin{bmatrix} A^T & C^T \\ B^T & D^T \end{bmatrix} \begin{bmatrix} H^T & 0 \\ 0 & J_2^T \end{bmatrix} \begin{bmatrix} A^T & C^T \\ B^T & D^T \end{bmatrix}^* = \begin{bmatrix} H^T & 0 \\ 0 & J_1^T \end{bmatrix} \tag{4.5}$$

for some (unique) hermitian matrix H. Plainly, equations (4.4) and (4.5) are equivalent. □

If J_1 and J_2 are identity matrices and (4.2) holds, then, for each $z \in \mathbf{T}$ which is not a pole of W, $\|W(z)\|_2 = 1$. Hence $\|W(z)\|_2 = 1$ for each $z \in \mathbf{T}$ and, in particular, W is analytic on $\mathbf{T}$. If J_2 is indefinite, W which satisfies (4.2) may have a pole on $\mathbf{T}$. This is the case e.g. when

$$J_1 = [1], \quad J_2 = \begin{bmatrix} 1 & 0 & 0 \\ 0 & -1 & 0 \\ 0 & 0 & 1 \end{bmatrix}, \quad \text{and} \quad W(z) = \begin{bmatrix} \frac{1}{z^2-z} \\ \frac{1}{z^2-z} \\ 1 \end{bmatrix}. \tag{4.6}$$

Similarly, in contrast to a (J_2, J_1)-inner function, an inner function is analytic in the unit disk. Consequently, Theorem 4.1 provides the following characterization of inner functions.

Corollary 4.2 *Suppose a function $W \in \mathcal{R}^{m\times n}$ has neither a pole nor zero at infinity and let (A, B, C, D) be a minimal realization of W. Then W is inner if and only if all eigenvalues of A are outside the closed unit disk and equation*

$$\begin{bmatrix} A & B \\ C & D \end{bmatrix}^* \begin{bmatrix} X & 0 \\ 0 & I_m \end{bmatrix} \begin{bmatrix} A & B \\ C & D \end{bmatrix} = \begin{bmatrix} X & 0 \\ 0 & I_n \end{bmatrix} \tag{4.7}$$

has a hermitian solution. Moreover, the hermitian solution of equation (4.7), if it exists, is unique.

Proof If W is inner, then, by Theorem 4.1, equation (4.7) has a unique hermitian solution. Also, (4.1) holds with $J_2 = I_m$ and $J_1 = I_n$, and W is analytic in the closed unit disk. Since the realization (A, B, C, D) is minimal, all eigenvalues of A are outside the closed unit disk.

Conversely, suppose (4.7) holds and W is analytic in the closed unit disk. By Theorem 4.1, (4.2) holds (with $J_1 = I_n$ and $J_2 = I_m$). By the Maximum Modulus Principle, (4.1) holds. Since $W\left(\frac{1}{\bar{z}}\right)^* W(z) = I_n$, $\|W(z)\|_2 \geq 1$ if $\|z\| > 1$ and (4.3) holds. □

Suppose a function $W \in \mathcal{R}^{m\times n}$ takes isometric values on the unit circle. Then $W(\frac{1}{\bar{z}})W(z) = I$, and W has a pole (resp. zero) at infinity if and only if W has a zero (resp. pole) at the origin. Note that if an inner function does have a zero at the origin, it cannot be proper and so it cannot have a realization of the form (A, B, C, D).

We wish to characterize inner rational matrix functions without a zero at the origin in terms of hermitian matrices associated with minimal realizations. First, we consider functions which are isometrically valued on the unit circle. The hermitian matrix associated with a realization (3.23) of the function $W(z) = \frac{1}{\sqrt{2}}[1 \quad \frac{1}{z}]$ is nonsingular. This is a rule when J_2 is definite.

Proposition 4.3 *If (A, B, C, D) is a minimal realization of a function $W \in \mathcal{R}^{m\times n}$ which is isometrically valued on the unit circle, then the associated hermitian matrix is nonsingular.*

Proof Let $\tilde{A}$ and $\tilde{B}$ be as in (3.12) and (3.13). Reversing the computation which lead to (3.17), we obtain

$$G\tilde{A}^* + \tilde{A}G = -\tilde{B}\tilde{B}^*. \tag{4.8}$$

Let U be a unitary matrix such that

$$UGU^* = \begin{bmatrix} G_{11} & 0 \\ 0 & 0 \end{bmatrix} \tag{4.9}$$

with G_{11} nonsingular, let $\hat{A} = U\tilde{A}U^*$, and let $\hat{B} = U\tilde{B}$. Then, by (4.8),

$$\begin{bmatrix} G_{11} & 0 \\ 0 & 0 \end{bmatrix}\begin{bmatrix} \hat{A}_{11}^* & \hat{A}_{21}^* \\ \hat{A}_{12}^* & \hat{A}_{22}^* \end{bmatrix} + \begin{bmatrix} A_{11} & A_{12} \\ A_{21} & A_{22} \end{bmatrix}\begin{bmatrix} G_{11} & 0 \\ 0 & 0 \end{bmatrix} = -\begin{bmatrix} \hat{B}_1 \\ \hat{B}_2 \end{bmatrix}[\hat{B}_1^* \quad \hat{B}_2^*] \tag{4.10}$$

where all matrices are partitioned conformally with (4.9). Comparing the lower right blocks in both sides of (4.10), we obtain $\hat{B}_2 = 0$. Then, equality of the lower left blocks in both sides of (4.10) gives $\hat{A}_{21} = 0$. Thus, for $j = 0, 1, 2, \ldots$, the matrix $\hat{A}^j\hat{B}$ has zero lower block of the size of $\hat{B}_2$. But since the McMillan degrees of $W(z)$ and $R(z) = W(\alpha\frac{z-1}{z+1})$ coincide, $(\hat{A}, \hat{B})$ is a controllable pair and $\sum \mathrm{im}\,\hat{A}^j\hat{B}$ fills out the whole space. It follows that G_{11} and G have the same size. □

The eigenvalues of the associated hermitian matrix can be used to determine the number of poles of W in the unit disk.

Lemma 4.4 *Suppose a proper $W \in \mathcal{R}^{m\times n}$ takes isometric values on the unit circle, let (A, B, C, D) be a minimal realization of W, and let G be the associated hermitian matrix. Then the number of poles of W inside the unit disk, counting multiplicities, is equal to the number of negative eigenvalues of G, counting multiplicities.*

Proof Let $\alpha \in \mathbb{T}\backslash\sigma(A)$. The number of poles of W in the unit disk (counting multiplicities) is equal to the number of poles in the right half-plane of the function $R(z) = W(\alpha\frac{z-1}{z+1})$ with a minimal realization (3.12)-(3.15). It follows from the proof of Theorem 3.1 that G is the associated hermitian matrix and $G\tilde{A}^*+\tilde{A}G = -\tilde{B}\tilde{B}^*$. Hence the real part of the matrix $-\tilde{A}G$, $-\frac{1}{2}(\tilde{A}G + G\tilde{A}^*)$, is positive semidefinite. Denote by $\pi(*)$ (resp. $\nu(*)$) the number of eigenvalues of a matrix $*$ in the open right (resp. left) half-plane. By Proposition 4.4, G is nonsingular and it follows from Corollary 4 in [OS] that

$$\nu(\tilde{A}) = \pi(-\tilde{A}) = \pi(-\tilde{A}GG^{-1}) \leq \pi(G^{-1}) = \pi(G)$$

and

$$\pi(\tilde{A}) = \nu(-\tilde{A}) = \nu(-\tilde{A}GG^{-1}) \leq \nu(G^{-1}) = \nu(G).$$

Since R does not have a pole on the imaginary axis, $\tilde{A}$ is nonsingular and

$$\nu(\tilde{A}) + \pi(\tilde{A}) = \nu(G) + \pi(G).$$

It follows that $\nu(G) = \pi(\tilde{A})$. □

We can now prove the following.

Theorem 4.5 *If $W \in \mathcal{R}^{m\times n}$, TFAE*

(i) *W has a minimal realization (A, B, C, D) such that equation*

$$\begin{bmatrix} A & B \\ C & D \end{bmatrix}^* \begin{bmatrix} X & 0 \\ 0 & -I_m \end{bmatrix} \begin{bmatrix} A & B \\ C & D \end{bmatrix} = \begin{bmatrix} X & 0 \\ 0 & -I_n \end{bmatrix} \tag{4.11}$$

has a positive semidefinite hermitian solution,

(ii) *W is proper and for each minimal realization (A, B, C, D) of W equation* (4.11) *has a unique hermitian solution; moreover, this unique hermitian solution is positive definite,*

(iii) *W is an inner function without a zero at the origin.*

Proof Suppose (i) holds. One can do computations similar to those in (3.10) to show that W takes isometric values on the unit circle. Hence, in particular, A has no eigenvalues on the unit circle. By Proposition 4.3, the associated hermitian matrix G is positive definite. By Lemma 4.4, all eigenvalues of A are outside the unit disk and, by Corollary 4.2, W is inner. Since W is proper, identity (3.2) implies that W does not have a zero at the origin. Thus, (iii) holds. Suppose (iii) holds. By (3.2), W has neither a pole nor zero at infinity. Pick a minimal realization (A, B, C, D) of W. By Corollary 4.2, equation (4.11) has a unique hermitian solution, the associated hermitian matrix G. By Proposition 4.3 and Lemma 4.4, G is positive definite. Thus, (ii) holds. Finally, implication (ii) $\Rightarrow$ (i) is clear. □

Note that in contrast to Corollary 4.2, conditions (i)-(ii) in theorem 4.5 do not involve eigenvalues of a state-space matrix A.

References

[AG] D. Alpay and I. Gohberg, Unitary Rational Matrix Functions, *in* "Topics in Interpolation Theory of Rational Matrix Valued Functions," ed. I. Gohberg, OT 33, Birkhäuser Verlag, Basel, 175-222, 1988.

[AR] D. Alpay and M. Rakowski, Rational Matrix Functions with Coisometric Values on the Imaginary Line, preprint.

[BG] A. Ben-Israel and T. N. E. Greville, "Generalized Inverses: Theory and Applications," John Wiley & Sons, New York/London/Sydney/Toronto, 1974.

[BGK] H. Bart, I. Gohberg and M. A. Kaashoek, "Minimal Factorization of Matrix and Operator Functions," OT 1, Birkhäuser Verlag, Basel/Boston/Stuttgart, 1979.

[BGR] J. A. Ball, I. Gohberg and L. Rodman, "Interpolation of Rational Matrix Functions," OT 45, Birkhäuser Verlag, Basel/Boston/Berlin, 1990.

[F] G. D. Forney, Jr., Minimal bases of rational vector spaces, with applications to multivariable linear systems, *SIAM J. Control* **13** (1975), 493-520.

[GDKDM] Y. Genin, P. Van Dooren, T. Kailath, J.M. Delosme, M. Morf, On Σ-Lossless Transfer Functions and Related Questions, *Linear Algebra and Its Applications* **50** (1983), 251-275.

[Gl] K. Glover, All optimal Hankel-norm approximations of linear multivariable systems and their L^∞-error bounds, *International J. Control* **39** (1984), 1115-1193.

[K] T. Kailath, "Linear Systems," Prentice-Hall, Englewood Cliffs, 1980.

[OS] A. Ostrowski and H. Schneider, Some Theorems on the Inertia of General Matrices, *J. Mathematical Analysis and Applications* **4** (1962), 72-84.

[R] M. Rakowski, Generalized Pseudoinverses of Matrix Valued Functions, *Integral Equations and Operator Theory* **14** (1991), 564-585.

[VDK] G. Verghese, P. Van Dooren and T. Kailath, Properties of the system matrix of a generalized state-space system, *International J. Control* **30** (1979), 235-243.

[WSCP] B. F. Wyman, M. K. Sain, G. Conte and A. M. Perdon, On the Zeros and Poles of a Transfer Function, *Linear Algebra and Its Applications* **122/123/124** (1989), 123-144.

Daniel Alpay
Department of Mathematics
Ben-Gurion University of the Negev
Post Office Box 653
84105 Beer-Sheva, Israel

Marek Rakowski
Department of Mathematics
The Ohio State University
231 West 18th Avenue
Columbus, OH 43210

AMS classification: 15A24, 47A57, 93B10

Operator Theory:
Advances and Applications, Vol. 80

ON SOME ASPECTS OF V.E. KATSNELSON'S INVESTIGATIONS ON INTERRELATIONS BETWEEN LEFT AND RIGHT BLASCHKE–POTAPOV PRODUCTS

D.Z. Arov, B. Fritzsche, B. Kirstein

0 INTRODUCTION

The study of a wide class of matricial versions of classical interpolation problems shows that the solution set of such a problem can be parametrized by a linear fractional transformation of matrices which is generated by some j_{pq}–inner function, where $j_{pq} := \text{diag}\,(I_p, -I_q)$, and where the Schur class $\mathcal{S}_{p\times q}(\mathbb{D})$ is used as the set of parameters (see, e.g., [BGR], [Dy] and [DFK]). Conversely, there arises the following inverse problem: Given a j_{pq}–inner function W, one has to construct an interpolation problem such that the image of $\mathcal{S}_{p\times q}(\mathbb{D})$ under the linear fractional transformation generated by W coincides with its solution set. For the case of an arbitrary A–regular j_{pq}–inner function W, there is always a Generalized Bitangential Schur-Nevanlinna-Pick Interpolation Problem which is associated with W in the sense described above (see [A2]-[A6]). The first main goal of this paper is to treat the analogous problem for the case of a given A–singular j_{11}–inner function. Our method is essentially based on recent results of V.E. Katsnelson [Ka1]–[Ka4] on interrelations between A–singular j_{pq}–inner functions, left and right Blaschke-Potapov products. The second aim of this paper is the study of some connections between the Potapov factorization of j_{pq}–inner functions and that factorization which was obtained by the first author [A2]-[A6] in the context of generalized bitangential Schur–Nevanlinna-Pick interpolation. In this way, we will obtain an alternate proof of a theorem due to V.E. Katsnelson on the Potapov factorization of Blaschke-Potapov products.

1 SOME NOTATION AND PRELIMINARIES

In this first section, we will give a summary on some facts on several classes of meromorphic functions. For a detailed treatment, we refer the reader to the monographs of R. Nevanlinna [Ne2] and P.L. Duren [Du]. We will start with some notations. Throughout this paper, let p and q be positive integers. We will use $\mathbb{N}, \mathbb{C}, \mathbb{D}, \mathbb{T}, \mathbb{C}_0$ and $\mathbb{E}$ to denote the set of positive integers, the set of complex numbers, the open unit disc, the unit circle, the extended complex plane and the exterior of the closed unit disc, respectively:

$$\mathbb{D} := \{z \in \mathbb{C} : |z| < 1\}, \mathbb{T} := \{z \in \mathbb{C} : |z| = 1\}, \mathbb{C}_0 := \mathbb{C} \cup \{\infty\}, \mathbb{E} := \mathbb{C}_0 \setminus (\mathbb{D} \cup \mathbb{T}).$$

The notation $\mathbb{C}^{p\times q}$ stands for the set of all $p \times q$ matrices with complex entries. For the null matrix that belongs to $\mathbb{C}^{p\times q}$ we will write $0_{p\times q}$, whereas the identity matrix that belongs to $\mathbb{C}^{q\times q}$ will be denoted by I_q. The linear Lebesgue-Borel measure on $\mathbb{T}$ will be designated by $\underline{\lambda}$. Assume that G is a simply connected domain of $\mathbb{C}_0$. Then let $\mathcal{NM}(G)$ be the Nevanlinna class of all functions which are meromorphic in G and which can be represented as quotient of two bounded holomorphic functions in G. If $g \in \mathcal{NM}(\mathbb{D})$ (respectively, $g \in \mathcal{NM}(\mathbb{E})$), then a well-known theorem due to Fatou implies that there exist a Borelian subset $\mathfrak{B}_0$ of the unit circle $\mathbb{T}$ with $\underline{\lambda}(\mathfrak{B}_0) = 0$ and a Borel measurable function $\underline{g} : \mathbb{T} \to \mathbb{C}$ such that

$$\lim_{r\to 1-0} g(rz) = \underline{g}(z) \quad \text{(respectively, } \lim_{r\to 1+0} g(rz) = \underline{g}(z))$$

for all $z \in \mathbb{T}\setminus\mathfrak{B}_0$. In the following, we will continue to use the symbol $\underline{g}$ to denote a boundary function of a function g which belongs to $\mathcal{NM}(\mathbb{D})$ or $\mathcal{NM}(\mathbb{E})$. The subalgebra of all $g \in \mathcal{NM}(G)$ which are holomorphic in G will be denoted by $\mathcal{N}(G)$. The class $\mathcal{N}(\mathbb{D})$ can be described as the set of all functions g which are holomorphic in $\mathbb{D}$ and which fulfill

$$\sup_{r\in[0,1)} \frac{1}{2\pi} \int_{\mathbb{T}} \log^+ | g(rz) | \underline{\lambda}(dz) < +\infty$$

where $\log^+ x := \max(\log x, 0)$ for each $x \in [0,\infty)$. If $g : \mathbb{D} \to \mathbb{C}$ admits a representation

$$g(w) = \alpha \exp\left(\frac{1}{2\pi} \int_{\mathbb{T}} \frac{z+w}{z-w} \log k(z) \underline{\lambda}(dz) \right) \quad , \quad w \in \mathbb{D} \;,$$

with some $\alpha \in \mathbb{T}$ and some Borel measurable function $k : \mathbb{T} \longrightarrow [0,\infty)$ which satisfies $\frac{1}{2\pi}\int_{\mathbb{T}} | \log k | \, d\underline{\lambda} < +\infty$, then g belongs to $\mathcal{N}(\mathbb{D})$. Such functions g are called *outer*. For all $g \in \mathcal{N}(\mathbb{D})$, the inequality

$$\frac{1}{2\pi} \int_{\mathbb{T}} \log^+ | \underline{g}(z) | \underline{\lambda}(dz) \leq \sup_{r\in[0,1)} \frac{1}{2\pi} \int_{\mathbb{T}} \log^+ | g(rz) | \underline{\lambda}(dz). \tag{1}$$

holds true. By the *Smirnov class* $\mathcal{N}_+(\mathbb{D})$ we will mean the set of all $g \in \mathcal{N}(\mathbb{D})$ for which equality holds true in (1). The class $\mathcal{N}_+(\mathbb{D})$ proves to be a subalgebra of $\mathcal{N}(\mathbb{D})$. If g is outer in $\mathcal{N}(\mathbb{D})$, then g necessarily belongs to $\mathcal{N}_+(\mathbb{D})$. Note that, for each $t \in (0,\infty]$, the Hardy class $H^t(\mathbb{D})$ is a subset of $\mathcal{N}_+(\mathbb{D})$.

If $\mathfrak{X}$ is one of the classes $\mathcal{NM}(G)$, $\mathcal{N}(G)$, $\mathcal{N}_+(\mathbb{D})$ or $H^s(\mathbb{D})$, where $s \in (0,\infty]$, then $\mathfrak{X}^{p\times q}$ designates the class of all $p \times q$ matrix-valued functions each entry of which belongs to $\mathfrak{X}$. A function $\Phi \in [\mathcal{N}_+(\mathbb{D})]^{q\times q}$ is called *outer* (in $[\mathcal{N}_+(\mathbb{D})]^{q\times q}$) if $\det \Phi$ is outer in $\mathcal{N}(\mathbb{D})$. If Φ is an outer function in $[\mathcal{N}_+(\mathbb{D})]^{q\times q}$, then $\det \Phi(z) \neq 0$ for all $z \in \mathbb{D}$, and Φ^{-1} is also an outer function in $[\mathcal{N}_+(\mathbb{D})]^{q\times q}$. Conversely, if $\Phi \in [\mathcal{N}_+(\mathbb{D})]^{q\times q}$ satisfies $\det \Phi(z) \neq 0$ for all $z \in \mathbb{D}$ and if $\Phi^{-1} \in [\mathcal{N}_+(\mathbb{D})]^{q\times q}$, then Φ and Φ^{-1} are necessarily outer functions in $[\mathcal{N}_+(\mathbb{D})]^{q\times q}$. If $\Phi \in [\mathcal{N}_+(\mathbb{D})]^{q\times q}$ and $\Psi \in [\mathcal{N}_+(\mathbb{D})]^{q\times q}$ are outer, then the product $\Phi\Psi$ is also an outer function in $[\mathcal{N}_+(\mathbb{D})]^{q\times q}$. An outer function $\Phi \in [\mathcal{N}_+(\mathbb{D})]^{q\times q}$ is called *normalized* if $\Phi(0)$ is nonnegative Hermitian. Let $\mathbb{C}_{\geq}^{q\times q}$ be the set of all $q \times q$ nonnegative Hermitian matrices. If $\Lambda : \mathbb{T} \to \mathbb{C}_{\geq}^{q\times q}$ is Lebesgue integrable on $\mathbb{T}$ and if Λ satisfies

$$\frac{1}{2\pi} \int_{\mathbb{T}} \log \det \Lambda d\underline{\lambda} > -\infty ,$$

then there exist unique normalized outer functions Φ and Ψ which belong to $[H^2(\mathbb{D})]^{q\times q}$ such that $\Lambda = \underline{\Phi}\ \underline{\Phi}^*\ \underline{\lambda}$–a.e. and $\Lambda = \underline{\Psi}^*\underline{\Psi}\ \underline{\lambda}$–a.e. on $\mathbb{T}$ (see, e.g., Wiener and Masani [WM]). A matrix-valued function f defined on a nonempty set $\mathfrak{X}$ is called contractive if $I_p - f(z)f^*(z) \in \mathbb{C}_{\geq}^{p\times p}$ for all $z \in \mathfrak{X}$. A function $f : \mathbb{D} \to \mathbb{C}^{p\times q}$ is said to be a $p \times q$ *Schur function* if f is both holomorphic and contractive in $\mathbb{D}$. The class of all $p \times q$ Schur functions will be denoted by $\mathcal{S}_{p\times q}(\mathbb{D})$. A function $f \in \mathcal{S}_{q\times q}(\mathbb{D})$ is called *inner* if f has unitary boundary values $\underline{\lambda}$–a.e. on $\mathbb{T}$. An inner function $f \in \mathcal{S}_{q\times q}(\mathbb{D})$ is said to be *singular* if $\det f$ nowhere vanishes in $\mathbb{D}$. Let $\mathfrak{X}$ be a nonempty subset of the extended complex plane $\mathbb{C}_0$, and let $f : \mathfrak{X} \to \mathbb{C}^{p\times q}$. Then we will use the symbol $\check{f}$ for the function $\check{f} : \mathfrak{Y} \to \mathbb{C}^{q\times p}$ which is given by $\mathfrak{Y} := \{z \in \mathbb{C}_0 : \overline{z} \in \mathfrak{X}\}$ and $\check{f}(z) := [f(\overline{z})]^*$. Furthermore, if $\mathcal{M}$ is a nonempty subset of $\mathfrak{X}$, then $\mathrm{Rstr.}_{\mathcal{M}} f$ designates the restriction of the function f onto $\mathcal{M}$.

Now we will turn our attention to linear fractional transformations. Let $A \in \mathbb{C}^{(p+q)\times(p+q)}$ and $B \in \mathbb{C}^{(p+q)\times(p+q)}$ be partitioned into blocks via

$$A = \begin{pmatrix} A_{11} & A_{12} \\ A_{21} & A_{22} \end{pmatrix} \quad \text{and} \quad B = \begin{pmatrix} B_{11} & B_{12} \\ B_{21} & B_{22} \end{pmatrix}$$

where $A_{11} \in \mathbb{C}^{p\times p}$ and $B_{11} \in \mathbb{C}^{p\times p}$. The set

$$\mathcal{D}_{A_{21},A_{22}} := \{X \in \mathbb{C}^{p\times q} : \det(A_{21}X + A_{22}) \neq 0\}$$

is nonempty if and only if $\mathrm{rank}(A_{21}, A_{22}) = q$. In this case, the right linear fractional transformation $\mathcal{S}_A^{(p,q)} : \mathcal{D}_{A_{21},A_{22}} \longrightarrow \mathbb{C}^{p\times q}$ is defined by

$$\mathcal{S}_A^{(p,q)} := (A_{11}X + A_{12})(A_{21}X + A_{22})^{-1}.$$

If $\mathrm{rank}(A_{21}, A_{22}) = q$, $\mathrm{rank}(B_{21}, B_{22}) = q$, and if the set $\mathcal{D} := \{X \in \mathcal{D}_{A_{21},A_{22}} : \mathcal{S}_A^{(p,q)}(X) \in \mathcal{D}_{B_{21},B_{22}}\}$ is nonempty, then

$$\mathcal{S}_{BA}^{(p,q)}(X) = \mathcal{S}_B^{(p,q)}\left(\mathcal{S}_A^{(p,q)}(X)\right)$$

for all $X \in \mathcal{D}$. The set

$$\mathcal{E}_{A_{12},A_{22}} := \{X \in \mathbb{C}^{q\times p} : \det(XA_{12} + A_{22}) \neq 0\}$$

is nonempty if and only if $\mathrm{rank}(A_{12}^*, A_{22}^*) = q$. In this case, the left linear fractional transformation $\mathcal{T}_A^{(p,q)} := \mathcal{E}_{A_{12},A_{22}} \longrightarrow \mathbb{C}^{q\times p}$ is defined by

$$\mathcal{T}_A^{(p,q)}(X) := (XA_{12} + A_{22})^{-1}(XA_{11} + A_{21}).$$

REMARK 1 *Let A be a j_{pq}–contractive matrix, and let $\mathbb{K}_{p\times q} := \{X \in \mathbb{C}^{p\times q} : I - XX^* \in \mathbb{C}_{\geq}^{p\times q}\}$. Then $\mathbb{K}_{p\times q} \subseteq \mathcal{D}_{A_{21},A_{22}}$ and $\mathbb{K}_{q\times p} \subseteq \mathcal{E}_{A_{12},A_{22}}$. Moreover, $\mathcal{S}_A^{(p,q)}(\mathbb{K}_{p\times q}) \subseteq \mathbb{K}_{p\times q}$ and $\mathcal{T}_A^{(p,q)}(\mathbb{K}_{q\times p}) \subseteq \mathbb{K}_{q\times p}$ (see, e.g., [P2] or [DFK, Theorem 1.6.1]).*

2 ON J-INNER FUNCTIONS AND THEIR FACTORIZATIONS

Throughout this section, let m be a positive integer, and let J be an $m \times m$ signature matrix, i.e., an $m \times m$ complex matrix with $J = J^*$ and $J^2 = I_m$. If $A \in \mathbb{C}^{m\times m}$ satisfies $J - A^*JA \in \mathbb{C}_{\geq}^{m\times m}$ (respectively, $A^*JA = J$), then A is called *J–contractive* (respectively, *J–unitary*).

REMARK 2 *If $J \neq \pm I_m$, then the multiplicities p and q of the eigenvalues $+1$ and -1 of J are positive integers with $p + q = m$, and there is an $m \times m$ unitary matrix U with $U^*JU = j_{pq}$ where $j_{pq} :=$ diag $(I_p, -I_q)$.*

If G is a simply connected domain of the extended complex plane $\mathbb{C}_0$, then we will use $\mathfrak{P}_J(G)$ for the Potapov class, i.e., the set of all $m \times m$ matrix-valued functions W which satisfy the following three conditions:

(i) W is meromorphic in G.

(ii) $\det W$ does not identically vanish in G.

(iii) $W(z)$ is J-contractive for all z which belong to the set $\mathbb{H}(W)$ of all points of analyticity of W.

Obviously, the class $\mathfrak{P}_J(G)$ is multiplicative, i.e., if W_1 and W_2 belong to $\mathfrak{P}_J(G)$, then the product W_1W_2 belongs to $\mathfrak{P}_J(\mathbb{D})$ as well. In the follwing, we will often deal with the class $\widetilde{\mathfrak{P}}_J(\mathbb{D})$ which consists of all $W \in \mathfrak{P}_J(\mathbb{D})$ for which both functions W and W^{-1} are holomorphic in $\mathbb{D}$. Obviously, the class $\widetilde{\mathfrak{P}}_J(\mathbb{D})$ is multiplicative. The Potapov class $\mathfrak{P}_J(\mathbb{D})$ is a subclass of $[\mathcal{NM}(\mathbb{D})]^{m\times m}$ (see, e.g., Dym [Dy, Corollary 2]). The radial boundary function $\underline{W}$ of $W \in \mathfrak{P}_J(\mathbb{D})$ is J-contractive, i.e., $J - \underline{W}^*(z)J\underline{W}(z) \in \mathbb{C}_{\geq}^{m\times m}$ holds for $\underline{\lambda}$–almost all $z \in \mathbb{T}$. A function $W \in \mathfrak{P}_J(\mathbb{D})$ is called *J–inner* if $\underline{W}$ is J–unitary $\underline{\lambda}$–a.e., i.e., if $\underline{W}^*J\underline{W} = J$ is fulfilled $\underline{\lambda}$–a.e. on $\mathbb{T}$. A J–inner function that belongs to $\widetilde{\mathfrak{P}}_J(\mathbb{D})$ is said to be *singular*. Recent investigations of the first author (see [A2]-[A6]) indicated that there is an important subclass of singular J–inner functions which satisfy some growth conditions. This led him to the following objects. A J–inner function $W \in \mathfrak{P}_J(\mathbb{D})$ is said to be *A-singular* if both functions W and W^{-1} belong to the Smirnov class $[\mathcal{N}_+(\mathbb{D})]^{m\times m}$. Obviously, every A-singular J–inner function that belongs to $\mathfrak{P}_J(\mathbb{D})$ even belongs to $\widetilde{\mathfrak{P}}_J(\mathbb{D})$. Note that a J–inner function $W \in \mathfrak{P}_J(\mathbb{D})$ is A-singular if and only if W is an outer function in $[\mathcal{N}_+(\mathbb{D})]^{m\times m}$ (see [A6]). The subclass of all A-singular J–inner functions is multiplicative.

A J-inner function W is called *left A–regular*, if W has the following property: If $W = W_1W_2$ is an arbitrary representation of W with some J-inner function W_1 and some

A-singular J-inner function W_2, then W_2 is necessarily constant. A J-inner function W is said to be *right* A-*regular*, if W fulfills the following condition: If $W = W_1 W_2$ is an arbitrary representation of W with some A-singular J-inner function W_1 and some J-inner function W_2, then W_1 is necessarily constant.

To define Blaschke–Potapov J-elementary factors, we need the usual Blaschke factors. If $w \in \mathbb{D} \setminus \{0\}$, then the Blaschke factor $\beta_w : \mathbb{C}_0 \setminus \{\frac{1}{\overline{w}}\} \to \mathbb{C}$ is defined by

$$\beta_w(z) := \begin{cases} \frac{\overline{w}}{|w|} \cdot \frac{w-z}{1-\overline{w}z} & , \quad z \in \mathbb{C}\setminus\{\frac{1}{\overline{w}}\} \\ \frac{1}{|w|} & , \quad z = +\infty \end{cases} , \tag{2}$$

whereas $\beta_0 : \mathbb{C} \to \mathbb{C}$ is given by

$$\beta_0(z) := z \quad , \quad z \in \mathbb{C}. \tag{3}$$

Now let $w \in \mathbb{D}$. If $P \in \mathbb{C}^{m\times m}\setminus\{0_{m\times m}\}$ satisfies $P^2 = P$ and $JP \in \mathbb{C}_{\geq}^{m\times m}$, then the rational matrix-valued function

$$B_{w,P} := I_m + (\beta_w - 1)P \tag{4}$$

is called *a Blaschke-Potapov J-elementary factor of first kind.* If $Q \in \mathbb{C}^{m\times m}\setminus\{0_{m\times m}\}$ fulfills $Q^2 = Q$ and $-JQ \in \mathbb{C}_{\geq}^{m\times m}$, then

$$C_{w,Q} := I_m + (\frac{1}{\beta_w} - 1)Q \tag{5}$$

is said to be *a Blaschke–Potapov J-elementary factor of second kind.*

It is readily checked that if $P \in \mathbb{C}^{m\times m}$ satisfies $P^2 = P$, then

$$(I_m + (\alpha - 1)P)\left(I_m - (\frac{1}{\alpha} - 1)P\right) = I_m$$

for all $\alpha \in \mathbb{C}\setminus\{0\}$. Consequently, if $B_{w,P}$ (respectively, $C_{w,Q}$) is a Blaschke-Potapov J-elementary factor of first (respectively, second) kind, then $B_{w,P}^{-1}$ (respectively, $C_{w,Q}^{-1}$) is a Blaschke-Potapov $(-J)$-elementary factor of second (respectively, first) kind.

A matrix-valued function B_l (respectively, B_r) is called a *left* (respectively, *right*) *Blaschke–Potapov product* with respect to J if B_l (respectively, B_r) is a constant matrix-valued function in $\mathbb{C}_0$ with some J-unitary value or if B_l (respectively, B_r) admits a product representation

$$B_l = \left(\prod_{k\in\mathcal{I}}^{\longrightarrow} D_k\right) U \qquad \left(\text{respectively, } B_r = U\left(\prod_{k\in\mathcal{I}}^{\longleftarrow} D_k\right)\right)$$

where $\mathcal{I}$ is some nonempty subset of $\mathbb{N}$, where U is some (constant) J-unitary matrix, and where, for each $k \in \mathcal{I}$, D_k is a Blaschke–Potapov J-elementary factor of first or second kind. If all factors D_k, $k \in \mathcal{I}$, are Blaschke-Potapov factors of first (respectively, second) kind, then B_l and B_r are called *Blaschke-Potapov products with respect to J of first* (respectively, *second*) *kind.* For convergence aspects of Blaschke-Potapov products with respect to J, we refer the reader to Potapov [P1] and Ginzburg [G]. Simakova and the first author [AS], [A1] proved that if B is a left or right Blaschke–Potapov product with respect to J, then its restriction onto $\mathbb{H}(B) \cap \mathbb{D}$ is a J–inner function. Moreover, the restriction of an arbitrary left or right Blaschke–Potapov product with respect to J onto $\mathbb{H}(B) \cap \mathbb{D}$ is even a left (respectively, right) A–regular J–inner function (see [A2], [A5]).

Observe that if $J = I_m$ (respectively, $J = -I_m$), then there is no matrix $P \in \mathbb{C}^{m\times m}\setminus\{0_{m\times m}\}$ such that $P^2 = P$ and $-JP \in \mathbb{C}_{\geq}^{m\times m}$ (respectively, $P^2 = P$ and $JP \in \mathbb{C}_{\geq}^{m\times m}$). Consequently, if $J = I_m$ (respectively, $J = -I_m$), then there is no Blaschke-Potapov J–elementary factor of second (respectively, first) kind.

REMARK 3 *Let $\rho \in \mathbb{D}$, let $P \in \mathbb{C}^{m\times m}\setminus\{0_{m\times m}\}$ be such that $P^2 = P$ and $JP \in \mathbb{C}_{\geq}^{m\times m}$, and let the functions b_ρ and $B_{\rho;P}$ be defined by (2)–(4). Then $\det B_{\rho;P} = (b_\rho)^r$ where $r := \operatorname{rank} P$.*

REMARK 4 *Let J be an $m \times m$ signature matrix. If W is a left (respectively, right) Blaschke–Potapov product with respect to J, then $\check{W}$ is a right (respectively, left) Blaschke–Potapov product with respect to J.*

Let us still recall the notion of a Blaschke-Potapov J–elementary factor of the third kind. Let $z_0 \in \mathbb{T}$, and let $P \in \mathbb{C}^{m\times m}\setminus\{0_{m\times m}\}$ be such that $JP \in \mathbb{C}_{\geq}^{m\times m}$ and $P^2 = 0$. Then the rational matrix-valued function $B_{z_0;P} : \mathbb{C}_0\setminus\{z_0\} \longrightarrow \mathbb{C}^{m\times m}$ given by

$$B_{z_0;P}(z) := \begin{cases} I - \frac{z_0+z}{z_0-z}P & , \quad z \in \mathbb{C}\setminus\{z_0\} \\ I + P & , \quad z = \infty \end{cases}$$

is called *Blaschke-Potapov J–elementary factor of third kind.* It is readily checked that $\mathrm{Rstr.}_{\mathbb{D}} B_{z_0;P}$ is an A–singular J–inner function.

REMARK 5 *Let J be an $m \times m$ signature matrix, and let $W \in [\mathcal{NM}(\mathbb{D})]^{m\times m}$.*
(a) If $W \in \mathfrak{P}_J(\mathbb{D})$, then $\check{W} \in \mathfrak{P}_J(\mathbb{D})$.
(b) If $W \in \widetilde{\mathfrak{P}}_J(\mathbb{D})$, then $\check{W} \in \widetilde{\mathfrak{P}}_J(\mathbb{D})$.
(c) If W is J–inner, then $\check{W}$ is J–inner.
(d) If $W \in \mathfrak{P}_J(\mathbb{D})$ is an A–singular J–inner function, then $\check{W}$ is also an A–singular J–inner function.
(e) If W is a left (respectively, right) A–regular J–inner function, then $\check{W}$ is a right (respectively, left) A–regular J–inner function.

Since Potapov's [P1] fundamental generalization of the F. Riesz– R. Nevanlinna– V. I. Smirnov factorization for bounded holomorphic functions in $\mathbb{D}$ will play a key role in our further considerations, we are going to recall this fundamental theorem.

THEOREM 1 *Let J be an $m \times m$ signature matrix, and let $W \in \mathfrak{P}_J(\mathbb{D})$.*

(a) Then there are a left Blaschke-Potapov product B_l with respect to J and a function Σ_{ls} belonging to $\widetilde{\mathfrak{P}}_J(\mathbb{D})$ such that $W = \left(\mathrm{Rstr.}_{\mathbb{D}\cap\mathbb{H}(B_l)} B_l\right) \cdot \Sigma_{ls}$.

(b) *If* $W = \left(\mathrm{Rstr.}_{\mathbb{D}\cap\mathbf{H}(B_l)}\widetilde{B_l}\right) \cdot \widetilde{\Sigma_{ls}}$ *is a factorization of* W *with some left Blaschke–Potapov product* $\widetilde{B_l}$ *with respect to* J *and some function* $\widetilde{\Sigma_{ls}} \in \widetilde{\mathfrak{P}}_J(\mathbb{D})$*, then there is a (constant)* J*-unitary matrix* U *such that* $\widetilde{B_l} = B_l U$ *and* $\widetilde{\Sigma_{ls}} = U^{-1}\Sigma_{ls}$*. If* W *is* J*-inner, then* $\widetilde{\Sigma_{ls}}$ *is* J*-inner.*

(c) *If* U *is an arbitrary (constant)* J*-unitary matrix, then* $\widetilde{B_l} := B_l U$ *and* $\widetilde{\Sigma_{ls}} := U^{-1}\Sigma_{ls}$ *are a left Blaschke–Potapov product with respect to* J *and a function belonging to* $\widetilde{\mathfrak{P}}_J(\mathbb{D})$*, respectively, such that* $W = \left(\mathrm{Rstr.}_{\mathbb{D}\cap\mathbf{H}(B_l)}\widetilde{B_l}\right) \cdot \widetilde{\Sigma_{ls}}$.

Because our investigations are mainly concerned with interrelations between left and right Blaschke-Potapov products we will formulate the right version of Theorem 1, which follows immediately from Theorem 1 and Remark 5.

THEOREM 2 *Let* J *be an* $m \times m$ *signature matrix, and let* $W \in \mathfrak{P}_J(\mathbb{D})$.

(a) *Then there are a right Blaschke–Potapov product* B_r *with respect to* J *and a function* Σ_{rs} *belonging to* $\widetilde{\mathfrak{P}}_J(\mathbb{D})$ *such that* $W = \Sigma_{rs} \cdot \left(\mathrm{Rstr.}_{\mathbb{D}\cap\mathbf{H}(B_r)}B_r\right)$.

(b) *If* $W = \widetilde{\Sigma_{rs}} \cdot \left(\mathrm{Rstr.}_{\mathbb{D}\cap\mathbf{H}(B_r)}\widetilde{B_r}\right)$ *is a factorization of* W *with some right Blaschke–Potapov product* $\widetilde{B_r}$ *with respect to* J *and some function* $\widetilde{\Sigma_{rs}} \in \widetilde{\mathfrak{P}}_J(\mathbb{D})$*, then there is a (constant)* J*-unitary matrix* U *such that* $\widetilde{B_r} = UB_r$ *and* $\widetilde{\Sigma_{rs}} = \Sigma_{rs}U^{-1}$*. If* W *is* J*-inner, then* $\widetilde{\Sigma_{rs}}$ *is* J*-inner.*

(c) *If* U *is an arbitrary (constant)* J*-unitary matrix, then* $\widetilde{B_r} := UB_r$ *and* $\widetilde{\Sigma_{rs}} := \Sigma_{rs}U^{-1}$ *are a right Blaschke–Potapov product with respect to* J *and a function belonging to* $\widetilde{\mathfrak{P}}_J(\mathbb{D})$*, respectively, such that* $W = \widetilde{\Sigma_{rs}} \cdot \left(\mathrm{Rstr.}_{\mathbb{D}\cap\mathbf{H}(B_r)}\widetilde{B_r}\right)$.

Now we will show that the class of inner matrix-valued functions and some distinguished subclasses of it can be completely characterized by the determinants of these functions. In particular, we will obtain that the set of all left Blaschke-Potapov products with respect to I_m and the set of all right Blaschke-Potapov products with respect to I_m coincide.

REMARK 6 *Let* $A \in \mathbb{K}_{q\times q}$*. Then* $|\det A| \leq 1$*. Moreover,* $|\det A| = 1$ *if and only if* A *is unitary.*

LEMMA 1 *Let* $f \in \mathcal{S}_{m\times m}(\mathbb{D})$*. Then:*

(a) *The function* $\det f$ *belongs to* $\mathcal{S}_{1\times 1}(\mathbb{D})$.

(b) *The function* f *is inner if and only if* $\det f$ *is inner. If* f *is inner, then* $\det f$ *does not identically vanish.*

(c) f *is a singular inner function if and only if* $\det f$ *is a singular inner function.*

(d) *The following statements are equivalent:*

(i) *f is the restriction of a left Blaschke-Potapov product with respect to I_m onto $\mathbb{D}$.*

(ii) *f is the restriction of a right Blaschke-Potapov product with respect to I_m onto $\mathbb{D}$.*

(iii) $\det f$ *is the restriction of a Blaschke product onto $\mathbb{D}$.*

PROOF The assertions stated in parts (a) and (b) are immediate consequences of Remark 6. Part (c) follows from part (a) and the definition of a singular inner function.

It remains to prove part (d). In view of Remark 3, it is readily checked that each of the conditions (i) and (ii) implies (iii). Now suppose that (iii) holds. By virtue of part (b), we see that f is an inner function. Applying Theorem 1 (with $J = I_m$) we get that there exist a left Blaschke-Potapov product B_l with respect to I_m and a singular inner function C_l such that $f = B_l^{\square} C_l$ where $B_l^{\square} := \mathrm{Rstr}._{\mathbb{D}} B_l$. Hence, $\det f = \det B_l^{\square} \cdot \det C_l$. The implication "(i) $\Rightarrow$ (iii)", which is already verified, shows that $\det B_l$ is a Blaschke product. Part (c) yields that $\det C_l$ is a singular inner function. Therefore, the uniqueness of the F. Riesz-R. Nevanlinna–V. I. Smirnov factorization of inner functions and condition (iii) provide that $\det C_l$ is a constant (inner) function with unimodular value. Hence, we obtain from Remark 6 that, for each $z \in \mathbb{D}$, the matrix $C_l(z)$ is unitary. Since C_l belongs to $\mathcal{S}_{m\times m}(\mathbb{D})$, the maximum modulus principle for matrix-valued Schur functions (see, e.g., [DFK, Corollary 2.3.2]) implies that C_l is a constant function. From $f = B_l^{\square} C_l$ we then infer that (i) holds.

We finally verify that (ii) is a consequence of (iii). If we suppose (iii), then Remark 4 shows that $(\det f)^{\check{}}$ is a Blaschke product. Because of $\det \check{f} = (\det f)^{\check{}}$ and the implication "(iii) $\Rightarrow$ (i)", which is already proved, we obtain that $\check{f}$ is a left Blaschke-Potapov product with respect to I_m. Thus, Remark 4 provides (ii). ■

In view of Lemma 1, one can speak of Blaschke–Potapov products with respect to I_m instead of left or right Blaschke–Potapov products with respect to I_m.

Studying a certain inverse problem in the context of generalized bitangential Schur–Nevanlinna–Pick interpolation, the first author [A2], [A4], [A5] discovered a further kind of factorization for the class of J–inner functions, which is of different nature in comparison with Potapov's factorizations given in Theorems 1 and 2. This interpolation problem is intimately connected with the $(p+q) \times (p+q)$ signature matrix $j_{pq} := \mathrm{diag}\,(I_p, -I_q)$. In this way, the first author [A2], [A5] proved originally the following Theorem 3 for $J = j_{pq}$. However, in the general case of an arbitrary $m \times m$ signature matrix, Theorem 3 can be immediately obtained from the special case $J = j_{pq}$ and the following fact:

REMARK 7 *Let J be an $m \times m$ signature matrix with $J \neq \pm I_m$. According to Remark 2 , let U be an $m \times m$ unitary matrix with $U^* J U = j_{pq}$. Let $W \in [\mathcal{NM}(\mathbb{D})]^{m\times m}$, and let $W_{\blacksquare} := U^* W U$. Then:*

(a) *$W \in \mathfrak{P}_J(\mathbb{D})$ if and only if $W_{\blacksquare} \in \mathfrak{P}_{j_{pq}}(\mathbb{D})$.*

(b) *$W \in \widetilde{\mathfrak{P}}_J(\mathbb{D})$ if and only if $W_{\blacksquare} \in \widetilde{\mathfrak{P}}_{j_{pq}}(\mathbb{D})$.*

(c) *W is a J-inner function if and only if $W_{\blacksquare}$ is a j_{pq}–inner function.*

(d) *W is an A-singular J–inner function that belongs to $\mathfrak{P}_J(\mathbb{D})$ if and only if $W_{\blacksquare}$ is an A–singular j_{pq}–inner function that belongs to $\mathfrak{P}_{j_{pq}}(\mathbb{D})$.*

(e) *W is a left (respectively, right) A–regular J–inner function if and only if $W_{\blacksquare}$ is a left (respectively, right) A–regular j_{pq}–inner function.*

THEOREM 3 *Let J be an $m \times m$ signature matrix, and let W be a J–inner function.*

(a) *There are a left A–regular J–inner function W_l and an A–singular J–inner function W_{ls} such that $W = W_l W_{ls}$.*

(b) *If $W = \widetilde{W_l}\widetilde{W_{ls}}$ with some left A–regular J–inner function $\widetilde{W_l}$ and an A–singular J–inner function $\widetilde{W_{ls}}$, then there is a (constant) J–unitary matrix U such that $\widetilde{W_l} = W_l U$ and $\widetilde{W_{ls}} = U^{-1} W_{ls}$.*

(c) *If U is an arbitrary (constant) J–unitary matrix, then $\widetilde{W_l} := W_l U$ and $\widetilde{W_{ls}} := U^{-1} W_{ls}$ are a left A–regular J–inner function and an A–singular J–inner function, respectively, such that $W = \widetilde{W_l}\widetilde{W_{ls}}$.*

Using Theorem 3 and Remark 5, one can easily infer the following right version of the foregoing result.

THEOREM 4 *Let J be an $m \times m$ signature matrix, and let W be a J–inner function.*

(a) *There are a right A–regular J–inner function W_r and an A–singular J–inner function W_{rs} such that $W = W_{rs} W_r$.*

(b) *If $W = \widetilde{W_{rs}}\widetilde{W_r}$ with some right A–regular J–inner function $\widetilde{W_r}$ and an A–singular J–inner function $\widetilde{W_{rs}}$, then there is a (constant) J–unitary matrix U such that $\widetilde{W_r} = U W_r$ and $\widetilde{W_{rs}} = W_{rs} U^{-1}$.*

(c) *If U is an arbitrary (constant) J–unitary matrix, then $\widetilde{W_r} := U W_r$ and $\widetilde{W_{rs}} := W_{rs} U^{-1}$ are a right A–regular J–inner function and an A–singular J–inner function, respectively, such that $W = \widetilde{W_{rs}}\widetilde{W_r}$.*

3 ON j_{pq}–INNER FUNCTIONS AND GENERALIZED BITANGENTIAL SCHUR–NEVANLINNA–PICK INTERPOLATION

In this section, we will describe the image of the Schur class $\mathcal{S}_{p\times q}(\mathbb{D})$ under the linear fractional transformation generated by an A–regular j_{pq}–inner function. We will recognize that it admits a representation as the solution set of an appropriately constructed Generalized Bitangential Schur–Nevanlinna–Pick Interpolation Problem. To explain this we summarize some essential results which were proved in [A2], [A5].

Let $W \in \mathfrak{P}_{j_{pq}}(\mathbb{D})$, and let

$$W = \begin{pmatrix} W_{11} & W_{12} \\ W_{21} & W_{22} \end{pmatrix} \tag{6}$$

be the block partition of W with $p \times p$ block W_{11}. Then $\det W_{22} \neq 0$ for each $z \in \mathbb{D}$. Moreover, both functions $S_{11} := W_{11} - W_{12}W_{22}^{-1}W_{21}$ and $S_{22} := W_{22}^{-1}$ are matrix-valued Schur functions with nonidentically vanishing determinant (see, e.g., [DD1] or [A2], [A5]). In particular, S_{11} and S_{22} admit inner–outer factorizations. If $b_1 \in \mathcal{S}_{p\times p}(\mathbb{D})$ and $b_2 \in \mathcal{S}_{q\times q}(\mathbb{D})$ are inner functions such that the functions $b_1^{-1}S_{11}$ and $S_{22}b_2^{-1}$ (respectively, $S_{11}b_1^{-1}$ and $b_2^{-1}S_{22}$) are outer, then $[b_1, b_2]$ is called a *left* (respectively, *right*) *pair of inner functions associated with* W.

In [A2], [A5], the first author proved the following fact.

LEMMA 2 *Let $W_\square$ be a j_{pq}–inner function, and let $W_\blacksquare$ be an A-singular j_{pq}–inner function.*

(a) If $[b_1, b_2]$ is a left pair of inner functions associated with $W_\square$, then $[b_1, b_2]$ is also a left pair of inner functions associated with $W_\square W_\blacksquare$.

(b) If $[c_1, c_2]$ is a right pair of inner functions associated with $W_\square$, then $[c_1, c_2]$ is also a right pair of inner functions associated with $W_\blacksquare W_\square$.

Now we characterize the case that a given j_{pq}–inner function is a left or right Blaschke-Potapov product with respect to j_{pq}.

THEOREM 5 *Let W be a j_{pq}–inner function.*

(a) Let $[b_1, b_2]$ be a left pair of inner functions associated with W. Then the following statements are equivalent:

(i) W is the restriction of some left Blaschke–Potapov product B_l with respect to j_{pq} onto $\mathbb{H}(B_l) \cap \mathbb{D}$.

(ii) W is left A–regular, b_1 is the restriction of some Blaschke–Potapov product β_1 with respect to I_p onto $\mathbb{H}(\beta_1) \cap \mathbb{D}$, and b_2 is the restriction of some Blaschke-Potapov product β_2 with respect to I_q onto $\mathbb{H}(\beta_2) \cap \mathbb{D}$.

(b) Let $[c_1, c_2]$ be a right pair of inner functions associated with W. Then the following statements are equivalent:

(iii) W is the restriction of some right Blaschke–Potapov product B_r with respect to j_{pq} onto $\mathbb{H}(B_r) \cap \mathbb{D}$.

(iv) W is right A–regular, c_1 is the restriction of some Blaschke–Potapov product γ_1 with respect to I_p onto $\mathbb{H}(\gamma_1) \cap \mathbb{D}$, and c_2 is the restriction of some Blaschke-Potapov product γ_2 with respect to I_q onto $\mathbb{H}(\gamma_2) \cap \mathbb{D}$.

A proof of Theorem 5 is given in [A2], [A5].

There is a useful sufficient condition for A-regularity of j_{pq}–inner functions in terms of two distinguished blocks of the given j_{pq}–inner function. To formulate this criterion, we use the spectral norm for complex matrices: If $A \in \mathbb{C}^{p\times q}$, then let $\|A\| := \sqrt{l_1(AA^*)}$ where $l_1(AA^*)$ is the largest eigenvalue of AA^*.

LEMMA 3 *Let W be a j_{pq}–unitary matrix, and let (6) be the block partition of W with $p \times p$ block W_{11}. Then* $\det W_{22} \neq 0$ *and* $\|W_{12}W_{22}^{-1}\| = \|W_{22}^{-1}W_{21}\|$.

PROOF Obviously, $W_{22}^*W_{22} = I_q + W_{12}^*W_{12}$. This implies $\det W_{22} \neq 0$ and

$$(W_{12}W_{22}^{-1})^*W_{12}W_{22}^{-1} = I_q - (W_{22}^{-1})^*W_{22}^{-1} \ .$$

On the other hand, from the identity

$$W_{22}W_{22}^* - W_{21}W_{21}^* = I_q$$

it follows

$$W_{22}^{-1}W_{21}(W_{22}^{-1}W_{21})^* = I_q - W_{22}^{-1}(W_{22}^{-1})^* \ .$$

Thus the assertion easily follows. ■

THEOREM 6 *Let W be a j_{pq}–inner function, and let (6) be the block partition of W where W_{11} is a $p \times p$ block. The functions $S_0 := W_{12}W_{22}^{-1}$ and $T_0 := W_{22}^{-1}W_{21}$ belong to $\mathcal{S}_{p\times q}(\mathbb{D})$ and $\mathcal{S}_{q\times p}(\mathbb{D})$, respectively. Moreover,*

$$\frac{1}{2\pi}\int\limits_{\mathbb{T}} \frac{1}{1-\|\underline{S_0}(z)\|}\lambda(dz) < +\infty \ , \tag{7}$$

if and only if

$$\frac{1}{2\pi}\int\limits_{\mathbb{T}} \frac{1}{1-\|\underline{T_0}(z)\|}\lambda(dz) < +\infty \ .$$

If (7) is satisfied, then W is both left A–regular and right A–regular.

PROOF Use Theorem 11 in [A2], Lemma 3 and Remark 5. ■

Theorem 6 enables us to construct an example for a j_{pq}–inner function which shows that there are singular j_{pq}–inner functions which are also A-regular.

Let $f \in \mathcal{S}_{p\times p}(\mathbb{D})$ and $g \in \mathcal{S}_{q\times q}(\mathbb{D})$ be inner functions. Then $W := \operatorname{diag}(f, g^{-1})$ is obviously a j_{pq}–inner function. Moreover, Theorem 6 shows that W is both left A-regular and right A-regular. If f and g are even singular inner functions, then W is obviously a singular j_{pq}–inner function, i.e., W is an example for a singular j_{pq}–inner function which is also left A-regular and right A-regular. On the other hand, this function W is not A-singular (see, e.g., Duren [Du, Section 2.5]).

Now we will indicate that j_{pq}–inner functions turn out to be intimately connected to the following interpolation problem.

Generalized Bitangential Schur–Nevanlinna–Pick Problem. *Let $S_0 \in \mathcal{S}_{p\times q}(\mathbb{D})$. Further, let b_1 and b_2 be inner functions which belong to $\mathcal{S}_{p\times p}(\mathbb{D})$ and $\mathcal{S}_{q\times q}(\mathbb{D})$, respectively. Describe the set $\mathcal{F}_{S_0;b_1,b_2}$ of all $S \in \mathcal{S}_{p\times q}(\mathbb{D})$ such that $b_1^{-1}(S - S_0)b_2^{-1} \in [H^\infty(\mathbb{D})]^{p\times q}$.*

In the sequel, this problem will be designated by GSNP$[S_0; b_1, b_2]$. Now we consider a function $S_0 \in \mathcal{S}_{p\times q}(\mathbb{D})$ and inner functions $b_1 \in \mathcal{S}_{p\times p}(\mathbb{D})$ and $b_2 \in \mathcal{S}_{q\times q}(\mathbb{D})$. Let $\mathcal{Z}_{b_1,b_2} := \{z \in \mathbb{D} : \det b_1(z) \det b_2(z) \neq 0\}$. Since b_1 and b_2 are inner, the set $\mathbb{D} \setminus \mathcal{Z}_{b_1,b_2}$ is a discrete subset of $\mathbb{D}$. For all $z \in \mathcal{Z}_{b_1,b_2}$, the set

$$\mathcal{K}_{S_0;b_1,b_2}(z) := \{S(z) : S \in \mathcal{F}_{S_0;b_1,b_2}\}$$

admits a representation as a matrix ball, i.e., there are matrices $M(z) \in \mathbb{C}^{p\times q}$, $L(z) \in \mathbb{C}^{p\times p}$ and $R(z) \in \mathbb{C}^{q\times q}$ such that

$$\mathcal{K}_{S_0;b_1,b_2}(z) := \{M(z) + L(z)KR(z) : K \in \mathbb{K}_{p\times q}\}.$$

If there is a $z_0 \in \mathcal{Z}_{b_1,b_2}$ such that $\det L(z_0) \neq 0$, then $\det L(z) \det R(z) \neq 0$ for all $z \in \mathcal{Z}_{b_1,b_2}$ (see [A2], [A5]). This statement does not depend on the special representation of $\mathcal{K}_{S_0;b_1,b_2}$ as matrix ball (see Smuljan [Sm]). The Problem GSNP$[S_0; b_1, b_2]$ is called *completely indeterminate* if there exists a $z_0 \in \mathcal{Z}_{b_1,b_2}$ such that $\det L(z_0) \neq 0$.

If $W \in \mathfrak{P}_{j_{pq}}(\mathbb{D})$ is partitioned via (6) where W_{11} is a $p \times p$ block, then one can easily see from Remark 1 that, for each $g \in \mathcal{S}_{p\times q}(\mathbb{D})$, the function $\det(W_{21}g + W_{22})$ does not identically vanish and

$$\mathcal{S}_{[W]}^{(p,q)}(g) := (W_{11}g + W_{12})(W_{21}g + W_{22})^{-1}$$

belongs to $\mathcal{S}_{p\times q}(\mathbb{D})$. Analogously, for each $h \in \mathcal{S}_{q\times p}(\mathbb{D})$, the function $\det(hW_{12} + W_{22})$ does not identically vanish and

$$\mathcal{T}_{[W]}^{(p,q)}(h) := (hW_{12} + W_{22})^{-1}(hW_{11} + W_{21})$$

belongs to $\mathcal{S}_{q\times p}(\mathbb{D})$. We set

$$\mathcal{S}_{[W]}^{(p,q)}(\mathcal{S}_{p\times q}(\mathbb{D})) := \{\mathcal{S}_{[W]}^{(p,q)}(g) : g \in \mathcal{S}_{p\times q}(\mathbb{D})\}$$

and

$$\mathcal{T}_{[W]}^{(p,q)}(\mathcal{S}_{q\times p}(\mathbb{D})) := \{\mathcal{T}_{[W]}^{(p,q)}(h) : h \in \mathcal{S}_{q\times p}(\mathbb{D})\}.$$

The following two theorems which are due to the first author [A2], [A5] provide a complete description of the interrelations between generalized bitangential Schur–Nevanlinna–Pick interpolation and j_{pq}–inner functions.

THEOREM 7 *Let $S_0 \in \mathcal{S}_{p\times q}(\mathbb{D})$. Further, let b_1 and b_2 be inner functions which belong to $\mathcal{S}_{p\times p}(\mathbb{D})$ and $\mathcal{S}_{q\times q}(\mathbb{D})$, respectively. Suppose that Problem GSNP$[S_0; b_1, b_2]$ is completely indeterminate. Then:*

(a) *There is a left A-regular j_{pq}-inner function W_l such that $\mathcal{S}^{(p,q)}_{[W_l]}(\mathcal{S}_{p\times q}(\mathbb{D})) = \mathcal{F}_{S_0;b_1,b_2}$. If V_l is an arbitrary j_{pq}-inner function such that $\mathcal{S}^{(p,q)}_{[V_l]}(\mathcal{S}_{p\times q}(\mathbb{D})) = \mathcal{F}_{S_0;b_1,b_2}$ and such that $[b_1,b_2]$ is a left pair of inner functions associated with V_l, then there is a (constant) j_{pq}-unitary matrix U such that $V_l = W_l U$. In particular, V_l is then left A-regular.*

(b) *There is a right A-regular j_{qp}-inner function W_r such that $\mathcal{T}^{(q,p)}_{[W_r]}(\mathcal{S}_{p\times q}(\mathbb{D})) = \mathcal{F}_{S_0;b_1,b_2}$. If V_r is an arbitrary j_{qp}-inner function such that $\mathcal{T}^{(q,p)}_{[V_r]}(\mathcal{S}_{p\times q}(\mathbb{D})) = \mathcal{F}_{S_0;b_1,b_2}$ and such that $[b_1,b_2]$ is a right pair of inner functions associated with V_r, then there is a (constant) j_{qp}-unitary matrix Q such that $V_r = QW_r$. In particular, V_r is then right A-regular.*

Conversely, a given j_{pq}-inner function W generates naturally two types of generalized bitangential Schur-Nevanlinna-Pick problems:

THEOREM 8 *(a) Let W be a j_{pq}-inner function, and let W be partitioned into blocks via (6) where W_{11} is a $p\times p$ block. Let $S_0 := W_{12}W_{22}^{-1}$, let $[b_1,b_2]$ be a left pair of inner functions associated with W. Then Problem $GSNP[S_0;b_1,b_2]$ is completely indeterminate and*

$$\mathcal{S}^{(p,q)}_{[W]}(\mathcal{S}_{p\times q}(\mathbb{D})) \subseteq \mathcal{F}_{S_0;b_1,b_2}$$

where equality holds true if and only if W is left A-regular.

(b) *Let W be a j_{qp}-inner function. Let $T_0 := W_{22}^{-1}W_{21}$, let $[c_1,c_2]$ be a right pair of inner functions associated with W. Then Problem $GSNP[T_0;c_1,c_2]$ is completely indeterminate and*

$$\mathcal{T}^{(q,p)}_{[W]}(\mathcal{S}_{p\times q}(\mathbb{D})) \subseteq \mathcal{F}_{T_0;c_1,c_2}$$

where equality holds true if and only if W is right A-regular.

4 ON V.E. KATSNELSON'S REFINEMENT OF THE FACTORIZATION THEORY OF J-INNER FUNCTIONS

Recall that the restriction of an arbitrary left or right Blaschke-Potapov product B with respect to J onto $\mathbb{H}(B)\cap\mathbb{D}$ is a J-inner function.

REMARK 8 *Let J be an $m\times m$ signature matrix with $J \neq \pm I_m$. According to Remark 2 , let U be an $m\times m$ unitary matrix with $U^*JU = j_{pq}$. Then B is a left (respectively, right) Blaschke-Potapov product with respect to J if and only if $B_\blacksquare := U^*BU$ is a left (respectively, right) Blaschke-Potapov product with respect to j_{pq}.*

Now we are going to study the factorizations given in Theorems 3 and 4 for the particular case that the J-inner function in question is a left or right Blaschke-Potapov product with respect to J.

THEOREM 9 *Suppose that J is an $m \times m$ signature matrix with $J \neq \pm I_m$.*

(a) Let B_l be a left Blaschke-Potapov product with respect to J, and let $B_l^\square := \mathrm{Rstr.}_{\mathbf{H}(B_l)\cap\mathbb{D}} B_l$. Let W_s be an $A-$singular $J-$inner function, and let W_r be a right $A-$regular $J-$inner function such that $B_l^\square = W_s W_r$. Then there is a right Blaschke-Potapov product $\widetilde{B_r}$ with respect to J such that $W_r = \mathrm{Rstr.}_{\mathbf{H}(\widetilde{B_r})\cap\mathbb{D}} \widetilde{B_r}$.

(b) Let B_r be a right Blaschke-Potapov product with respect to J, and let $B_r^\square := \mathrm{Rstr.}_{\mathbf{H}(B_r)\cap\mathbb{D}} B_r$. Let V_s be an $A-$singular $J-$inner function, and let V_l be a left $A-$regular $J-$inner function such that $B_r^\square = V_l V_s$. Then there is a left Blaschke-Potapov product $\widetilde{B_l}$ with respect to J such that $V_l = \mathrm{Rstr.}_{\mathbf{H}(\widetilde{B_l})\cap\mathbb{D}} \widetilde{B_l}$.

PROOF (a) First we consider the case $J = j_{pq}$. Let $[b_1, b_2]$ be a left pair of inner functions associated with the $j_{pq}-$inner function $B_l^\square$. By virtue of part (a) of Theorem 5, the functions b_1 and b_2 are restrictions of Blaschke-Potapov products with respect to I_p and I_q, respectively. Hence, in view of Lemma 1, $\det b_1$ and $\det b_2$ are restrictions of some Blaschke products β_1 and β_2, respectively. Let $[c_1, c_2]$ be a right pair of inner functions associated with the $j_{pq}-$inner function W_r. From Lemma 2 we infer that $[c_1, c_2]$ is also a right pair of inner functions associated with $B_l^\square$. Let B_l be partitioned into blocks via

$$B_l^\square = \begin{pmatrix} B_{11}^\square & B_{12}^\square \\ B_{21}^\square & B_{22}^\square \end{pmatrix}$$

where $B_{11}^\square$ is a $p \times p$ block. Then $\Phi_2 := (B_{22}^\square)^{-1} b_2^{-1}$ and $\Psi_2 := c_2^{-1} (B_{22}^\square)^{-1}$ are outer functions, i.e., $\det \Phi_2$ and $\det \Psi_2$ are outer functions. Since b_2 and c_2 are inner functions in $\mathcal{S}_{q\times q}(\mathbb{D})$, part (b) of Lemma 1 shows that $\det b_2$ and $\det c_2$ are inner functions in $\mathcal{S}_{1\times 1}(\mathbb{D})$. In view of $(B_{22}^\square)^{-1} = \Phi_2 b_2$ and $(B_{22}^\square)^{-1} = c_2 \Psi_2$, it follows that

$$\det((B_{22}^\square)^{-1}) = \det b_2 \cdot \det \Phi_2 \quad \text{and} \quad \det((B_{22}^\square)^{-1}) = \det c_2 \cdot \det \Psi_2$$

are inner-outer factorizations of the Schur function $\det((B_{22}^\square)^{-1})$. Hence, there is a $\xi \in \mathbb{T}$ such that $\det c_2 = \xi \cdot \det b_2$. Since $\det b_2$ is the restriction of some Blaschke product, we thus see that $\det c_2$ is the restriction of some Blaschke product as well. Consequently, using part (d) of Lemma 1, we obtain that c_2 is the restriction of some Blaschke-Potapov product with respect to I_q. Similarly, considering the $p \times p$ Schur function $B_{11}^\square - B_{12}^\square (B_{22}^\square)^{-1} B_{21}^\square$ instead of the $q \times q$ Schur function $(B_{22}^\square)^{-1}$ one can see that c_1 is the restriction of some Blaschke-Potapov product with respect to I_p. Since W_r is a right A-regular $j_{pq}-$inner function, part (b) of Theorem 5 yields that W_r is the restriction of some right Blaschke-Potapov product with respect to j_{pq}. Thus, the theorem is proved for $J = j_{pq}$. Applying Remark 8, we immediately get the assertion in the general case that J is an arbitrary $m \times m$ signature matrix with $J \neq \pm I_m$.
(b) Because of Remark 4, the function $\check{B}_r$ is a left Blaschke-Potapov product with respect to J. In view of Remark 4 and parts (c)–(e) of Remark 5, then we obtain the assertion stated in part (b) immediately from part (a). ■

The following theorem which is due to V.E. Katsnelson [Ka1], [Ka2], [Ka3], [Ka4] is a refinement of Potapov's Factorization Theorems 1 and 2 for the special case that the

$J-$inner function W is a left or right Blaschke–Potapov product. The proof of V.E. Katsnelson is based on growth estimates for meromorphic matrix-valued functions. We will present a completely different approach to Katsnelson's result. It is based on Theorem 9, the proof of which essentially uses Theorem 5 that has its origin in generalized bitangential Schur–Nevanlinna–Pick interpolation.

THEOREM 10 *Suppose that J is an $m \times m$ signature matrix with $J \neq \pm I_m$.*

(a) Let B_l be a left Blaschke-Potapov product with respect to J, and let $B_l^{\square} := \mathrm{Rstr.}_{\mathbb{H}(B_l)\cap\mathbb{D}} B_l$. If Σ_{rs} is a function that belongs to $\widetilde{\mathfrak{P}}_J(\mathbb{D})$, and if $\widetilde{B_r}$ is a right Blaschke-Potapov product with respect to J such that

$$B_l^{\square} = \Sigma_{rs} \widetilde{B_r^{\square}} \tag{8}$$

where $\widetilde{B_r^{\square}} := \mathrm{Rstr.}_{\mathbb{H}(B_r)\cap\mathbb{D}} \widetilde{B_r}$, then Σ_{rs} is an A-singular $J-$inner function.

(b) Let B_r be a right Blaschke-Potapov product with respect to J, and let $B_r^{\square} := \mathrm{Rstr.}_{\mathbb{H}(B_r)\cap\mathbb{D}} B_r$. If Σ_{ls} is a function that belongs to $\widetilde{\mathfrak{P}}_J(\mathbb{D})$, and if $\widetilde{B_l}$ is a left Blaschke-Potapov product with respect to J such that

$$B_r^{\square} = \widetilde{B_l^{\square}} \Sigma_{ls}$$

where $\widetilde{B_l^{\square}} := \mathrm{Rstr.}_{\mathbb{H}(B_l)\cap\mathbb{D}} \widetilde{B_l}$, then Σ_{ls} is an A-singular $J-$inner function.

PROOF (a) Since $B_l^{\square}$ belongs to $\mathfrak{P}_J(\mathbb{D})$, we see from Theorem 2 that there are a function $\Sigma_{rs} \in \widetilde{\mathfrak{P}}_J(\mathbb{D})$ and a right Blaschke-Potapov product $\widetilde{B_r}$ with respect to J such that (8) holds. By virtue of Theorem 4, there is a right A-regular $J-$inner function W_r and an A-singular $J-$inner function W_{rs} such that

$$B_l^{\square} = W_{rs} W_r. \tag{9}$$

Using part (a) of Theorem 9, we obtain that $W_r = B_r^{\blacksquare}$, where $B_r^{\blacksquare}$ is the restriction of some right Blaschke-Potapov product $\widehat{B_r}$ onto $\mathbb{H}(\widehat{B_r}) \cap \mathbb{D}$. Hence, comparing (8) and (9) we infer from part (b) of Theorem 2 that there is a (constant) $J-$unitary matrix U such that $\widetilde{B_r} = U\widehat{B_r}$ and $\Sigma_{rs} = W_{rs}U^{-1}$. Since W_{rs} is A-singular, it follows immediately that Σ_{rs} is also A-singular.
(b) Use part (a), Remark 5 and Remark 4. ∎

5 AN INVERSE PROBLEM FOR A-SINGULAR $j_{pq}-$INNER FUNCTIONS

Let E be an A-singular $j_{pq}-$inner function. Then we want to look for an interpolation problem the solution set of which coincides exactly with $\mathcal{S}_{[E]}^{(p,q)}(\mathcal{S}_{p\times q}(\mathbb{D}))$. Since an A-singular $j_{pq}-$inner function has neither zeros nor singularities in the unit disc $\mathbb{D}$, such an interpolation problem will be a boundary interpolation problem. Interpolation problems of this type occur first in Nevanlinna's fundamental paper [Ne1]. For the case of a given A–singular j_{pq}–inner function which is a restriction of some rational matrix-valued function with exactly one pole which is located on the unit circle $\mathbb{T}$, I. V. Kovalishina [Ko2], [Ko3] constructed explicitly a boundary Nevanlinna–Pick interpolation problem which has the desired property. Her approach is based on V. P. Potapov's method of the so-called Fundamental Matrix Inequality. One way to handle wider subclasses of A–singular j_{pq}–inner functions could be based on a careful study of the boundary behaviour of matrix-valued functions which belong to the classes of Schur, Carathéodory or Potapov. Investigations on this topic can be found in [BGR], [C], [DD2], [Dy], [Ko1], [Ko3], [M1], [M2], [A6] and [A7].

In the following we will concentrate on the case $p = q = 1$. Our way to represent the set $\mathcal{S}_{[E]}^{(1,1)}(\mathcal{S}_{1\times 1}(\mathbb{D}))$ as solution set of an appropriate interpolation problem is via approximation along an A–regular path. In the heart of our construction lies a deep result of V.E. Katsnelson on interrelations between A–singular J–inner functions and Blaschke-Potapov products with respect to J. To state V.E. Katsnelson's theorem, we need a little preparation.

Let $(z_k)_{k\in\mathbb{N}}$ be a sequence of pairwise different points from $\mathbb{D}$, and let $(\Sigma_k)_{k\in\mathbb{N}}$ be a sequence of points which belong to $\mathbb{D}\cup\mathbb{T}$. Then the Schur-Nevanlinna-Pick Problem $(SNP(z_k, \Sigma_k)_{k\in\mathbb{N}})$ consists of describing the set $\mathcal{G}\left[(z_k, \Sigma_k)_{k\in\mathbb{N}}\right]$ of all functions $f \in \mathcal{S}_{1\times 1}(\mathbb{D})$ satisfying $f(z_k) = \Sigma_k$ for all $k \in \mathbb{N}$. Now we are able to formulate V.E. Katsnelson's [Ka3] result.

THEOREM 11 *Let E be an A–singular j_{11}–inner function. Then there is a sequence $(B_{r,k})_{k\in\mathbb{N}}$ of Blaschke-Potapov j_{11}–elementary factors of first kind with the following properties:*

(i) The product $\overleftarrow{\prod_{k\in\mathbb{N}}} B_{r,k}(z)$ converges for all $z \in \mathbb{D}$.

(ii) If $B_r := \overleftarrow{\prod_{k\in\mathbb{N}}} B_{r,k}$, then $B_l := EB_r$ is a (convergent) infinite left Blaschke-Potapov j_{11}–product of first kind.

(iii) There are a sequence $(z_k)_{k\in\mathbb{N}}$ of pairwise different points from $\mathbb{D}$ and a sequence $(\Sigma_k)_{k\in\mathbb{N}}$ of points from $\mathbb{D}$ such that, for each $m \in \mathbb{N}$, the function

$$B_m^{\blacksquare} := E \cdot \overleftarrow{\prod_{k\in\mathbb{N}}} B_{r,k+m-1} \tag{10}$$

satisfies

$$\mathcal{S}_{[B_m^{\blacksquare}]}^{(1,1)}(\mathcal{S}_{1\times 1}(\mathbb{D})) = \mathcal{G}\left[(z_{k+m-1}, \Sigma_{k+m-1})_{k\in\mathbb{N}}\right]. \tag{11}$$

Observe that the proof of Theorem 11 given in [Ka3] is a constructive one. The most difficult part of this construction consists of the concrete choice of the sequence $(z_k)_{k\in\mathbb{N}}$. V.E. Katsnelson obtained this sequence by considering an appropriate weighted approximation problem for pseudocontinuable functions. Moreover, he used essentially V. P. Potapov's results on Nevanlinna–Pick interpolation and splitting-off of Blaschke-Potapov J–elementary factors, which he reconsidered, in some sense, as a backward Schur type algorithm.

DEFINITION 1 *Let $(z_k)_{k\in\mathbb{N}}$ be a sequence of pairwise different points that belong to $\mathbb{D}$, and let $(\Sigma_k)_{k\in\mathbb{N}}$ be a sequence from $\mathbb{D}$.*

(a) A Schur function $S \in \mathcal{S}_{1\times 1}(\mathbb{D})$ is called a primitive singular element associated with $(z_k, \Sigma_k)_{k\in\mathbb{N}}$ if there is an $m \in \mathbb{N}$ such that S belongs to $\mathcal{G}\left[(z_{k+m-1}, \Sigma_{k+m-1})_{k\in\mathbb{N}}\right]$.

(b) A Schur function $S \in \mathcal{S}_{1\times 1}(\mathbb{D})$ is said to be a singular element associated with $(z_k, \Sigma_k)_{k\in\mathbb{N}}$ if there is a sequence $(S_n)_{n\in\mathbb{N}}$ of primitive singular elements associated with $(z_k, \Sigma_k)_{k\in\mathbb{N}}$ such that

$$\lim_{n\to\infty} S_n(z) = S(z)$$

for all $z \in \mathbb{D}$.

In the following, we will use $\sigma\left[(z_k, \Sigma_k)_{k\in\mathbb{N}}\right]$ to denote the set of all singular elements associated with $(z_k, \Sigma_k)_{k\in\mathbb{N}}$.

REMARK 9 *Let $(f_n)_{n\in\mathbb{N}}$ be a sequence from $\mathcal{S}_{1\times 1}(\mathbb{D})$. Then, in view of a theorem due to Montel (see, e.g., Burckel [B, p. 220]), there are a Schur function $f \in \mathcal{S}_{1\times 1}(\mathbb{D})$ and a subsequence $(f_{n_k})_{k\in\mathbb{N}}$ of $(f_n)_{n\in\mathbb{N}}$ such that*

$$\lim_{k\to\infty} f_{n_k}(z) = f(z)$$

for all $z \in \mathbb{D}$. This convergence is uniform for all compact subsets of $\mathbb{D}$.

Now we are able to formulate the main result of this section.

THEOREM 12 *Let E be an A–singular j_{11}–inner function. Let $(B_{r,k})_{k\in\mathbb{N}}$ be a sequence of Blaschke-Potapov j_{11}–elementary factors of first kind, and let $(z_k)_{k\in\mathbb{N}}$ be a sequence of pairwise different points from $\mathbb{D}$, and let $(\Sigma_k)_{k\in\mathbb{N}}$ be a sequence of points from $\mathbb{D}$ which are associated with E via Theorem 11. Then*

$$\sigma\left[(z_k, \Sigma_k)_{k\in\mathbb{N}}\right] = \mathcal{S}_{[E]}^{(1,1)}\left(\mathcal{S}_{1\times 1}(\mathbb{D})\right). \tag{12}$$

PROOF Assume $S \in \sigma\left[(z_k, \Sigma_k)_{k\in\mathbb{N}}\right]$. First we consider the case that S is a primitive singular element associated with $(z_k, \Sigma_k)_{k\in\mathbb{N}}$. Then there is an $m \in \mathbb{N}$ such that S belongs to $\mathcal{G}\left[(z_{k+m-1}, \Sigma_{k+m-1})_{k\in\mathbb{N}}\right]$. Because of (11), we have

$$S \in \mathcal{S}_{[B_m^\blacksquare]}^{(1,1)}\left(\mathcal{S}_{1\times 1}(\mathbb{D})\right).$$

Using Remark 1 and (10), we obtain then

$$S \in \mathcal{S}_{[E]}^{(1,1)}\left(\mathcal{S}_{1\times 1}(\mathbb{D})\right). \tag{13}$$

Now assume that S is an arbitrary singular element associated with $(z_k, \Sigma_k)_{k\in\mathbb{N}}$. Then there is a sequence $(S_n)_{n\in\mathbb{N}}$ of primitive singular elements associated with $(z_k, \Sigma_k)_{k\in\mathbb{N}}$ such that

$$\lim_{n\to\infty} S_n(z) = S(z) \tag{14}$$

for all $z \in \mathbb{D}$. In view of the considerations above, then, for all $n \in \mathbb{N}$,

$$S_n = \mathcal{S}_{[E]}^{(1,1)}(f_n) \tag{15}$$

with some function f_n belonging to $\mathcal{S}_{1\times 1}(\mathbb{D})$. By virtue of Remark 9, there are a function $f \in \mathcal{S}_{1\times 1}(\mathbb{D})$ and a subsequence $(f_{n_k})_{k\in\mathbb{N}}$ of $(f_n)_{n\in\mathbb{N}}$ such that

$$\lim_{k\to\infty} f_{n_k}(z) = f(z) \tag{16}$$

for all $z \in \mathbb{D}$. If

$$E = \begin{pmatrix} E_{11} & E_{12} \\ E_{21} & E_{22} \end{pmatrix}, \tag{17}$$

then we see from (14), (15) and (16) that

$$\begin{aligned} S(z) &= \lim_{k\to\infty} \left(E_{11}(z) f_{n_k}(z) + E_{12}(z)\right)\left(E_{21}(z) f_{n_k}(z) + E_{22}(z)\right)^{-1} \\ &= \left(E_{11}(z) f(z) + E_{12}(z)\right)\left(E_{21}(z) f(z) + E_{22}(z)\right)^{-1} \end{aligned}$$

for all $z \in \mathbb{D}$, i.e.,

$$S = \mathcal{S}_{[E]}^{(1,1)}(f). \tag{18}$$

Hence,

$$\sigma\left[(z_k, \Sigma_k)_{k\in\mathbb{N}}\right] \subseteq \mathcal{S}_{[E]}^{(1,1)}\left(\mathcal{S}_{1\times 1}(\mathbb{D})\right). \tag{19}$$

Conversely, now let $S \in \mathcal{S}_{[E]}^{(1,1)}\left(\mathcal{S}_{1\times 1}(\mathbb{D})\right)$ be given. Then there is an $f \in \mathcal{S}_{1\times 1}(\mathbb{D})$ such that (18) holds. Because of Theorem 11, the infinite right Blaschke-Potapov products of first kind

$$B_{r,n}^{\square} := \overleftarrow{\prod_{k\in\mathbb{N}}} B_{r,k+n-1}, \quad n \in \mathbb{N},$$

converge. In particular, the functions $B_{r,n}^{\blacksquare} := E \cdot B_{r,n}^{\square}$, $n \in \mathbb{N}$, are j_{11}−inner. For all $n \in \mathbb{N}$, the function

$$S_n := \mathcal{S}_{[B_{r,n}^{\blacksquare}]}^{(1,1)}(f) \tag{20}$$

is a (well-defined) 1×1 Schur function. From Theorem 11 we see that, for all $n \in \mathbb{N}$, the function S_n belongs to $\mathcal{G}\left[(z_{k+n-1}, \Sigma_{k+n-1})_{k\in\mathbb{N}}\right]$. Since $B_{r,1}^{\square}$ is a j_{11}−inner function, the function $\det B_{r,1}^{\square}$ does not identically vanish in $\mathbb{D}$. Because the infinite product

$$\overleftarrow{\prod_{k\in\mathbb{N}}} B_{r,k}$$

converges, there is a discrete subset $\mathfrak{N}$ of $\mathbb{D}$ such that

$$\lim_{n\to\infty} B_{r,n}^{\square}(z) = I_2 \tag{21}$$

for all $z \in \mathbb{D}\setminus\mathfrak{N}$. Thus, using (18), (20) and (21) it follows

$$\begin{aligned} S(z) &= \mathcal{S}_{E(z)}^{(1,1)}(f(z)) = \mathcal{S}_{E(z)}^{(1,1)}\left(\mathcal{S}_{I_2}^{(1,1)}(f(z))\right) \\ &= \lim_{n\to\infty} \mathcal{S}_{E(z)}^{(1,1)}\left(\mathcal{S}_{B_{r,n}^{\square}(z)}^{(1,1)}(f(z))\right) = \lim_{n\to\infty} \mathcal{S}_{B_{r,n}^{\blacksquare}(z)}^{(1,1)}(f(z)) = \lim_{n\to\infty} S_n(z) \end{aligned} \tag{22}$$

for all $z \in \mathbb{D}\setminus\mathfrak{N}$. On the other hand, we know from Remark 9 that there are a Schur function $\widetilde{S} \in \mathcal{S}_{1\times 1}(\mathbb{D})$ and a subsequence $(S_{n_k})_{k\in\mathbb{N}}$ of $(S_n)_{n\in\mathbb{N}}$ such that

$$\lim_{k\to\infty} S_{n_k}(z) = \widetilde{S}(z) \tag{23}$$

for all $z \in \mathbb{D}$. Comparing (22) and (23), we get from the Identity Theorem for holomorphic functions that $S = \widetilde{S}$, i.e., the sequence $(S_{n_k})_{k\in\mathbb{N}}$ of primitive singular elements associated with $(z_k, \Sigma_k)_{k\in\mathbb{N}}$ converges pointwise to S. Therefore, the function S is a singular element associated with $(z_k, \Sigma_k)_{k\in\mathbb{N}}$. Hence

$$\mathcal{S}^{(1,1)}_{[E]}(\mathcal{S}_{1\times 1}(\mathbb{D})) \subseteq \sigma\left[(z_k, \Sigma_k)_{k\in\mathbb{N}}\right]. \tag{24}$$

From (19) and (24) it follows finally (12). ∎

The description of all sequences $(z_k)_{k\in\mathbb{N}}$ of pairwise different points from $\mathbb{D}$ which can be taken in Theorem 12 seems to be very difficult. The construction of V. E. Katsnelson in [Ka3] provides only sufficient conditions.

References

[A1] Arov, D.Z.: On the boundary values of convergent sequences of meromorphic matrix–valued functions (Russian), Mat. Zametki 25 (1979), 335-339.

[A2] Arov, D.Z.: γ-generating matrices, J–inner matrix–functions and related extrapolation problems (Russian), deposited in Ukr. NIINTI, no. 726-Uk. 86, 1986.

[A3] Arov, D.Z.: Regular J–inner matrix–functions and related continuation problems (Russian), deposited in Ukr. NIINTI, no. 406-Uk. 87, 1987.

[A4] Arov, D.Z.: On regular and singular J–inner matrix–functions and related extrapolation problems (Russian), Functional Analysis Appl. 22 (1988), 57-59.

[A5] Arov, D.Z.: γ-generating matrices, J–inner matrix–functions and related extrapolation problems (Russian), Theory of Functions, Functional Analysis and their Applications (Kharkov), Part I: 51 (1989), 61-67; Part II: 52 (1989), 103-109; Part III: 53 (1990), 57-64.

[A6] Arov, D.Z.: Regular J–inner matrix–functions and related continuation problems, in: Linear Operators in Function Spaces (G. Arsene et al., eds.), Operator Theory: Advances and Applications, Vol. 43, Birkhäuser, Basel 1990, p. 63-87.

[A7] Arov, D.Z.: A theorem of Carathéodory for matrix-valued functions and the maximal jump of spectral functions in extension problems (Russian), Mat. Zametki 48 (1990), Issue 3, 3-11.

[AS] Arov, D.Z. and Simakova, L.A.: On the boundary values of convergent sequences of J–contractive matrix–valued functions (Russian), Mat. Zametki 19 (1976), 491-499.

[BGR] Ball, J., Gohberg, I. and Rodman, L.: Interpolation of Rational Matrix Functions, Operator Theory: Advances and Applications, Vol. 45, Birkhäuser, Basel 1990.

[B] Burckel, R.B.: An Introduction to Complex Analysis, Volume I, Birkhäuser, Basel 1979.

[C] Carathéodory, C.: Über die Winkelderivierten von beschränkten Funktionen, Sitzungsber. Preuß. Akad. Wiss. 1929, 39-52.

[DD1] Dewilde, P. and Dym, H.: Lossless chain scattering matrices and optimum linear prediction, Circuit Theory and Applications 9 (1981), 135-175.

[DD2] Dewilde, P. and Dym, H.: Lossless inverse scattering for digital filters, IEEE Trans. Inf. Theory 30 (1984), 644-662.

[DFK] Dubovoj, V.K., Fritzsche, B. and Kirstein, B.: Matricial Version of the Classical Schur Problem, Teubner-Texte zur Mathematik, Band 129, B.G. Teubner Stuttgart-Leipzig 1992.

[Du] Duren, P.: Theory of H^p Spaces, Academic Press, New York 1970.

[Dy] Dym, H.: J Contractive Matrix Functions, Reproducing Kernel Hilbert Spaces and Interpolation, CBMS Lecture Notes, No. 71, Amer. Math. Soc., Providence, R.I. 1989.

[G] Ginzburg, J.P.: On multiplicative representations of J–contractive matrix–valued functions (Russian), Mat. Issled. (Kishinev), part I: 2, 2 (1967), 52-83; part II: 2, 3 (1967), 20-51.

[Ka1] Katsnelson, V.E.: A left Blaschke–Potapov product is not necessarily a right Blaschke–Potapov product (Russian), Dokl. Akad. Nauk Ukrainian SSR, Series A, 10 (1989), 15-17.

[Ka2] Katsnelson, V.E.: Left and right Blaschke–Potapov products and Arov–singular matrix–valued functions, Integral Equations and Operator Theory 13 (1990), 836-848.

[Ka3] Katsnelson, V.E.: Weighted spaces of pseudocontinuable functions and approximation by rational functions with prescribed poles, Zeitschr. für Analysis und ihre Anwendungen 12 (1993), 27-67.

[Ka4] Katsnelson, V.E.: Left and right Blaschke–Potapov products and Arov-singular J–inner functions, to appear.

[Ko1] Kovalishina, I.V.: The Carathéodory–Julia theorem for matrix–valued functions (Russian), Theory of Functions, Functional Analysis and their Applications (Kharkov) 43 (1985), 70-82.

[Ko2] Kovalishina, I.V.: The multiple Nevanlinna–Pick boundary interpolation problem for matrix–valued functions which are contractive in the unit circle (Russian), Deposited in VINITI, No. 95–B86 (1986).

[Ko3] Kovalishina, I.V.: The theory of j–elementary factors with a multiple pole on the unit circle (Russian), Theory of Functions, Functional Analysis and their Applications (Kharkov) 50 (1988), 62-74.

[M1] Melamud, E.J.: The boundary Nevanlinna–Pick problem for J–contractive matrix–valued functions (Russian), Izvestija Vuzov, Series Mathematics, Issue 6 (1984), 36-43.

[M2] Melamud, E.J.: A theorem of Carathédory and a Nevanlinna–Pick boundary interpolation problem for J–contractive matrix–valued functions (Russian), Dokl. Akad. Nauk Armj. SSR, Series Mathematics 80 (1985), 12-16.

[Ne1] Nevanlinna, R.: Über beschränkte analytische Funktionen, Ann. Acad. Sci. Fenn. A 32 (1929), 1-75.

[Ne2] Nevanlinna, R.: Eindeutige analytische Funktionen, Springer, Berlin 1953.

[P1] Potapov, V.P.: The multiplicative structure of J–contractive matrix functions (Russian), Trudy Moskov. Mat. Obžč. 4 (1955), 125-236. English transl. in Amer. Math. Soc. Transl. (2), Vol. 15 (1960), 131-243.

[P2] Potapov, V.P.: Linear fractional transformations of matrices (Russian), in: Studies in the Theory of Operators and Their Applications (V.A. Marcenko, ed.), Naukova Dumka, Kiev (1975), 75-97. English transl. in Amer. Math. Soc. Transl. (2), Vol. 138 (1988), 21-35.

[Sa] Sarason, D.: Angular derivatives via Hilbert space, Complex Variables 10 (1988), 1-10.

[Šm] Šmuljan, J.L.: Operator balls (Russian), Theory of Functions, Functional Analysis and their Applications (Kharkov) 6 (1968), 68-81; English translation in: Integral Equations and Operator Theory 13 (1990), 864-882.

[WM] Wiener, N. and Masani, P.R.: The prediction theory of multivariate stochastic processes, Acta Math., Part I: 98 (1957), 111-150, Part II: 99 (1958), 93-137.

D.Z. Arov
Department of Mathematics
State Pedagogical Institute "K.D. Ushinskii"
270 020 Odessa
Ukraine

B. Fritzsche, B. Kirstein
Fakultät für Mathematik und Informatik
Universität Leipzig
Augustusplatz 10
04109 Leipzig
Germany

MSC: 47A57

Operator Theory:
Advances and Applications, Vol. 80

On Some Development of the S. Krein Pencil Theory

T.Ya. Azizov and L.I. Sukhocheva

1. Let H be an infinite dimensional Hilbert space with inner product (x, y) and let A, B, C be bounded selfadjoint operators on H. An operator function L:

$$L(\lambda) = \lambda^2 A + \lambda B + C \tag{1}$$

is called a selfadjoint operator pencil. The point $\lambda \in C$ is said to be a regular point of L ($\lambda \in \rho(L)$) if $0 \in \rho(L(\lambda))$. Analogously $\lambda \in C$ is a point of the spectrum of L (an eigenvalue of L, respectively) if $0 \in \sigma(L(\lambda))$ ($0 \in \sigma_p(L(\lambda))$, respectively). The point infinity is said to belong to $\sigma(L)$: $\infty \in \sigma(L)$ ($\infty \in \sigma_p(L)$, respectively) if $0 \in \sigma(L_1)$ ($0 \in \sigma_p(L_1)$, respectively) where $L_1(\lambda) = A + \lambda B + \lambda^2 C$.

A vector $x_0 \neq 0$ is an eigenvector of L if there exists a $\lambda_0 \in \sigma_p(L)$ such that $L(\lambda_0)x_0 = 0$ if $\lambda_0 \neq \infty$ and $Ax_0 = 0$ if $\lambda_0 = \infty$. The set $\{x_0,\ x_1, \dots,\ x_p\}$ is called a Jordan chain of L if

$$\begin{aligned}&(2\lambda_0 A + B)x_0 + L(\lambda_0)x_1 = 0,\\ &Ax_{i-2} + (2\lambda_0 A + B)x_{i-1} + L(\lambda_0)x_i = 0, \quad (i = 2,\ 3, \dots, p)\end{aligned}$$

for $\lambda_0 \neq \infty$ and

$$Bx_0 + Ax_1 = 0,\ Cx_{i-2} + Bx_{i-1} + Ax_i = 0 \quad (i = 2,\ 3, \dots, p)$$

for $\lambda_0 = \infty$. The set of all Jordan chains is said to be doubly complete in H if the system

$$\begin{aligned}&x_0 \oplus \lambda_0 x_0,\ x_1 \oplus (\lambda_0 x_1 + x_0), \dots, x_p \oplus (\lambda_0 x_p + x_{p-1}) \quad (\lambda_0 \in \sigma_p(L))\\ &0 \oplus x_0,\ x_0 \oplus x_1, \dots, x_{p-1} \oplus x_p \quad (\infty \in \sigma_p(L))\end{aligned} \tag{2}$$

is complete in the space $\tilde{H} = H \oplus H$, and this set is said to be a double basis of H if there exists a basis in $\tilde{H}$ formed by vectors of the system (2).

We shall say that an operator $\Phi :\ \tilde{H} \to \tilde{H}$ reflects the spectral properties of the pencil L (in short: Φ is a linearization of L) if the following holds:

1) There exists an injective function f, which is analytic on $\sigma(L)$ and such that $f(\sigma_p(L)) = \sigma_p(\Phi)$, $f(\sigma(L)) = \sigma(\Phi)$;

2) In $\tilde{H}$ there exists a bounded and boundedly invertible operator which maps the system (2) onto a Jordan chain system of the operator Φ.

We study the pencil (1) with a compact operator A ($A \in \mathcal{S}_\infty$) and $B = B_1 + B_2$, where $\pm B_1$ is a uniformly positive operator and $B_2 \in \mathcal{S}_\infty$. Such pencils with a positive operator A ($A > 0$), $B = I$ and $C \geq 0$, $C \in \mathcal{S}_\infty$ were first studied by S. Krein (for the references see, for example, [1], [2]). We shall say that a pencil L is a modified selfadjoint S. Krein pencil (m.s. S.Krein p.), if $A \in \mathcal{S}_\infty$, $B = B_1 + B_2$, $\pm B_1 \gg 0$, $B_2 \in \mathcal{S}_\infty$.

In this paper it is proved that every m.s. S.Krein p. L with $\rho(L) \neq \emptyset$ has a linearization Φ which is a selfadjoint operator in some Pontryagin space Π_κ (with κ negative squares, see [1], p.64) and an estimate for κ is obtained. We mention that the ideas of the transition from a pencil L to an operator Φ can be found also in [3] (see also [4]).

2. It is known that not each pair A, B of selfadjoint matrices can be simultaneously reduced to diagonal form. This problem has a solution if, for example, the matrix B is nondegenerate and the following equivalent conditions are fulfilled:

a) There exists a positive matrix S such that the matrices $S^{-1}A$ and $S^{-1}B$ commute.

b) The matrix $B^{-1}A$ is similar to a selfadjoint one.

In particular, if A is a positive matrix these conditions are fulfilled. Consider a generalization of this result to an infinite dimensional Hilbert space H.

Let $B = B^*$ be a bounded and boundedly invertible operator, $[x, y] = (Bx, y)$. Then the space $\{H, [\cdot, \cdot]\}$ is a Krein space [1].

Theorem 1. *Let A, B be bounded selfadjoint operators in the Hilbert space H, $0 \in \rho(B)$. Then the following conditions are equivalent:*

a) The operator $B^{-1}A$ is similar to a selfadjoint operator;

b) the operator $B^{-1}A$ has a maximal uniformly B-positive and a maximal uniformly B-negative invariant subspace N_+ and N_-, such that N_+ is B-orthogonal to N_- and

$$H = N_+ \, [+] \, N_-; \tag{3}$$

c) there exists a uniformly positive, bounded operator S such that $S^{-1}A$ and $S^{-1}B$ commute.

If these conditions are satisfied S can be choosen as $S = BJ$ with $J = P_+ - P_-$, where $P_\pm$ is the B-orthogonal projection onto $N_\pm$.

Proof. The equivalence of the conditions a) and b) follows immediately from the well-known theorem of Phillips [5] (see [1], Corollary II.5.20).

b) $\to$ c). Consider the new Hilbert inner product $(x, y)_1 = [x_+, y_+] - [x_-, y_-]$, where $x = x_+ + x_-$, $y = y_+ + y_-$, $x_\pm$, $y_\pm \in N_\pm$. Then there exists a bounded and uniformly positive operator S such that $(x, y)_1 = (Sx, y)$. Since $(BJx, y) = [Jx, y] = (J^2x, y)_1 = (x, y)_1 = (Sx, y)$ we have $S = BJ$. The operator $B^{-1}A$ is S-selfadjoint and it commutes with J. Therefore $S^{-1}B \cdot S^{-1}A = S^{-1}B \cdot S^{-1}B \cdot B^{-1}A = S^{-1}B \cdot B^{-1}A \cdot S^{-1}B = S^{-1}A \cdot S^{-1}B$. From this it follows that

$$S^{-1}AN_\pm \subset N_\pm. \tag{4}$$

c) → a). The operator $B^{-1}A$ is S-selfadjoint and therefore the operator $S^{1/2}(B^{-1}A)S^{-1/2}$ is selfadjoint and similar to $B^{-1}A$.

We say that operators A and B can be simultaneously reduced to diagonal form if the equivalent conditions a) - c) are fulfilled.

Corollary 1. Let A be a bounded positive operator, let $B = B_1 + B_2$ be a bounded selfadjoint operator with $0 \in \rho(B)$, $B_1 \gg 0$ and $B_2 \in \mathcal{S}_\infty$. Then the operators A and B can be simultaneously reduced to diagonal form.

In fact, the space $\{H, [\cdot,\cdot]\}$ with $[x,y] = (Bx,y)$ is a Pontryagin space and $B^{-1}A$ is a B-positive operator. From the theorem of Langer [6] it follows the existence of invariant subspaces $N_\pm$ and of the decomposition (4).

3. Let L be a m.s. S.Krein p. with $\rho(L) \neq \emptyset$: $L(\lambda) = (\lambda^2 A + \lambda B_2) + (\lambda B_1 + C)$. Let $B_1 \gg 0$. For $|\lambda|$ large enough we have $\lambda B_1 + C \gg 0$ if $\lambda > 0$ and $\lambda B_1 + C \ll 0$ if $\lambda < 0$. Since $\lambda^2 A + \lambda B_2 \in \mathcal{S}_\infty$, the accumulation points of $\sigma(L(\lambda))$ are positive in the first case and negative in the second case.

Let Λ_- (Λ_+, respectively) be the set of all real regular points λ of L for which $\sigma(L(\lambda))$ has only positive (negative, respectively) accumulation points. Let $\kappa_-(\lambda)$ ($\kappa_+(\lambda)$, respectively) be the number of the negative (positive, respectively) eigenvalues (counting multiplicities) of the operator $L(\lambda)$. We denote

$$\begin{aligned}\kappa_-(L) &= \min\{\kappa_-(\lambda) : \ \lambda \in \Lambda_-\},\\ \kappa_+(L) &= \min\{\kappa_+(\lambda) : \ \lambda \in \Lambda_+\}.\end{aligned}$$

The functions $\kappa_+(\lambda)$ and $\kappa_-(\lambda)$ are discrete and $\Lambda_\pm \subset \rho(L)$. Therefore there are numbers a and b such that $a \neq b$, $\kappa_-(L) = \kappa_-(a)$ and $\kappa_+(L) = \kappa_+(b)$.

Theorem 2. *Let A, B and C be bounded selfadjoint operators in H and let $A \in \mathcal{S}_\infty$, $B = B_1 + B_2$ with $B_1 \gg 0$ and $B_2 \in \mathcal{S}_\infty$. Then the m.s. S.Krein p. L has a linearization Φ in a Pontryagin space Π_κ with $\kappa = \kappa_+(L) + \kappa_-(L)$.*

First we shall prove a Lemma. Let L be a selfadjoint pencil, $a, b \in \rho(L) \cap R$ and $a \neq b$. Let

$$\lambda = (a + b\mu)/(\mu + 1) \quad (\mu = (a - \lambda)/(\lambda - b)). \tag{5}$$

Consider the selfadjoint pencil

$$M(\mu) = \mu^2 D + \mu F + G \tag{6}$$

with $D = L(b)$, $F = 2abA + (a+b)B + 2C$, $G = L(a)$.

The eigenvalues of the pencils L and M are connected by (5) and the corresponding Jordan chains $\{x_0, x_1, \ldots, x_n\}$ and $\{y_0, y_1, \ldots, y_n\}$ by the relations

$$\begin{aligned} y_k &= \frac{(-1)^k}{(\mu+1)^k} \sum_{j=0}^{k} \frac{(-1)^j C k^j (b-a)^j x_j}{(\mu+1)^j} \quad \text{if } \lambda \neq \infty \quad (\mu \neq -1),\\ y_k &= \frac{(-1)^k b^k}{(a-b)^k} \sum_{j=0}^{k} \frac{(-1)^j C k^j x_j}{(b)^j} \quad \text{if } \lambda = \infty \quad (\mu = -1),\end{aligned} \tag{7}$$

$k = 0, 1, \ldots, n$.

Let $E(L)$ denote the closed linear span of the vectors (2) and let $E_0(L)$ denote the closed linear span of the first vectors of (2). From (7) it follows that

$$\dim E(L)/E_0(L) = \dim E(M)/E_0(M). \tag{8}$$

Lemma. *If the number given by (8) is finite then the system of the Jordan chains of the pencil L is doubly complete (a double basis, respectively) if and only if the system of the Jordan chains of M has this property.*

Proof. Since the number given by (8) is finite it is sufficient to prove the statement for the case $E(L) = E_0(L)$ and $E(M) = E_0(M)$. The bounded and boundedly invertible operator

$$R = \begin{pmatrix} bI & -aI \\ -I & I \end{pmatrix} : \quad \tilde{H} \to \tilde{H}$$

maps $E(M)$ bijectively onto $E(L)$. Therefore $E(L) = \tilde{H}$ if and only if $E(M) = \tilde{H}$. Let x_i be an eigenvector of the pencil L corresponding to $\lambda_i \in \sigma_p(L)$. Without loss of generality we assume that $\infty \notin \sigma_p(L)$. Then $y_i = x_i$ is an eigenvector of M corresponding to $\mu_i = (a-\lambda_i)/(\lambda_i - b)$. Since the vectors $\sum_i c_i^{(n)}[x_i \oplus \lambda_i x_i]$ and $\sum_i c_i^{(n)}[(\lambda_i - b)x_i \oplus (a-\lambda_i)x_i]$ converge to 0 only simultaneously, the linear operator $T : \tilde{H} \to \tilde{H}$ defined by the relation $T(x_i \oplus \lambda_i x_i) = ((\lambda_i - b)x_i \oplus (a - \lambda_i)x_i)$ is bounded and boundedly invertible. Thus the system of the Jordan chains of L is a double basis if and only if the system of the Jordan chains of M has this property.

Proof of Theorem 2. Assume that in (5) the numbers a and b are such that $\kappa_-(L) = \kappa_-(a)$ and $\kappa_+(L) = \kappa_+(b)$. Using the Lemma we conclude that it is sufficient (and necessary) to find a linearization Φ for the pencil M. We introduce in $\tilde{H}$ the inner product $[x, y] = (Kx, y)$ with

$$K = \begin{pmatrix} G & 0 \\ 0 & -D \end{pmatrix}. \tag{9}$$

By the construction of G and D the negative spectrum of the operator K consists of $\kappa = \kappa_+(L) + \kappa_-(L) < \infty$ eigenvalues, counted according to their multiplicities. Therefore $\{\tilde{H}, [\cdot,\cdot]\}$ is a Pontryagin space Π_κ.

Consider the operator

$$\Phi = \begin{pmatrix} 0 & I \\ -D^{-1}G & -D^{-1}F \end{pmatrix}. \tag{10}$$

It can be verified that this operator is K-selfadjoint, $\sigma(\Phi) = \sigma(M)$ and that the systems of the Jordan chains of the pencil M and of the operator Φ coincide. Thus the operator Φ has the desired properties.

Remark 1. If in Theorem 2 the operator A (or C) is positive and $0 \in \rho(B)$ then $\kappa_-(L) = 0$.

In fact, the equality $\kappa_-(L) = 0$ is equivalent to the existence of a number a such that $L(a) \gg 0$. It follows from Corollary 1 that the operators A and B can be simultaneously reduced to diagonal form. Using Theorem 1 we can assume that $B = J$, where $J = P_+ - P_-$

with $\dim P_- < \infty$, and $JA = AJ$. With respect to the decomposition (3) the operator $L(a)$ has the following matrix form (observe (5)):

$$L(a) = \begin{pmatrix} a^2 A_{11} + aI_1 + C_{11} & C_{12} \\ C_{12}^* & a^2 A_{22} - aI_2 + C_{22} \end{pmatrix}.$$

For a sufficiently large positive number a we have $a^2 A_{11} + aI_1 + C_{11} \gg \|C_{12}\| I_1$ and, using that N_- is of finite dimension, $a^2 A_{22} - aI_2 + C_{22} \gg \|C_{12}\| I_2$. From [1], Lemma II.3.21 it follows that $L(a) \gg 0$ and therefore $\kappa_-(L) = 0$.

Remark 2. The condition $B = B_1 + B_2$ with $\pm B_1 \gg 0$ and $B_2 \in \mathcal{S}_\infty$ is essential for the existence of a linearization in Π_κ. Indeed, let $B = B^*$ with $0 \in \rho(B)$ and $B \neq B_1 + B_2$ with $\pm B_1 \gg 0$ and $B_2 \in \mathcal{S}_\infty$. Then the space H with the inner product $[x, y] = (Bx, y)$ is a Krein space. Let H be a separable space. Then there exists a Volterra B-selfadjoint operator X with $0 \notin \sigma_p(X)$. Let $A = BX$, $C = BX$. Then $\sigma_p(L) = \emptyset$. Since selfadjoint operators on Π_κ with $0 < \kappa < \infty$ have at least one eigenvalue the pencil L has no linearization in any Pontryagin space.

4. Consider the problem of completeness and the basisness of the Jordan chain of a m.s. S.Krein p.. Without loss of generality we assume again $\infty \notin \sigma_p(L)$. Let the operators B and C in Theorem 2 be such that the spectrum of the operator $B_1^{-1}C$ has at most countably many accumulation points. In this case the system of the Jordan chains of the linearization Φ is complete if and only if it is a basis. This holds if and only if the root subspaces of the operator Φ, corresponding to the points $\mu = (a - \lambda)/(\lambda - b)$, where λ runs through the set of all accumulation points of $\sigma(-B_1^{-1}C)$, are nondegenerate (see [1], Remark IV.2.13).

We denote by $E(\Phi)$ and $E_0(\Phi)$ the closed linear span of the root vectors and eigenvectors, respectively, of the operator Φ. Let

$$\Omega(\lambda) = \text{Ker } L(\bar{\lambda}) \cup (L'(\lambda)\text{Ker } L(\lambda))^\perp,$$

$$\omega(\lambda) = \text{Ker } L(\bar{\lambda}) \cup (\text{l.s. } \{\frac{1}{2}L''(\lambda)x_{k-1} + L'(\lambda)x_k \mid 0 \leq k \leq m\})^\perp,$$

where the linear span runs over all Jordan chains $\{x_0, x_1, \ldots, x_m\}$ of the pencil L, corresponding to $\lambda \in \sigma_p(L)$, and $x_{-1} = 0$. We mention that $\omega(\lambda) \subset \Omega(\lambda)$.

Theorem 3. *Let A, B and C be as in Theorem 2 and suppose that the set of accumulation points of $\sigma(-B_1^{-1}C)$ is at most countable. Then:*

a) $\dim \tilde{H}/E(\Phi) \leq \dim \tilde{H}/E_0(\Phi) < \infty$;

b) $E_0(\Phi) = \tilde{H}$ *if and only if* $\Omega(\lambda) = \{0\}$ *for all* $\lambda \neq \bar{\lambda}$ *and* $\lambda \in \Lambda$;

c) $E(\Phi) = \tilde{H}$ *if and only if* $\omega(\lambda) = \{0\}$ *for all* $\lambda \in \Lambda$;

d) if $E_0(\Phi) = \tilde{H}$ ($E(\Phi) = \tilde{H}$, respectively) then there exists in $\tilde{H}$ an almost K-orthonormal ([1], p.76) Riesz basis consisting of eigenvectors of the operator Φ if and only if $\sigma_p(L) \subset R$.

The proof follows immediately from the connection between the root vectors of the pencil L and of the operator Φ (cf. [1] Theorem IV.3.7). If, in Theorem 3, $C \in \mathcal{S}_\infty$ then $\Lambda = \{0\}$.

Therefore $\Omega(0) = \{0\}$ if and only if Ker C is B-nondegenerate. Let $A > 0$. From Remark 1 there follows the existence of a and b such that $L(a) \gg 0$ and $L(b) \gg 0$, $a \neq b$.

An eigenvector x_0 of the pencil L corresponding to $\lambda_0 \in \sigma_p(L)$ is called (a,b)-positive ((a,b)-negative and (a,b)-neutral, respectively) if

$$|\lambda_0 - b|^2(L(a)x_0, x_0) > |\lambda_0 - a|^2(L(b)x_0, x_0)$$

$$(|\lambda_0 - b|^2(L(a)x_0, x_0) < |\lambda_0 - a|^2(L(b)x_0, x_0)$$

and

$$|\lambda_0 - b|^2(L(a)x_0, x_0) = |\lambda_0 - a|^2(L(b)x_0, x_0),$$

respectively).

Theorem 4. *If in Theorem 3 the operator A is positive then from $\omega(\lambda) = \{0\}$ ($\Omega(\lambda) = \{0\}$, respectively) it follows the existence of two Riesz bases ϕ^+ and ϕ^- in H which consist of root vectors (eigenvectors respectively) of the pencil L. The basis ϕ^+ (ϕ^-, respectively) consists of (a,b)-positive ((a,b)-negative, respectively) eigenvectors with the possible exception of a finite number of associated vectors and (a,b)-neutral eigenvectors.*

Proof. The space $\tilde{H}$ with the K-metric (see (9)) is a Krein space and the operator Φ (see (10)) is selfadjoint. It follows from the Lemma that there exists a basis in $\tilde{H}$ consisting of root vectors of Φ. Therefore (cf. [1]. Theorem III.4.14) the operator Φ has an invariant maximal dual pair $\{N^+, N^-\}$ such that $N^+ = N_0 + N_1^+$ and $N^- = N_0 + N_1^-$ with $N_0 = N^+ \cup N^-$, $\dim N_0 < \infty$ and $N_1^\pm$ are uniformly positive and negative subspaces, respectively. From [7] it follows the existence of two Riesz bases in N^+ and N^- consisting of root vectors of the operator Φ. The projections of the basis in N^+ onto $H \oplus 0$ and of the basis in N^- onto $0 \oplus H$ give us the desired bases in H.

The proof of the case $\Omega(\lambda) = \{0\}$ is analogous.

References

[1] Azizov, T.Ya., I.S. Iohvidov, Linear operators in spaces with an indefinite metric. John Wiley and Sons, New York, 1989.

[2] Kopachevskii, N.D., S.G. Krein, Kan Ngo Zui, Operator methods in linear hydrodynamics. Nauka, Moscow, 1989.

[3] Markus, A.S., Introduction to the spectral theory of polynomial operator pencils. AMS Translations of Mathematical Monographs, Vol. 71, Providence, Rhode Island, 1988.

[4] Shkalikov, A.A., V.T. Pliev, Compact perturbation of strongly damped operator pencils. Matemat. Zametki, 45, 2 (1989), 118-129.

[5] Phillips, R.S., The extension of dual subspaces invariant under an algebra. Proc. Intern. Symp. Linear Spaces. Jerusalem, 1960. Jerusalem Press, 1961, 366-398.

[6] Langer, H., Invariante Teilräume definisierbarer J-selbstadjungierter Operatoren. Ann. Acad. Sci. Fenn., Ser. A1, No 475 (1971).

[7] Azizov, T.Ya., On compact operators which are selfadjoint under a degenerate metric. Matem. Issled., 4 (1972) 237-240.

[8] Askerov, N.K., S.G. Krein, G.I. Laptev, A problem on oscillations of a viscous liquid and the operator equations connected with it. Funktional Analysis and its Applications, 2,1 (1968) 21-31.

[9] Gohberg, I.Ts., M.G., Krein Introduction to the theory of linear non-selfadjoint operators in a Hilbert space. AMS Translations of Mathematical Monographs, Vol. 18, Providence, Rhode Island, 1969.

Research Institute of Mathematics
Voronezh State University
Universitetskaya pl. 1
Voronezh 394693
Russia.

AMS Subject Classification: 47 A 56, 47 B 37

Operator Theory:
Advances and Applications, Vol. 80

DISCRETE NONSTATIONARY BOUNDED REAL LEMMA IN INDEFINITE METRICS, THE STRICT CONTRACTIVE CASE

A. Ben-Artzi, I. Gohberg and **M.A. Kaashoek**

Necessary and sufficient conditions for an input-output operator of a linear time variant finite dimensional system to be a strict contraction are suggested. This is a generalization of the well known bounded real lemma. Both the definite and indefinite cases are considered. The results are presented in state space form.

1. INTRODUCTION

In this paper, we consider input-output operators of systems of the form

$$(1.1)\qquad \begin{cases} x_{n+1} = A_n x_n + B_n u_n & (n = 0, \pm 1, \dots)\ , \\ v_n = C_n x_n + D_n u_n & (n = 0, \pm 1, \dots)\ . \end{cases}$$

We will always assume that $(A_n)_{n=-\infty}^{\infty}$, $(B_n)_{n=-\infty}^{\infty}$, $(C_n)_{n=-\infty}^{\infty}$ and $(D_n)_{n=-\infty}^{\infty}$ are bounded sequences of matrices of sizes $r'' \times r''$, $r'' \times r$, $r' \times r''$ and $r' \times r$, respectively. We use the concept of a dichotomy of a system, and refer to the next section for its definition and properties. Assume that the system

$$(1.2)\qquad x_{n+1} = A_n x_n \qquad (n = 0, \pm 1, \dots)$$

admits a dichotomy. For $k = 1, 2, \dots$ we denote by $\ell_k^2(\mathbb{Z})$ the space of two sided square summable sequences $u = (u_n)_{n=-\infty}^{\infty}$ with entries u_n in $\mathbb{C}^k$. It follows from the dichotomy assumption (see Corollary 2.3) that for each input sequence $u = (u_n)_{n=-\infty}^{\infty} \in \ell_r^2(\mathbb{Z})$, there exists a unique sequence $x = (x_n)_{n=-\infty}^{\infty} \in \ell_{r''}^2(\mathbb{Z})$ satisfying the first row of (1.1). The output $v = (v_n)_{n=-\infty}^{\infty} \in \ell_{r'}^2(\mathbb{Z})$ is then given by the second row of (1.1). We define an operator $T : \ell_r^2(\mathbb{Z}) \to \ell_{r'}^2(\mathbb{Z})$ via $Tu = v$, and call T the *input-output operator* of the system (1.1). The operator T is linear and bounded.

The next result gives conditions for T to be a strict contraction. For a self-adjoint matrix M we denote the inertia of M by $\operatorname{In} M = (\nu_+, \nu_0, \nu_-)$, where $\nu_0 =$

dim Ker$M$, and ν_+ and ν_- denote the number of positive and negative eigenvalues of M, counting multiplicities. We say that a sequence of self-adjoint matrices $(M_n)_{n=-\infty}^{\infty}$ is of *constant inertia* if In $M_n =$ In M_{n+1} $(n = 0, \pm 1, \ldots)$. Let us remark that in Theorems 1.1 and 1.2 below we do not make any assumption on the system (1.1) of the type: stability, stabilizability, controllability or observability.

THEOREM 1.1. *Consider the system*

$$\text{(1.3)} \qquad \begin{cases} x_{n+1} = A_n x_n + B_n u_n & (n = 0, \pm 1, \ldots) , \\ v_n = C_n x_n + D_n u_n & (n = 0, \pm 1, \ldots) , \end{cases}$$

where $(A_n)_{n=-\infty}^{\infty}, (B_n)_{n=-\infty}^{\infty}, (C_n)_{n=-\infty}^{\infty}$ *and* $(D_n)_{n=-\infty}^{\infty}$ *are bounded sequences of matrices of sizes* $r'' \times r'', r'' \times r, r' \times r''$ *and* $r' \times r$ *respectively. Then the following conditions are equivalent:*
I) The system

$$\text{(1.4)} \qquad x_{n+1} = A_n x_n \qquad (n = 0, \pm 1, \ldots)$$

admits a dichotomy, and the input-output operator T *of* (1.3) *is a strict contraction* $(\|T\| < 1)$.
II) There exists a bounded sequence of self-adjoint $r'' \times r''$ *matrices* $(M_n)_{n=-\infty}^{\infty}$ *of constant inertia, such that the following inequalities hold for some positive number* ε

$$\text{(1.5)} \qquad I - B_n^* M_{n+1} B_n - D_n^* D_n \geq \varepsilon I \qquad (n = 0, \pm 1, \ldots) ,$$

and
(1.6)

$$\begin{gathered} M_n - A_n^* M_{n+1} A_n - C_n^* C_n \\ -(A_n^* M_{n+1} B_n + C_n^* D_n)(I - B_n^* M_{n+1} B_n - D_n^* D_n)^{-1}(B_n^* M_{n+1} A_n + D_n^* C_n) \geq \varepsilon I \end{gathered}$$

for $n = 0, \pm 1, \ldots$.
Moreover, if the system (1.4) *admits a dichotomy and* $(M_n)_{n=-\infty}^{\infty}$ *is a bounded sequence of self-adjoint matrices which satisfies the inequalities* (1.5) *and* (1.6), *then* M_n *is invertible,* $\sup_n \|M_n^{-1}\| < \infty$, *and* $\text{In}(M_n) = (p, 0, r'' - p)$ $(n = 0, \pm 1, \ldots)$, *where* p *is the rank of the dichotomy of* (1.4). *Finally, if the sequences* $(A_n)_{n=-\infty}^{\infty}, (B_n)_{n=-\infty}^{\infty}, (C_n)_{n=-\infty}^{\infty}$, *and* $(D_n)_{n=-\infty}^{\infty}$ *are constant or periodic with period* m *and I) holds, then* $(M_n)_{n=-\infty}^{\infty}$ *in II) can be chosen to be constant or periodic with period* m, *respectively.*

Assume now that the system (1.3) is stationary, namely $A_n = A, B_n = B, C_n = C$, and $D_n = D$ $(n = 0, \pm 1, \ldots)$ for some matrices A, B, C, and D. Then the system

(1.4) admits a dichotomy if and only if A does not have eigenvalues on the unit circle, and the rank of the dichotomy is equal to the number of eigenvalues of A in the unit disc, counting multiplicities. See Proposition 2.1. In this case, the input-output operator T of (1.3) is a Toeplitz operator with symbol $C(\lambda I - A)^{-1}B + D$ ($|\lambda| = 1$). We now apply Theorem 1.1 to this situation. Note that the sequence $(M_n)_{n=-\infty}^{\infty}$ can be chosen to be constant, say $M_n = M$ ($n = 0, \pm 1, \ldots$), by the last part of the theorem. Thus, we have the following result.

COROLLARY 1.2. *Let A, B, C, D be matrices of sizes $r'' \times r'', r'' \times r, r' \times r''$ and $r' \times r$ respectively. The following conditions are equivalent:*
I) A does not have eigenvalues on the unit circle and $\sup_{|\lambda|=1} \|C(\lambda I - A)^{-1}B + D\| < 1$.
II) There exists a self-adjoint matrix M such that $I - B^*MB - D^*D > 0$ and

$$M - A^*MA - C^*C - (A^*MB + C^*D)(I - B^*MB - D^*D)^{-1}(B^*MA + D^*C) > 0 .$$

Moreover, if these conditions hold, then M is invertible and the number of positive (respectively negative) eigenvalues of M is equal to the number of eigenvalues of A inside (respectively outside) the unit circle, counting multiplicities.

Let us also remark that the case when A is stable corresponds to $M > 0$. This stable case appears as the equivalence of conditions (a) and (b) in Theorem 2.2 of [SX] (after replacing B by $\gamma^{-1}B$, and D by $\gamma^{-1}D$ in Theorem 2.2 of [SX]).

Next, we consider (J, J')-strict contractions. Let

$$(J, J') = ((J_n)_{n=-\infty}^{\infty}, (J'_n)_{n=-\infty}^{\infty}) \tag{1.7}$$

be a pair of bounded sequences of self-adjoint matrices of orders r and r', respectively. We say that T is a *(J, J')-strict contraction* if there exists a positive number δ such that

$$\sum_{n=-\infty}^{\infty} (u_n^* J_n u_n - v_n^* J'_n v_n) \geq \delta \sum_{n=-\infty}^{\infty} \|u_n\|^2 \tag{1.8}$$

whenever $u = Tv$, where $u = (u_n)_{n=-\infty}^{\infty} \in \ell_r^2(\mathbb{Z})$ and $v = (v_n)_{n=-\infty}^{\infty} \in \ell_{r'}^2(\mathbb{Z})$. The norms $\|u_n\|^2$ on the right hand side denote the usual positive definite Euclidean norms of $u_n \in \mathbb{C}^r$. Note that if $J_n = I_r$ and $J_n = I_{r'}$ ($n = 0, \pm 1, \ldots$), then T is a (J, J')-strict contraction if and only if $\|T\| < 1$, where $\|T\|$ is the usual norm of $T : \ell_r^2(\mathbb{Z}) \to \ell_{r'}^2(\mathbb{Z})$.

THEOREM 1.3. *Let T be the input-output operator of the system (1.1), and assume the system (1.2) admits a dichotomy. Put $\Delta_n = J_n - B_n^* M_{n+1} B_n - D_n^* J'_n D_n$ ($n = 0, \pm 1, \ldots$). Then T is a (J, J')-strict contraction if and only if there exists a* *bounded sequence* $(M_n)_{n=-\infty}^{\infty}$ *of self-adjoint matrices of order* r'' *such that the following inequalities hold for some* $\varepsilon > 0$

$$\Delta_n \geq \varepsilon I \qquad (n = 0, \pm 1, \ldots) , \tag{1.9}$$

and

$$\begin{aligned} M_n - A_n^* M_{n+1} A_n - C_n^* J_n' C_n \\ -(A_n^* M_{n+1} B_n + C_n^* J_n' D_n)\Delta_n^{-1}(B_n^* M_{n+1} A_n + D_n^* J_n' C_n) \geq \varepsilon I \end{aligned} \tag{1.10}$$

for $n = 0, \pm 1, \ldots$.
Moreover, if all the sequences $(A_n)_{n=-\infty}^{\infty}$, $(B_n)_{n=-\infty}^{\infty}$, $(C_n)_{n=-\infty}^{\infty}$, $(D_n)_{n=-\infty}^{\infty}$, $(J_n)_{n=-\infty}^{\infty}$, *and* $(J_n')_{n=-\infty}^{\infty}$ *are constant or periodic with period* m, *then* $(M_n)_{n=-\infty}^{\infty}$ *can be chosen to be constant or periodic with period* m, *respectively.*

The proofs of Theorem 1.1 and 1.3 appear in Section 5. The next Section 2 contains some preliminary definitions and results. Section 3 deals with a special case when B_n is right invertible. This special case is applied in Section 4 in order to obtain necessary conditions for T to be a (J, J')-strict contraction.

Related results appear in Chapter 3 of the recent monograph of A. Halanay and V. Ionescu [HI], where the conditions on the Riccati equations lead to positive semi-definite solutions, while the cases treated here give solutions of arbitrary inertia.

Finally, let us remark that the methods of proof in this paper can also be applied to the study of extremal and stabilizing solutions of time-varying Riccati equations. These results will appear in another publication.

2. PRELIMINARIES

Let us begin by defining the notion of dichotomy. Let $(A_n)_{n=-\infty}^{\infty}$ be a sequence of $r'' \times r''$ matrices, and let $(R_n)_{n=-\infty}^{\infty}$ be a bounded sequence of projections in $\mathbb{C}^{r''}$, such that rank R_n $(n = 0, \pm 1, \ldots)$ is constant, and the equalities $A_n R_n = R_{n+1} A_n$ $(n = 0, \pm 1, \ldots)$ hold. We say that the system

$$x_{n+1} = A_n x_n \qquad (n = 0, \pm 1, \ldots) , \tag{2.1}$$

admits the *dichotomy* $(R_n)_{n=-\infty}^{\infty}$ if there exist two positive constants a and b, with $a < 1$, such that the following inequalities hold

$$\|A_{n+j-1} \cdots A_n x\| \leq b a^j \|x\| \quad (x \in \operatorname{Im} R_n) , \tag{2.2}$$

$$\|A_{n+j-1} \cdots A_n y\| \geq \frac{1}{b a^j} \|y\| \quad (y \in \operatorname{Ker} R_n) , \tag{2.3}$$

for $n = 0, \pm 1, \ldots; j = 1, 2, \ldots$. The constant integer rank R_n $(n = 0, \pm 1, \ldots)$ is called the rank of the dichotomy.

For the definition of the dichotomy see [BG1], [BG3], [CS], [GKvS], [S1] and [S2]. A definition of dichotomy for more general systems appears in [BG2], and [BGK1] and [BGK2].

It follows from Theorem 2.4 of [BG1] that the system (2.1) admits at most one dichotomy. Moreover, in case of time-invariant systems the following proposition is contained in Proposition 2.2 of [BG1], and Theorem 3.3 of [BGK2]. We call $\mathbb{T} = \{\lambda \in \mathbb{C} : |\lambda| = 1\} \subset \mathbb{C}$ the unit circle.

PROPOSITION 2.1. *Let A be a $r'' \times r''$ matrix. Then the time-invariant system*

$$x_{n+1} = Ax_n \qquad (n = 0, \pm 1, \ldots) \tag{2.4}$$

admits a dichotomy if and only if A has no eigenvalues on the unit circle. Moreover, if (2.4) admits a dichotomy $(R_n)_{n=-\infty}^{\infty}$, then the projections R_n $(n = 0, \pm 1, \ldots)$ are equal to the Riesz projection of A corresponding to the unit disc $\{\lambda \in \mathbb{C} : |\lambda| < 1\}$, namely

$$R_n = \frac{1}{2\pi i} \int_{\mathbb{T}} (\lambda I - A)^{-1} d\lambda \qquad (n = 0, \pm 1, \ldots) . \tag{2.5}$$

The notion of dichotomy is related to operator theory in the following way. We denote by $\ell_k^2(\mathbb{Z})$ $(k = 1, 2, \ldots)$ the Hilbert space of two sided sequences $y = (y_n)_{n=-\infty}^{\infty}$ with $y_n \in \mathbb{C}^k$ $(n = 0, \pm 1, \ldots)$ and such that $\|y\|^2 = \sum_{n=-\infty}^{\infty} \|y_n\|^2 < \infty$. Assume that the sequence $(A_n)_{n=-\infty}^{\infty}$ is bounded. Then the mapping V defined via

$$V(\cdots, x_{-1}, x_0, x_1, \cdots) = (\cdots, A_{-1}x_{-1}, A_0 x_0, Ax_1, \cdots)$$

defines a bounded linear operator in $\ell_{r''}^2$. The block matrix representing V is given by $V = (\delta_{ij} A)_{ij=-\infty}^{\infty}$. With V we also consider the two-sided block shift $S : \ell_{r''}^2(\mathbb{Z}) \to \ell_{r''}^2(\mathbb{Z})$ given by $S = (\delta_{i,j+1} I_{r''})_{ij=-\infty}^{\infty}$ and its inverse $S^{-1} = (\delta_{i+1,j} I_{r''})_{ij=-\infty}^{\infty}$. The following result is contained in Theorem 2.1 of [BG2].

THEOREM 2.2. *Let $(A_n)_{n=-\infty}^{\infty}$ be a bounded sequence of $r'' \times r''$ matrices, and define operators $V = (\delta_{ij} A_j)_{ij=-\infty}^{\infty}$ and $S^{-1} = (\delta_{i+1,j} I_{r''})_{ij=-\infty}^{\infty}$ in $\ell_{r''}^2(\mathbb{Z})$. Then the operator $S^{-1} - V$ is invertible in $\ell_{r''}^2(\mathbb{Z})$ if and only if the system (2.1) admits a dichotomy.*

Assume now that $(A_n)_{n=-\infty}^{\infty}$ is bounded and that the system (2.1) admits a dichotomy, and let V and S^{-1} be as above. Let $y = (y_n)_{n=-\infty}^{\infty} \in \ell^2_{r''}(\mathbb{Z})$ be arbitrary. Then the equation

$$S^{-1}x = Vx + y \tag{2.6}$$

admits the unique solution $x = (S^{-1} - V)^{-1}y$ in $\ell^2_{r''}(\mathbb{Z})$. Put $x = (x_n)_{n=-\infty}^{\infty}$. Then (2.6) is equivalent to

$$x_{n+1} = A_n x_n + y_n \qquad (n = 0, \pm 1, \ldots)\ . \tag{2.7}$$

We summarize this as follows.

COROLLARY 2.3. *Let $(A_n)_{n=-\infty}^{\infty}$ be a bounded sequence of $r'' \times r''$ matrices such that the system (2.1) admits a dichotomy. Then for each vector $y = (y_n)_{n=-\infty}^{\infty} \in \ell^2_{r''}(\mathbb{Z})$ there exists a unique vector $x = (x_n)_{n=-\infty}^{\infty} \in \ell^2_{r''}(\mathbb{Z})$ such that the equalities (2.7) hold , and the mapping in $\ell^2_{r''}(\mathbb{Z})$ sending y into x is linear and bounded.*

The existence of a dichotomy for a system can be inferred from some matrix inequalities. The following result is of this type, and is contained in Theorem 5.3 of [BG1].

THEOREM 2.4. *Let $(A_n)_{n=-\infty}^{\infty}$ and $(M_n)_{n=-\infty}^{\infty}$ be two sequences of $r'' \times r''$ matrices with $(M_n)_{n=-\infty}^{\infty}$ bounded, and such that M_n is self-adjoint $(n = 0, \pm 1, \ldots)$ and of constant inertia $M_n = (\nu_+, \nu_0, \nu_-)$ $(n = 0, \pm 1, \ldots)$. If the matrix inequalities*

$$M_n - A_n^* M_{n+1} A_n \geq \varepsilon I \qquad (n = 0, \pm 1, \ldots)$$

hold for some $\varepsilon > 0$, then $\nu_0 = 0$ and the system (2.1) admits a dichotomy of rank ν_+.

It turns out that the requirement that the inertia of M_n is constant in the preceding theorem is necessary. This is the contents of the following proposition which we prove here for completeness.

PROPOSITION 2.5. *Let $(A_n)_{n=-\infty}^{\infty}$ be a sequence of $r'' \times r''$ matrices such that the system (2.1) admits a dichotomy, and $(M_n)_{n=-\infty}^{\infty}$ be a bounded sequence of self-adjoint matrices of order $r'' \times r''$. If the inequalities*

$$M_n - A_n^* M_{n+1} A_n \geq \varepsilon I \qquad (n = 0, \pm 1, \ldots) \tag{2.8}$$

hold for some positive number ε, then

$$\mathrm{In}(M_n) = (p, 0, r'' - p) \qquad (n = 0, \pm 1, \ldots)\ , \tag{2.9}$$

where p is the rank of the dichotomy of (2.1), and if in addition the sequence $(A_n)_{n=-\infty}^{\infty}$ is bounded then

(2.10) $$\sup_n \|M_n^{-1}\| \leq \varepsilon^{-1} \max(1, \sup_n \|A_n\|^2) .$$

PROOF. Let $(R_n)_{n=-\infty}^{\infty}$ be the dichotomy of the system (2.1). Then rank $R_n = p$ $(n = 0, \pm 1, \ldots)$, the equalities

(2.11) $$A_n R_n = R_{n+1} A_n \qquad (n = 0, \pm 1, \ldots)$$

hold, and there are positive numbers a and b such that inequalities (2.2) and (2.3) are satisfied. Let k be an integer and $x \in \operatorname{Im} R_k$ a vector. Define the sequence $x_k, x_{k+1}, \ldots$ via the recursion

(2.12) $$x_k = x$$

and

(2.13) $$x_{n+1} = A_n x_n \qquad (n = k, k+1, \ldots) .$$

Then multiplying (2.8) by x_n on the right and x_n^* on the left, and taking into account (2.13) we obtain

$$x_n^* M_n x_n - x_{n+1}^* M_{n+1} x_{n+1} \geq \varepsilon \|x_n\|^2 \qquad (n = k, k+1, \ldots) .$$

Summing these inequalities for $n = k, \ldots, k+j-1$ and taking into account (2.12), we have

(2.14) $$x^* M_k x \geq \varepsilon \|x\|^2 + x_{k+j}^* M_{k+j} x_{k+j} \qquad (j = 1, 2, \ldots) .$$

Now note that (2.12) and (2.13) lead to $x_{k+j} = A_{k+j-1} \cdots A_k x$. Since $x \in \operatorname{Im} R_k$ it follows from the dichotomy inequality (2.2) that $\lim_{j\to\infty} x_{k+j} = 0$. Since the sequence $(M_n)_{n=-\infty}^{\infty}$ is bounded, this leads to $\lim_{j\to\infty} x_{k+j}^* M_{k+j} x_{k+j} = 0$. Thus, taking the limit in (2.14) yields

$$x^* M_k x \geq \varepsilon \|x\|^2 .$$

This inequality holds for each $x \in \operatorname{Im} R_k$. Since $p = \dim \operatorname{Im} R_k$, it follows that M_k has at least p eigenvalues greater or equal to ε, counting multiplicities. If $p = r''$, this implies $M_k \geq \varepsilon I$ $(k = 0, \pm 1, \ldots)$, whence the result follows.

Assume in the sequel that $p < r''$. Note that this implies $\mathrm{Ker} R_n \neq \{0\}$ $(n = 0, \pm 1, \dots)$. Hence, by (2.3)

$$A_n \neq 0 \qquad (n = 0, \pm 1, \dots) . \tag{2.15}$$

Let us now remark that inequality (2.3) implies in particular that

$$\mathrm{Ker} A_n \cap \mathrm{Ker} R_n = \{0\} \qquad (n = 0, \pm 1, \dots) .$$

In addition, the commutation relations (2.11) lead to $A_n(\mathrm{Ker} R_n) \subset \mathrm{Ker} R_{n+1}$ $(n = 0, \pm 1, \dots)$. It follows from these facts that the mappings

$$A_n\big|_{\mathrm{Ker} R_n} : \mathrm{Ker} R_n \longrightarrow \mathrm{Ker} R_{n+1} \qquad (n = 0, \pm 1, \dots)$$

are injective. Since $\dim \mathrm{Ker} R_n = r'' - p = \dim \mathrm{Ker} R_{n+1}$, the map $A_n\big|_{\mathrm{Ker} R_n}$ is invertible, and

$$(A_n\big|_{\mathrm{Ker} R_n})^{-1} : \mathrm{Ker} R_{n+1} \longrightarrow \mathrm{Ker} R_n \qquad (n = 0, \pm 1, \dots) .$$

Now let k be an integer and $y \in \mathrm{Ker} R_k$ a vector. Define the sequence $y_k, y_{k-1}, \dots$ via $y_k = y$ and

$$y_n = (A_n\big|_{\mathrm{Ker} R_n})^{-1} \cdots (A_{k-1}\big|_{\mathrm{Ker} R_{k-1}})^{-1} y \qquad (n = k-1, k-2, \dots) . \tag{2.16}$$

Since $y_n = A_{n-1} y_{n-1}$ $(n = k, k-1, \dots)$, we obtain from (2.8)

$$y_{n-1}^* M_{n-1} y_{n-1} - y_n^* M_n y_n \geq \varepsilon \|y_{n-1}\|^2 \qquad (n = k, k-1, \dots) .$$

Adding these inequalities for $n = k, \dots, k-j+1$, we obtain

$$y_{k-j}^* M_{k-j} y_{k-j} - y_k^* M_k y_k \geq \varepsilon \|y_{k-1}\|^2 \qquad (j = 1, 2, \dots) . \tag{2.17}$$

Note also that since $A_{k-1} y_{k-1} = y_k$, then $\|A_{k-1}\|\,\|y_{k-1}\| \geq \|y_k\|$. Taking into account (2.15), this inequality may be rewritten as $\|y_{k-1}\| \geq \|A_{k-1}\|^{-1}\|y_k\|$. Inserting this in (2.17), and taking into account $y_k = y$, we obtain

$$y^* M_k y \leq -\varepsilon \|A_{k-1}\|^{-2} \|y\|^2 + y_{k-j}^* M_{k-j} y_{k-j} \qquad (j = 1, 2 \dots) . \tag{2.18}$$

Moreover, definition (2.16) implies

$$y_{k-j} \in \mathrm{Im}\left(A_{k-j}\big|_{\mathrm{Ker} R_{k-j}}\right)^{-1} = \mathrm{Ker} R_{k-j} .$$

Hence, (2.3) leads to

$$\|A_{k-1}\cdots A_{k-j}y_{k-j}\| \geq \frac{1}{ba^j}\|y_{k-j}\| \qquad (j=1,2\ldots)\ .$$

However, by (2.16) again we have $A_{k-1}\cdots A_{k-j}y_{k-j} = y$. Combining this with the previous inequality we obtain $\|y_{k-j}\| \leq ba^j\|y\|$. Thus, $\lim_{j\to\infty} y_{k-j} = 0$, and since the sequence $(M_n)_{n=-\infty}^{\infty}$ is bounded, $\lim_{j\to\infty} y_{k-j}^* M_{k-j} y_{k-j} = 0$. Hence, taking the limit in (2.18) we obtain

$$y^* M_k y \leq -\varepsilon\|A_{k-1}\|^{-2}\|y\|^2 \qquad (k=0,\pm1,\ldots\ ;\ y\in \mathrm{Ker} R_k)\ .$$

Here k is an arbitrary integer and $y \in \mathrm{Ker}R_k$. Since $\dim \mathrm{Ker}R_k = r'' - p$, it follows that M_k has at least $r''-p$ eigenvalues less than or equal to $-\varepsilon\|A_{k-1}\|^{-2} < 0$. We have shown above that M_k has at least p eigenvalues greater or equal than ε. These two facts imply (2.9). In addition, we have

$$\|M_k^{-1}\| \leq \max\left(\varepsilon^{-1}, \varepsilon^{-1}\|A_{k-1}\|^2\right) \qquad (k=0,\pm1,\ldots)\ .$$

Thus, if the sequence $(A_n)_{n=-\infty}^{\infty}$ is bounded we obtain (2.10). □

Finally, we will also use the following result about Schur complements, which we give for completeness.

LEMMA 2.6. *Let* $\Lambda_n = \begin{pmatrix} a_n & b_n^* \\ b_n & c_n \end{pmatrix}$ $(n=0,\pm1\ldots)$ *be a bounded sequence of self-adjoint block matrices. Then, there exists a positive number* ε_1 *such that*

$$\Lambda_n \geq \varepsilon_1 I \qquad (n=0,\pm,1\ldots)\ , \tag{2.19}$$

if and only if there exists a positive number ε_2 *such that*

$$c_n \geq \varepsilon_2 I \quad \text{and} \quad a_n - b_n^* c_n^{-1} b_n \geq \varepsilon_2 I \quad (n=0,\pm1,\ldots)\ . \tag{2.20}$$

PROOF. Clearly (2.19) implies $c_n \geq \varepsilon_1 I$ $(n=0,\pm1,\ldots)$. Thus we may assume without loss of generality that

$$c_n \geq \varepsilon_3 I \qquad (n=0,\pm1,\ldots) \tag{2.21}$$

for some positive ε_3. In particular

$$\sup_n \|c_n^{-1}\| < \infty\ . \tag{2.22}$$

We now decompose for each n

$$\begin{pmatrix} a_n & b_n^* \\ b_n & c_n \end{pmatrix} = \begin{pmatrix} I & b_n^* c_n^{-1} \\ 0 & I \end{pmatrix} \begin{pmatrix} a_n - b_n^* c_n^{-1} b_n & 0 \\ 0 & c_n \end{pmatrix} \begin{pmatrix} I & 0 \\ c_n^{-1} b_n & I \end{pmatrix} . \tag{2.23}$$

Put

$$S_n = \begin{pmatrix} I & 0 \\ c_n^{-1} b_n & I \end{pmatrix} \qquad (n = 0, \pm 1 \ldots) .$$

From (2.22) and the boundedness of $(\Lambda_n)_{n=-\infty}^{\infty}$, it follows that $(S_n)_{n=-\infty}^{\infty}$ is a sequence of invertible matrices with

$$\sup_n \{\|S_n\|, \ \|S_n^{-1}\|\} < \infty . \tag{2.24}$$

Moreover, by (2.23)

$$\Lambda_n = S_n^* \begin{pmatrix} a_n - b_n^* c_n^{-1} b_n & 0 \\ 0 & c_n \end{pmatrix} S_n \quad (n = 0, \pm 1, \ldots) .$$

In view of (2.21) and (2.24), it is clear that (2.19) and (2.20) are equivalent. □

3. STRICT (J, J') CONTRACTIONS: NECESSARY CONDITIONS FOR A SPECIAL CASE

In this section we consider systems of the form

$$\begin{cases} x_{n+1} = A_n x_n + B_n u_n & (n = 0, \pm 1, \ldots) , \\ v_n = C_n x_n + D_n u_n & (n = 0, \pm 1, \ldots) , \end{cases} \tag{3.1}$$

where $(u_n)_{n=-\infty}^{\infty} \in \ell_r^2(\mathbb{Z})$, $(x_n)_{n=-\infty}^{\infty} \in \ell_{r''}^2(\mathbb{Z})$, $(u_n)_{n=-\infty}^{\infty} \in \ell_{r'}^2(\mathbb{Z})$, and $(A_n)_{n=-\infty}^{\infty}$, $(B_n)_{n=-\infty}^{\infty}$, $(C_n)_{n=-\infty}^{\infty}$ and $(D_n)_{n=-\infty}^{\infty}$ are bounded sequences of matrices of sizes $r'' \times r'', r'' \times r, r' \times r''$ and $r' \times r$ respectively. We assume that the system

$$x_{n+1} = A_n x_n \qquad (n = 0, \pm 1 \ldots) \tag{3.2}$$

admits a dichotomy. The definition of a dichotomy and its relevant properties have been given in Section 2. In particular, it follows from Corollary 2.3 and the boundedness of the sequence $(B_n)_{n=-\infty}^{\infty}$ that for each input sequence $u = (u_n)_{n=-\infty}^{\infty} \in \ell_r^2(\mathbb{Z})$, there exists a unique sequence $x = (x_n)_{n=-\infty}^{\infty} \in \ell_{r''}^2(\mathbb{Z})$, satisfying

$$x_{n+1} = A_n x_n + B_n u_n \qquad (n = 0, \pm 1 \ldots) , \tag{3.3}$$

and the mapping $K : \ell^2_r(\mathbb{Z}) \to \ell^2_{r''}(\mathbb{Z})$ defined by

$$Ku = x \tag{3.4}$$

is linear and bounded. Moreover, for u and $x = Ku$ as above, the second set of equalities in (3.1) defines a sequence $v = (v_n)_{n=-\infty}^{\infty} \in \ell^2_{r'}(\mathbb{Z})$, and the mapping $T : \ell^2_r(\mathbb{Z}) \to \ell^2_{r'}(\mathbb{Z})$ defined by

$$Tu = v \tag{3.5}$$

is called the *input-output operator of the system* (3.1). The boundedness of K and of the sequences $(C_n)_{n=-\infty}^{\infty}$ and $(D_n)_{n=-\infty}^{\infty}$ imply that T is a bounded linear operator. In this section we consider a special case by requiring that for each integer n the matrix B_n is right invertible, and that a right inverse $B_n^{[-r]}$ of B_n can be chosen $(n = 0, \pm 1, \ldots)$ so that

$$\sup_n \|B_n^{[-r]}\| < \infty \, . \tag{3.6}$$

Let $(J, J') = ((J_n)_{n=-\infty}^{\infty}, (J'_n)_{n=-\infty}^{\infty})$ be a pair of bounded sequences of self-adjoint matrices of orders r and r' respectively. We say that the input-output operator T of the system (3.1) is *a strict* (J, J') *contraction* if there exists a positive number δ such that the following inequality holds

$$\sum_{n=-\infty}^{\infty} (u_n^* J_n u_n - v_n^* J'_n v_n) \geq \delta \sum_{n=-\infty}^{\infty} \|u_n\|^2 \, , \tag{3.7}$$

for each $(u_n)_{n=-\infty}^{\infty} \in \ell^2_r(\mathbb{Z})$, where $(v_n)_{n=-\infty}^{\infty} = T((u_n)_{n=-\infty}^{\infty})$. Throughout this section, we assume that B_n is right invertible $(n = 0, \pm 1 \ldots)$ with (3.6), that T is a strict (J, J') contraction, and let $\delta > 0$ be such that (3.7) holds. We also denote

$$N = 1 + \|T\| + \|K\| + \sup_n\{\|A_n\|, \|B_n\|, \|C_n\|, \|D_n\|, \|B_n^{[-r]}\|, \|J_n\|, \|J'_n\|\} \, . \tag{3.8}$$

We now construct a certain optimization problem related to the system (3.1). Let s be an integer and $\xi \in \mathbb{C}^{r''}$ a vector. We denote by $\Sigma_{\xi,s}$ the set of all square summable sequences $(u_n)_{n=s}^{\infty}$ $(u_n \in \mathbb{C}^r : n = s, s+1 \ldots)$ such that the sequence $(x_n)_{n=s}^{\infty}$ is square summable, where $(x_n)_{n=s}^{\infty}$ is defined via the recursion

$$x_s = \xi \tag{3.9}$$

and

$$x_{n+1} = A_n x_n + B_n u_n \qquad (n = s, s+1, \dots) . \tag{3.10}$$

For each $u = (u_n)_{n=s}^{\infty} \in \Sigma_{\xi,s}$, define $(x_n)_{n=s}^{\infty}$ via (3.9) and (3.10), and define the sequence $(v_n)_{n=-\infty}^{\infty}$ via

$$v_n = C_n x_n + D_n u_n \qquad (n = s, s+1, \dots) . \tag{3.11}$$

For each $s = 0, \pm 1, \dots$ and $\xi \in \mathbb{C}^n$ we now define a function $f_{\xi,s} : \Sigma_{\xi,s} \to \mathbb{R}$ via

$$f_{\xi,s}(u) = \sum_{n=s}^{\infty} (u_n^* J_n u_n - v_n^* J_n' v_n) \qquad (u = (u_n)_{n=s}^{\infty} \in \Sigma_{\xi,s}) . \tag{3.12}$$

The optimization problem consists of finding the infimum of $f_{\xi,s}$. Thus, we define

$$\mu_{\xi,s} = \inf \left\{ f_{\xi,s}(u) : u \in \Sigma_{\xi,s} \right\} . \tag{3.13}$$

Before we proceed some remarks are in order. First, the set $\Sigma_{\xi,s}$ is not void. In fact, put $u_s = -B_s^{[-r]} A_s \xi$ and $u_n = 0$ $(n = s+1, s+2 \dots)$. Then (3.9)-(3.10) lead to $x_s = \xi$, $x_{s+1} = A_s \xi - B_s B_s^{[-r]} A_s \xi = 0$ and $x_n = 0$ $(n = s+2, s+3 \dots)$. Since $(u_n)_{n=s}^{\infty}$ and $(x_n)_{n=s}^{\infty}$ are square summable, $(u_n)_{n=s}^{\infty} \in \Sigma_{\xi,s}$. Thus, $\Sigma_{\xi,s}$ is not void. Next, we remark that the sequence $(v_n)_{n=s}^{\infty}$ is square summable by the boundedness of the sequences $(C_n)_{n=-\infty}^{\infty}$ and $(D_n)_{n=-\infty}^{\infty}$, and the fact that $(u_n)_{n=s}^{\infty} \in \Sigma_{\xi,s}$ and $(x_n)_{n=s}^{\infty}$ are square summable. Since $(J_n)_{n=-\infty}^{\infty}$ and $(J_n')_{n=-\infty}^{\infty}$ are also bounded, it follows that the function $f_{\xi,s} : \Sigma_{\xi,s} \to \mathbb{R}$ is well defined by formula (3.12). Consequently $\mu_{\xi,s}$ is well defined by (3.13), and $\mu_{\xi,s}$ is a real number or $-\infty$.

THEOREM 3.1. *There exists a bounded sequence $(M_s)_{s=-\infty}^{\infty}$ of $r'' \times r''$ self-adjoint matrices such that*

$$\mu_{\xi,s} = -\xi^* M_s \xi \qquad (\xi \in \mathbb{C}^{r''}; s = 0, \pm 1, \dots) . \tag{3.14}$$

The proof of this theorem will be given at the end of this section. We now draw some consequences. First, $\mu_{\xi,s}$ is a finite real number. Next, we prove a monotonicity property of $\mu_{\xi,s}$.

LEMMA 3.2. *Let s be an integer, and $\xi \in \mathbb{C}^{r''}$ and $u_s \in \mathbb{C}^r$ be vectors. Define $\xi' = A_s \xi + B_s u_s$ and $v_s = C_s \xi + D_s u_s$. Then*

$$\mu_{\xi,s} - \mu_{\xi',s+1} \le u_s^* J_s u_s - v_s^* J' v_s . \tag{3.15}$$

PROOF. Let ε be an arbitrary positive number. By the definition (3.13) of μ, there exists a sequence $u' = (u_n)_{n=s+1}^{\infty} \in \Sigma_{\xi',s+1}$ such that

$$f_{\xi',s+1}(u') \leq \mu_{\xi',s+1} + \varepsilon \ . \tag{3.16}$$

Further, by definition of $\Sigma_{\xi',s+1}$, the sequence $(x_n)_{n=s+1}^{\infty}$ defined via $x_{s+1} = \xi'$ and

$$x_{n+1} = A_n x_n + B_n u_n \qquad (n = s+1, s+2, \ldots) \tag{3.17}$$

is square summable, and we have

$$f_{\xi',s+1}(u') = \sum_{n=s+1}^{\infty} (u_n^* J_n u_n - v_n^* J_n' v_n) \ , \tag{3.18}$$

where $v_n = C_n x_n + D_n u_n$ $(n = s+1, s+2, \ldots)$. Now define $x_s = \xi$, let u_s be as in the statement, and consider the sequences $u = (u_n)_{n=s}^{\infty}$ and $x = (x_n)_{n=s}^{\infty}$. Both sequences are square summable. Moreover, $x_s = \xi$ by definition and the equalities $x_{n+1} = A_n x_n + B_n u_n$ $(n = s+1, s+2, \ldots)$ hold by the last paragraph. By the last paragraph we also have $x_{s+1} = \xi'$. Thus, $x_{s+1} = \xi' = A_s \xi + B_s u_s = A_s x_s + B_s u_s$, where the middle equality follows from the statement of the lemma. Hence $x_{n+1} = A_n x_n + B_n u_n$ holds for $n = s, s+1 \ldots$. This implies $u \in \Sigma_{\xi,s}$. Hence

$$f_{\xi,s}(u) \geq \mu_{\xi,s} \ . \tag{3.19}$$

Now note that

$$f_{\xi,s}(u) = \sum_{n=s}^{\infty} (u_n^* J_n u_n - v_n^* J_n' v_n) \ , \tag{3.20}$$

where v_n $(n = s+1, s+2 \ldots)$ are as above, and $v_s = C_s x_s + D_s u_s = C_s \xi + D_s u_s$ is as in the statement of the lemma. Comparing (3.20) with (3.18) we obtain

$$f_{\xi,s}(u) - f_{\xi',s+1}(u') = u_s^* J_s u_s - v_s^* J_s' v_s \ .$$

However, by (3.16) and (3.19) we have

$$f_{\xi,s}(u) - f_{\xi' s+1}(u') \geq \mu_{\xi,s} - u_{\xi',s+1} - \varepsilon \ .$$

These two relations lead to $\mu_{\xi,s} - \mu_{\xi',s+1} - \varepsilon \leq u_s^* J u_s - v_s^* J' v_s$. Since $\varepsilon > 0$ is arbitrary, we obtain (3.15). □

Translating this result into matrices we obtain the following.

LEMMA 3.3. *The following matrix inequalities hold*

$$(3.21)\qquad \begin{pmatrix} M_s & 0 \\ 0 & J_s \end{pmatrix} - \begin{pmatrix} A_s^* & B_s^* \\ C_s^* & D_s^* \end{pmatrix} \begin{pmatrix} M_{s+1} & 0 \\ 0 & J_s' \end{pmatrix} \begin{pmatrix} A_s & B_s \\ C_s & D_s \end{pmatrix} \geq 0 \quad (s = 0, \pm 1 \ldots),$$

where $(M_s)_{s=-\infty}^{\infty}$ *is the sequence of self-adjoint matrices given by Theorem 3.1.*

PROOF. Let s be an arbitrary integer, and $\xi \in \mathbb{C}^{r''}$ and $u_s \in \mathbb{C}^{r}$ vectors. Then (3.15) holds, where $\xi' = A_s\xi + B_s u_s$ and $v_s = C_s\xi + D_s u_s$. On the other hand, Theorem 3.1 leads to $\mu_{\xi,s} = -\xi^* M_s \xi$ and $\mu_{\xi',s+1} = -\xi'^* M_{s+1}\xi'$. Inserting this in (3.15), we obtain

$$-\xi^* M_s \xi + \xi'^* M_{s+1}\xi' \leq u_s^* J_s u_s - v_s^* J_s' v_s \ .$$

Thus

$$(\xi^* \ u_s^*) \begin{pmatrix} M_s & 0 \\ 0 & J_s \end{pmatrix} \begin{pmatrix} \xi \\ u_s \end{pmatrix} - (\xi'^* \ v_s^*) \begin{pmatrix} M_{s+1} & 0 \\ 0 & J_s' \end{pmatrix} \begin{pmatrix} \xi' \\ v_s \end{pmatrix} \geq 0 \ .$$

However, by the definitions of ξ' and v_s,

$$\begin{pmatrix} \xi' \\ v_s \end{pmatrix} = \begin{pmatrix} A_s & B_s \\ C_s & D_s \end{pmatrix} \begin{pmatrix} \xi \\ u_s \end{pmatrix} .$$

Therefore, the preceding inequality leads to

$$(\xi^* \ u_s^*) \left(\begin{pmatrix} M_s & 0 \\ 0 & J_s \end{pmatrix} - \begin{pmatrix} A_s^* & C_s^* \\ B_s^* & D_s^* \end{pmatrix} \begin{pmatrix} M_{s+1} & 0 \\ 0 & J_s' \end{pmatrix} \begin{pmatrix} A_s & B_s \\ C_s & D_s \end{pmatrix} \right) \begin{pmatrix} \xi \\ u_s \end{pmatrix} \geq 0 \ .$$

Since $\xi \in \mathbb{C}^{r''}$ and $u_s \in \mathbb{C}^{r}$ are arbitrary, this inequality implies the matrix inequality (3.21). □

We can now present the main result of this section.

THEOREM 3.4. *Assume that the input-output operator T of the system* (3.1) *is a strict* (J, J') *contraction, and that the matrix B_n admits a right inverse $B_n^{[-r]}$* ($n = 0, \pm 1, \ldots$) *with* $\sup_n \|B_n^{[-r]}\| < \infty$. *Then there exists a bounded sequence* $(M_n)_{n=-\infty}^{\infty}$ *of* $r'' \times r''$ *matrices and a positive number* γ *such that the following matrix inequality holds for each* $n = 0, \pm 1, \ldots$

$$(3.22)\qquad \begin{pmatrix} M_n & 0 \\ 0 & J_n \end{pmatrix} - \begin{pmatrix} A_n^* & C_n^* \\ B_n^* & D_n^* \end{pmatrix} \begin{pmatrix} M_{n+1} & 0 \\ 0 & J_n' \end{pmatrix} \begin{pmatrix} A_n & B_n \\ C_n & D_n \end{pmatrix} \geq \gamma I \ .$$

PROOF. Consider the system

$$(3.23)\qquad \begin{cases} x_{n+1} = A_n x_n + B_n u_n & (n = 0, \pm 1 \ldots), \\ w_n = \begin{pmatrix} C_n \\ I_{r''} \\ 0 \end{pmatrix} x_n + \begin{pmatrix} D_n \\ 0 \\ I_r \end{pmatrix} u_n & (n = 0, \pm 1 \ldots). \end{cases}$$

Here, $\begin{pmatrix} C_n \\ I_{r''} \\ 0 \end{pmatrix}$ is a matrix of order $(r' + r'' + r) \times r''$ where the lower block entry is the null $r \times r''$ matrix, and $\begin{pmatrix} D_n \\ 0 \\ I_r \end{pmatrix}$ is a matrix of order $(r' + r'' + r) \times r$ where the middle block entry is the null $r'' \times r$ matrix. We now show that the system (3.23) also satisfies the assumptions of this section. Clearly, all the sequences of matrix coefficients of (3.23) are bounded, and B_n admits a right inverse $B_n^{[-r]}$ satisfying $\sup_n \|B_n^{[-r]}\| < \infty$ by assumption. Since the system (3.2) admits a dichotomy, then (3.23) has a well defined input-output operator which we denote by T'. As in the beginning of this section, we denote by $K : \ell_r^2(\mathbb{Z}) \to \ell_{r''}^2(\mathbb{Z})$ the bounded operator defined by (3.4). Thus, for each $(u_n)_{n=-\infty}^{\infty} \in \ell_r^2(\mathbb{Z})$, $K((u_n)_{n=-\infty}^{\infty}) = (x_n)_{n=-\infty}^{\infty}$, where $(x_n)_{n=-\infty}^{\infty}$ is the unique sequence in $\ell_{r''}^2(\mathbb{Z})$ satisfying $x_{n+1} = A_n x_n + B_n u_n$ $(n = 0, \pm 1 \ldots)$. It is clear from the forms of the systems (3.1) and (3.23) that T and T' are related as follows. Let $u = (u_n)_{n=-\infty}^{\infty} \in \ell_r^2(\mathbb{Z})$ be arbitrary, and put

$$x = (x_n)_{n=-\infty}^{\infty} = Ku, \ v = (v_n)_{n=-\infty}^{\infty} = Tu, \ w = (w_n)_{n=-\infty}^{\infty} = T'u \, . \tag{3.24}$$

Then

$$w_n = \begin{pmatrix} v_n \\ x_n \\ u_n \end{pmatrix} \qquad (n = 0, \pm 1 \ldots) \, . \tag{3.25}$$

Now let J_n, J_n' and $\delta > 0$ be the self-adjoint matrices and positive number such that inequality (3.7) holds. Put $\gamma = \delta/(2(1 + \|K\|^2))$, and define a sequence $J'' = (J_n'')_{n=-\infty}^{\infty}$ of self-adjoint matrices of order $r' + r'' + r$ via

$$J_n'' = \begin{pmatrix} J_n' & 0 & 0 \\ 0 & \gamma I_{r''} & 0 \\ 0 & 0 & \gamma I_r \end{pmatrix} \qquad (n = 0, \pm 1, \ldots) \, . \tag{3.26}$$

Since $(J_n')_{n=-\infty}^{\infty}$ is a bounded sequence so is $(J_n'')_{n=-\infty}^{\infty}$. It follows from (3.25) and (3.26) that

$$w_n^* J_n'' w_n = v_n^* J_n' v_n + \gamma \|x_n\|^2 + \gamma \|u_n\|^2 \, ,$$

where u, x, v and w are related as in (3.24). This equality and $x = Ku$ lead to

$$\sum_{n=-\infty}^{\infty} (w_n^* J_n'' w_n - v_n^* J_n' v_n) = \gamma \sum_{n=-\infty}^{\infty} (\|x_n\|^2 + \|u_n\|^2) = \gamma(\|x\|^2 + \|u\|^2)$$

$$= \gamma(\|Ku\|^2 + \|u\|^2) \le \frac{\delta}{2(1 + \|K\|^2)} (\|K\|^2 + 1)\|u\|^2 = \frac{\delta}{2} \sum_{n=-\infty}^{\infty} \|u_n\|^2 \, .$$

Combining this inequality with (3.7) we obtain

$$\sum_{n=-\infty}^{\infty} (u_n^* J_n u_n - w_n^* J_n'' w_n) \geq \frac{\delta}{2} \sum_{n=-\infty}^{\infty} \|u_n\|^2 , \tag{3.27}$$

where $(w_n)_{n=-\infty}^{\infty} = T'((u_n)_{n=-\infty}^{\infty})$ by (3.24).

Inequality (3.27) means that T' is a strict (J, J'') contraction. Hence, we may apply Lemma 3.3 with T and J' replaced by T' and J''. Thus, there exists a bounded sequence $(M_n)_{n=-\infty}^{\infty}$ of $r'' \times r''$ matrices such that

$$\begin{pmatrix} M_n & 0 \\ 0 & J_n \end{pmatrix} - \begin{pmatrix} A_n^* & C_n^* & I_{r''} & 0 \\ B_n^* & D_n^* & 0 & I_r \end{pmatrix} \begin{pmatrix} M_{n+1} & 0 & 0 & 0 \\ 0 & J_n' & 0 & 0 \\ 0 & 0 & \gamma I_{r''} & 0 \\ 0 & 0 & 0 & \gamma I_r \end{pmatrix} \begin{pmatrix} A_n & B_n \\ C_n & D_n \\ I_{r''} & 0 \\ 0 & I_r \end{pmatrix} \geq 0$$

for $n = 0, \pm 1, \ldots$. This sequence of inequalities imply (3.22). □

We now turn to the proof of Theorem 3.1. We begin by proving some bounds on the infimum μ defined in (3.13). Recall that we deal with the input-output operator T of the system (3.1), where the matrices B_n admit right inverses $B_n^{[-r]}$ satisfying (3.6). We also assume that T is a strict (J, J') contraction and let $\delta > 0$ be a number such that (3.7) holds. Finally we will use the constant N defined by (3.8).

LEMMA 3.5. *For each integer s and vector $\xi \in \mathbb{C}^{r''}$*

$$|\mu_{\xi,s}| \leq N^7 \|\xi\|^2 \tag{3.28}$$

where N is given by (3.8). *Moreover, we have*

$$f_{\xi,s}(u) \geq -N^4 \|\xi\|^2 + \delta \sum_{n=s}^{\infty} \|u_n\|^2 \qquad (u = (u_n)_{n=-\infty}^{\infty} \in \Sigma_{\xi,s}) . \tag{3.29}$$

PROOF. We first obtain an upper bound for $\mu_{\xi,s}$. Define the sequences $u = (u_n)_{n=s}^{\infty}$ and $x = (x_n)_{n=s}^{\infty}$ via $u_s = -B_s^{[-r]} A_s \xi$, $u_n = 0$ $(n = s+1, s+2 \ldots)$, $x_s = \xi$, $x_n = 0$ $(n = s+1, s+2 \ldots)$. Then u and x are square summable and $x_{n+1} = A_n x_n + B_n u_n$ holds for $n = s, s+1 \ldots$. Thus $u \in \Sigma_{\xi,s}$. Consequently, the definition of $\mu_{\xi,s}$ in (3.13) leads to

$$\mu_{\xi,s} \leq f_{\xi,s}(u) . \tag{3.30}$$

Define $v_n = C_n x_n + D_n u_n$ $(n = s, s+1 \ldots)$. Then $v_s = (C_s - D_s B_s^{[-r]} A_s)\xi$ and $v_n = 0$ $(n = s+1, s+2, \ldots)$. Since $u_n = 0$ and $v_n = 0$ for $n = s+1, s+2, \ldots$, we obtain

$$f_{\xi,s}(u) = u_s^* J_s u_s - v_s^* J_s' v_s \leq N(\|u_s\|^2 + \|v_s\|^2) , \tag{3.31}$$

where N is defined in (3.8). From $u_s = -B_s^{[-r]}A_s\xi$ and $v_s = (C_s - D_sB_s^{[-r]}A_s)\xi$, we also obtain $\|u_s\|^2 + \|v_s\|^2 \leq N^6\|\xi\|^2$. Inserting this in (3.31) it follows $f_{\xi,s}(u) \leq N^7\|\xi\|^2$. Hence, (3.13) leads to

$$\mu_{\xi,s} \leq N^7\|\xi\|^2 . \tag{3.32}$$

We now proceed to obtain a lower bound for $\mu_{\xi,s}$. Let $u = (u_n)_{n=s}^{\infty} \in \Sigma_{\xi,s}$ be arbitrary. Define $(x_n)_{u=s}^{\infty}$ via the recursion (3.9) and (3.10). Then $(x_n)_{n=s}^{\infty}$ is square summable and $f_{\xi,s}(u)$ is given by (3.12), where v_n $(n = s, s+1, \ldots)$ is defined by (3.11). Put

$$u_{s-1} = B_{s-1}^{[-r]}\xi\ , \quad u_n = 0 \quad (n = s-2, s-3, \ldots)\ ,$$

and $x_n = 0$ $(n = s-1, s-2, \ldots)$, and denote $u' = (u_n)_{n=-\infty}^{\infty}$ and $x' = (x_n)_{n=-\infty}^{\infty}$. We have $x_s = \xi = A_{s-1}0 + B_{s-1}B_{s-1}^{[-r]}\xi = A_{s-1}x_{s-1} + B_{s-1}u_{s-1}$. In addition, it is clear that $x_{n+1} = 0 = A_nx_n + B_nu_n$ $(n = s-2, s-3, \ldots)$. Combining this with (3.10), it follows that $x_{n+1} = A_nx_n + B_nu_n$ $(n = 0, \pm1, \ldots)$. Since $u' \in \ell_r^2(\mathbb{Z})$ and $x' \in \ell_{r''}^2(\mathbb{Z})$, then $x' = Ku'$. Thus, we may apply (3.7) with $v_n = C_nx_n + D_nu_n$ $(n = 0, \pm1, \ldots)$. Note that v_n has the same meaning as above for $n = s, s+1, \ldots$, and furthermore, $v_n = 0$ for $n = s-2, s-3, \ldots$. Hence, (3.7) implies

$$\sum_{n=s-1}^{\infty} (u_n^*J_nu_n - v_n^*J_n'v_n) \geq \delta \sum_{n=-\infty}^{\infty} \|u_n\|^2 \geq \delta \sum_{n=s}^{\infty} \|u_n\|^2 .$$

By the definition (3.12) of $f_{\xi,s}(u)$ we obtain

$$f_{\xi,s}(u) + u_{s-1}^*J_{s-1}u_{s-1} - v_{s-1}^*J_{s-1}'v_{s-1} \geq \delta \sum_{n=s}^{\infty} \|u_n\|^2 .$$

Thus,

$$\begin{aligned} f_{\xi,s}(u) &\geq -\|u_{s-1}\|^2\|J_{s-1}\| - \|v_{s-1}\|^2\|J_{s-1}'\| + \delta \sum_{n=s}^{\infty} \|u_n\|^2 \\ &\geq -N(\|u_{s-1}\|^2 + \|v_{s-1}\|^2) + \delta \sum_{n=s}^{\infty} \|u_n\|^2 . \end{aligned}$$

By $u_{s-1} = B_{s-1}^{[-r]}\xi$ and $v_{s-1} = C_{s-1}x_{s-1} + D_{s-1}u_{s-1} = D_{s-1}u_{s-1}$ we have $\|u_{s-1}\|^2 + \|v_{s-1}\|^2 \leq \|B_{s-1}^{[-r]}\|^2(1 + \|D_{s-1}\|^2)\|\xi\|^2 \leq N^4\|\xi\|^2$. Hence, the above inequality for $f_{\xi,s}(u)$ implies

$$f_{\xi,s}(u) \geq -N^4\|\xi\|^2 + \delta \sum_{n=s}^{\infty} \|u_n\|^2 \quad (u = (u_n)_{n=s}^{\infty} \in \Sigma_{\xi,s}) .$$

Thus, (3.29) holds. In particular, $f_{\xi,s}(u) \geq -N^4\|\xi\|^2$ $(u \in \Sigma_{\xi,s})$. Hence the definition (3.13) leads to $\mu_{\xi,s} \geq -N^4\|\xi\|^2$. This inequality and (3.32) imply (3.28). □

We now introduce some notation. For each integer s, we denote by $\ell^2_{r,s}$ the Hilbert space of all square summable sequences $u = (u_n)_{n=s}^\infty$, with $u_n \in \mathbb{C}^r (n = s, s+1\ldots)$, endowed with the norm $\|u\| = (\sum_{n=s}^\infty \|u_n\|^2)^{1/2}$. The spaces $\ell^2_{r',s}$ and $\ell^2_{r'',s}$ are defined similarly. We will also use the Hilbert space $\ell^2_{r,s} \oplus \mathbb{C}^{r''}$ consisting of all pairs $((u_n)_{n=s}^\infty, \xi)$, where $(u_n)_{n=s}^\infty \in \ell^2_{r,s}$, $\xi \in \mathbb{C}^{r''}$ with the norm

$$\|((u_n)_{n=s}^\infty, \xi), \| = \left(\sum_{n=s}^\infty \|u_n\|^2 + \|\xi\|^2\right)^{1/2} .$$

Now fix an integer s. Denote by H_s the set of all pairs $((u_n)_{n=s}^\infty, \xi) \in \ell^2_{r,s} \oplus \mathbb{C}^{r''}$ such that $(u_n)_{n=s}^\infty \in \Sigma_{\xi,s}$. We define a mapping $\Pi_s : H_s \to \ell^2_{r',s}$ as follows. Let $((u_n)_{n=s}^\infty, \xi) \in H_s$. Since $(u_n)_{n=s}^\infty \in \Sigma_{\xi,s}$, the sequence $(x_n)_{n=s}^\infty$ defined by $x_s = \xi$ and $x_{n+1} = A_n x_n + B_n u_n$ $(n = s, s+1, \ldots)$ is square summable. Put $v_n = C_n x_n + D_n u_n$ $(n = s, s+1, \ldots)$. Since $(C_n)_{n=-\infty}^\infty$ and $(D_n)_{n=-\infty}^\infty$ are bounded, the sequence $(v_n)_{n=s}^\infty$ is square summable. We now define

$$\Pi_s((u_n)_{n=s}^\infty, \xi) = (v_n)_{n=s}^\infty \in \ell^2_{r',s} . \tag{3.33}$$

LEMMA 3.6. *The set H_s is a closed linear subspace of $\ell^2_{r,s} \oplus \mathbb{C}^{r''}$, and the mapping $\Pi_s : H_s \to \ell^2_{r',s}$ is linear and bounded.*

PROOF. We first define an operator $\Lambda : \ell^2_{r,s} \oplus \mathbb{C}^{r''} \to \ell^2_r(\mathbb{Z}) \oplus \mathbb{C}^{r''}$ as follows. Let $((u_n)_{n=s}^\infty, \xi) \in \ell^2_{r,s} \oplus \mathbb{C}^{r''}$. Put $u_{s-1} = B_{s-1}^{[-r]}\xi$, and $u_n = 0$ $(n = s-2, s-3, \ldots)$, and set $\Lambda((u_n)_{n=s}^\infty, \xi) = ((u_n)_{n=-\infty}^\infty, \xi)$. Clearly Λ is a bounded and linear operator. Next we define two operators

$$\Gamma : \ell^2_r(\mathbb{Z}) \oplus \mathbb{C}^{r'} \to \ell^2_{r',s}$$

and

$$\Omega : \ell^2_r(\mathbb{Z}) \oplus \mathbb{C}^{r''} \to \mathbb{C}^{r''}$$

as follows. Let $((u_n)_{n=-\infty}^\infty, \xi) \in \ell^2_r(\mathbb{Z}) \oplus \mathbb{C}^{r''}$. Denote $(x_n)_{n=-\infty}^\infty = K((u_n)_{n=-\infty}^\infty)$, and $(v_n)_{n=-\infty}^\infty = T((u_n)_{n=-\infty}^\infty)$. Then put

$$\Gamma((u_n)_{n=-\infty}^\infty, \xi) = (v_n)_{n=s}^\infty \tag{3.34}$$

and

$$\Omega((u_n)_{n=-\infty}^\infty, \xi) = x_s - \xi . \tag{3.35}$$

Since K and T are linear and bounded, also Γ and Ω are linear and bounded. The lemma follows from the equalities

$$H_s = \operatorname{Ker}(\Omega\Lambda) , \tag{3.36}$$

and

$$\Pi_s = \Gamma\Lambda\big|_{H_s} , \tag{3.37}$$

which we now prove. Indeed (3.36) shows that H_s is a closed linear subspace of $\ell^2_{r,s} \oplus \mathbb{C}^{r''}$, and (3.37) shows that Π_s is linear and bounded.

To prove (3.36) and (3.37) assume first that $((u_n)_{n=s}^{\infty}, \xi) \in H_s$. Then $(u_n)_{n=s}^{\infty} \in \Sigma_{\xi,s}$, whence the sequence $(x_n)_{n=s}^{\infty}$ defined via

$$x_s = \xi \tag{3.38}$$

and the recursion

$$x_{n+1} = A_n x_n + B_n u_n \qquad (n = s, s+1, \ldots) \tag{3.39}$$

is square summable. Put $u_{s-1} = B_{s-1}^{[-r]}\xi$, $u_n = 0$ $(n = s-2, s-3, \ldots)$, and $x_n = 0$ $(n = s-1, s-2, \ldots)$. Then

$$\Lambda((u_n)_{n=s}^{\infty}, \xi) = ((u_n)_{n=-\infty}^{\infty}, \xi) . \tag{3.40}$$

Moreover, note that $x_s = \xi = A_{s-1}0 + B_{s-1}B_{s-1}^{[-r]}\xi = A_{s-1}x_{s-1} + B_{s-1}u_{s-1}$, and in addition $x_{n+1} = 0 = A_n x_n + b_n u_n$ $(n = s-2, s-3, \ldots)$. Hence

$$x_{n+1} = A_n x_n + B_n u_n \qquad (n = 0, \pm 1, \ldots) .$$

Since $(x_n)_{n=s}^{\infty}$ and $(u_n)_{n=s}^{\infty}$ are square summable, also $(x_n)_{n=-\infty}^{\infty}$ and $(u_n)_{n=-\infty}^{\infty}$ are square summable. Hence the last equalities lead to

$$(x_n)_{n=-\infty}^{\infty} = K((u_n)_{n=-\infty}^{\infty}) . \tag{3.41}$$

Denote

$$v_n = C_n x_n + D_n u_n \qquad (n = 0, \pm 1, \ldots) . \tag{3.42}$$

Then, by the definition of the input-output operator T, we have

$$(v_n)_{n=-\infty}^{\infty} = T((u_n)_{n=-\infty}^{\infty}) . \tag{3.43}$$

By the equalities (3.41) and (3.43), and the definitions (3.34) and (3.35) of Γ and Ω we have

$$\Gamma((u_n)_{n=-\infty}^{\infty}, \xi) = (v_n)_{n=s}^{\infty} , \tag{3.44}$$

and

$$\Omega((u_n)_{n=-\infty}^{\infty}, \xi) = x_s - \xi .$$

Combining the last equality with (3.38) and (3.40) we obtain

$$\Omega\Lambda((u_n)_{n=s}^{\infty}, \xi) = \Omega((u_n)_{n=-\infty}^{\infty}, \xi) = x_s - \xi = 0 .$$

Since $((u_n)_{n=s}^{\infty}, \xi) \in H_s$ is arbitrary, this implies

$$H_s \subset \operatorname{Ker}(\Omega\Lambda) . \tag{3.45}$$

In addition, it follows from the definition (3.33) of the mapping Π_s, and the equalities (3.38), (3.39) and (3.42) that

$$\Pi_s((u_n)_{n=s}^{\infty}, \xi) = (v_n)_{n=s}^{\infty} .$$

By (3.40) and (3.44), this implies

$$\Pi_s((u_n)_{n=s}^{\infty}, \xi) = \Gamma((u_n)_{n=-\infty}^{\infty}, \xi) = \Gamma\Lambda\ ((u_n)_{n=s}^{\infty}, \xi) .$$

This equality implies (3.37).

Conversely, assume that $((u_n)_{n=s}^{\infty}, \xi) \in \operatorname{Ker}(\Omega\Lambda)$. Then $0 = \Omega\Lambda((u_n)_{n=s}^{\infty}, \xi) = \Omega((u_n)_{n=-\infty}^{\infty}, \xi)$, where $u_{s-1} = B_{s-1}^{[-r]}\xi$, and $u_n = 0$ $(n = s-2, s-3, \ldots)$. Denote $(x_n)_{n=-\infty}^{\infty} = K((u_n)_{n=-\infty}^{\infty}) \in \ell_{r''}^2(\mathbb{Z})$. Then by the definition (3.35) of Ω, $x_s - \xi = \Omega((u_n)_{n=-\infty}^{\infty}, \xi) = 0$. Thus, $x_s = \xi$. In addition, the definition of the operator K and $(x_n)_{n=-\infty}^{\infty} = K((u_n)_{n=-\infty}^{\infty})$ imply $x_{n+1} = A_n x_n + B_n u_n$ $(n = 0, \pm 1, \ldots)$. In particular, $x_{n+1} = A_n x_n + B_n u_n$ $(n = s, s+1, \ldots)$. Combining this with $x_s = \xi$, and the fact that $(x_n)_{n=s}^{\infty}$ and $(u_n)_{n=s}^{\infty}$ are square summable, we obtain $((u_n)_{n=s}^{\infty}, \xi) \in H_s$. Thus, $\operatorname{Ker}(\Omega\Lambda) \subset H_s$. This inclusion and (3.45) imply (3.36). Hence, (3.36) and (3.37) hold true. As remarked above, this implies the lemma. □

PROOF THEOREM 3.1. Let s be an integer. Recall that by Lemma 3.6, H_s is a closed linear subspace of $\ell_{r,s}^2 \oplus \mathbb{C}^{r''}$. In particular, H_s is a Hilbert space. We now define an operator

$$R_s : H_s \to \ell_{r,s}^2 \oplus \ell_{r',s}^2 \oplus \mathbb{C}^{r''}$$

as follows. Let $((u_n)_{n=s}^{\infty}, \xi) \in H_s$. By the definition of H_s, $(u_n)_{n=s}^{\infty} \in \Sigma_{\xi,s}$. Hence, the sequence $(x_n)_{n=s}^{\infty}$ defined via

$$x_s = \xi \quad \text{and} \quad x_{n+1} = A_n x_n + B_n u_n \ (n = s, s+1, \dots) , \tag{3.46}$$

is square summable. Denote

$$v_n = C_n x_n + D_n u_n \ (n = s, s+1, \dots) , \tag{3.47}$$

and put

$$R_s((u_n)_{n=s}^{\infty}, \xi) = ((u_n)_{n=s}^{\infty}, (v_n)_{n=s}^{\infty}, \xi) . \tag{3.48}$$

Clearly $(v_n)_{n=s}^{\infty}$ is square summable, whence R_s is well defined. Since the operator $\Pi_s : H_s \to \ell^2_{r',s}$ defined by $\Pi_s((u_n)_{n=s}^{\infty}, \xi) = (v_n)_{n=s}^{\infty}$ is linear and bounded by Lemma 3.6, the operator R_s is linear and bounded. Now let N be the constant defined by (3.8), and set $N' = N^4 + \delta$. We define a self-adjoint operator Γ_s in $\ell^2_{r,s} \oplus \ell^2_{r',s} \oplus \mathbb{C}^{r''}$ via

$$\Gamma_s((u_n)_{n=s}^{\infty}, (v_n)_{n=s}^{\infty}, \xi) = ((J_n u_n)_{n=s}^{\infty}, (-J'_n v_n)_{n=s}^{\infty}, N'\xi) . \tag{3.49}$$

Since $(J_n)_{n=-\infty}^{\infty}$ and $(J'_n)_{n=-\infty}^{\infty}$ are bounded, Γ is a bounded self-adjoint operator. Consider now the operator $L_s = R_s^* \Gamma_s R_s$. Then L_s is a bounded self-adjoint operator on H_s. Note that if $w = ((u_n)_{n=s}^{\infty}, \xi) \in H_s$, then by (3.48)-(3.49)

$$\langle L_s w, w \rangle = \langle \Gamma_s R_s w, R_s w \rangle = \sum_{n=s}^{\infty} (u_n^* J_n u_n - v_n^* J'_n v_n) + N' \|\xi\|^2 ,$$

where v_n $(n = s, s+1 \dots)$ is defined by (3.46)-(3.47). Comparing this with the definition (3.12) of $f_{\xi,s}(u)$, we obtain

$$\langle L_s w, w \rangle = f_{\xi,s}(u) + N' \|\xi\|^2 , \tag{3.50}$$

where $w = ((u_n)_{n=s}^{\infty}, \xi) \in H_s$ and $u = (u_n)_{n=s}^{\infty}$. Moreover, applying inequality (3.29) to (3.50) we obtain

$$\begin{aligned} \langle L_s w, w \rangle &\geq \delta \sum_{n=s}^{\infty} \|u_n\|^2 + (N' - N^4) \|\xi\|^2 \\ &= \delta \sum_{n=s}^{\infty} \|u_n\|^2 + \delta \|\xi\|^2 = \delta \|w\|^2 , \end{aligned}$$

where we used the definition $N' = N^4 + \delta$. This inequality means that

$$L_s \geq \delta I . \tag{3.51}$$

We now use the operator $Q_s : H_s \to \mathbb{C}^{r''}$ giving the projection on the second entry, namely $Q_s((u_n)_{n=s}^{\infty}, \xi) = \xi$. The operator Q_s is surjective. In fact, let $\xi \in \mathbb{C}^{r''}$ and define $u_s = -B_s^{[-r]} A_s \xi$ and $u_n = 0$ $(n = s, s+1, \ldots)$. Then $u = (u_n)_{n=s}^{\infty} \in \Sigma_{\xi,s}$. Indeed, put $x_s = \xi$, and define $x_{n+1} = A_n x_n + B_n u_n$ $(n = s, s+1, \ldots)$. Then $x_{s+1} = A_s \xi - B_s B_s^{[-r]} A_s \xi = 0$, and $x_n = 0$ $(n = s+2, s+3, \ldots)$. Thus $(x_n)_{n=s}^{\infty}$ is square summable, and hence $(u_n)_{n=s}^{\infty} \in \Sigma_{\xi,s}$. By definition of H_s, this leads to $((u_n)_{n=s}^{\infty}, \xi) \in H_s$. Since $Q_s((u_n)_{n=s}^{\infty}, \xi) = \xi$, we conclude that Q_s is surjective.

It follows from the surjectivity of $Q_s : H_s \to \mathbb{C}^{r''}$ and the inequality (3.51) for the bounded self-adjoint operator L_s in H_s, that there exists a self-adjoint matrix V_s of order $r'' \times r''$ such that for each $\xi \in \mathbb{C}^{r''}$ we have

$$(3.52) \qquad \min\{\langle L_s w, w\rangle : w \in H_s, Q_s w = \xi\} = \xi^* V_s \xi \qquad (s = 0, \pm 1, \ldots; \xi \in \mathbb{C}^{r''}).$$

This is an elementary fact of Hilbert spaces, which is given in Lemma 3.7 below, for completeness.

Now note that for given $s = 0, \pm 1, \ldots$ and $\xi \in \mathbb{C}^{r''}$, the set $\{w \in H_s : Q_s w = \xi\}$ consists of all $w = ((u_n)_{n=s}^{\infty}, \xi) \in H_s$. By definition of H_s, this set is $\{(u, \xi) : u \in \Sigma_{\xi,s}\}$. It follows from this description, and from formula (3.50), that (3.52) may be written as

$$\min\left\{ f_{\xi,s}(u) + N'\|\xi\|^2 : u \in \Sigma_{\xi,s} \right\} = \xi^* V_s \xi \qquad (s = 0, \pm 1, \ldots; \xi \in \mathbb{C}^{r''}).$$

Since the term $N'\|\xi\|^2$ is independent of u, we obtain

$$\min\left\{ f_{\xi,s}(u) : u \in \Sigma_{\xi,s} \right\} = \xi^* V_s \xi - N'\|\xi\|^2 \qquad (s = 0, \pm 1, \ldots; \xi \in \mathbb{C}^{r''}).$$

Now put $M_s = N'I - V_s$. Then

$$\min\left\{ f_{\xi,s}(u) : u \in \Sigma_{\xi,s} \right\} = -\xi^* M_s \xi \qquad (s = 0, \pm 1, \ldots; \xi \in \mathbb{C}^{r''}).$$

By the definition (3.13) of $\mu_{\xi,s}$ we obtain $\mu_{\xi,s} = -\xi^* M_s \xi$. Thus, (3.14) holds.

Finally, in order to show that the sequence M_s is bounded, note that by Lemma 3.5, we have $|\xi^* M_s \xi| = |\mu_{\xi,s}| \leq N^7 \|\xi\|^2$ $(\xi \in \mathbb{C}^{r''})$. Hence $\|M_s\| \leq N^7 (s = 0, \pm 1, \ldots)$. □

In the proof of Lemma 3.1 we used the following elementary result, whose proof we include for completeness. Since we use inner products in different spaces, we use the notation $\langle \cdot, \cdot \rangle_H$ to indicate the inner product in the Hilbert space H.

LEMMA 3.7. *Let H and H' be two Hilbert spaces, $Q : H \to H'$ a surjective bounded linear operator, and L a bounded self-adjoint operator in H such that $L \geq \delta I$*

for some $\delta > 0$. Then there exists a bounded self-adjoint operator V in H' such that for each $\xi \in H'$

$$\min\{\langle Lw, w\rangle_H : w \in H, Qw = \xi\} = \langle V\xi, \xi\rangle_{H'} .$$

PROOF. For $w \in H$, put $\|w\|_L^2 = \langle Lw, w\rangle_H$. Since L is self-adjoint bounded and $L \geq \delta I$, the function $w \longrightarrow \|w\|_L$ is a norm, and the space H endowed with this norm is a Hilbert space. The original norm and the L norm on H are topologically equivalent. Now let R be a right inverse of Q, and P be the L-orthogonal projection in H along Ker$Q$ and onto $(\mathrm{Ker}Q)^{\perp_L}$. Note that Ker$P$ = Ker$Q$ leads to $Q(I-P) = 0$, whence to $QP = Q$. Moreover, the vector of minimal L-norm in the affine subspace $\{w \in H : Qw = \xi\}$ is given by $w_0 = PR\xi$. In fact, note first that $Qw_0 = QPR\xi = QR\xi = \xi$, where we used $QP = Q$ and $QR = I$. In addition, assume $w \in H$ satisfies $Qw = \xi$. Then $w - w_0 \in \mathrm{Ker}Q = \mathrm{Ker}P$. Since $w_0 \in \mathrm{Im}\, P$ we obtain $w - w_0 \perp_L w_0$. Thus, $\|w\|_L^2 = \|w_0\|_L^2 + \|w - w_0\|_L^2 \geq \|w_0\|_L^2$ with equality if and only if $w = w_0$. This means that

$$\min\{\|w\|_L^2 : w \in H, Qw = \xi\} = \|PR\xi\|_L^2 = \langle LPR\xi, PR\xi\rangle_H .$$

Since $\|w\|_L^2 = \langle Lw, w\rangle_H$, this equality implies

$$\min\{\langle Lw, w\rangle_H : w \in H, Qw = \xi\} = \langle R^*P^*LPR\xi, \xi\rangle_{H'} .$$

Hence the lemma holds with $V = R^*P^*LPR$. □

We conclude this section with the following remark. Assume that the coefficients sequences $(A_n)_{n=-\infty}^{\infty}$, $(B_n)_{n=-\infty}^{\infty}$, $(C_n)_{n=-\infty}^{\infty}$, and $(D_n)_{n=-\infty}^{\infty}$ of the system (3.1), and the sequences $(J_n)_{n=-\infty}^{\infty}$, $(J'_n)_{n=-\infty}^{\infty}$ are periodic with period m, where m is a natural number. Then it is clear from the definition that $\mu_{\xi,s} = \mu_{\xi,s+m}$. Hence, Theorem 3.1 shows that $M_s = M_{s+m}$ $(s = 0, \pm 1, \ldots)$. Finally, the extended system (3.23) in Theorem 3.4 is also periodic with period m. Hence, the matrices M_n in Theorem 3.4 can be chosen to be periodic with period m.

4. STRICT (J, J') CONTRACTIONS: NECESSARY CONDITIONS IN THE GENERAL CASE

In this section we consider a system

$$\text{(4.1)} \qquad \begin{cases} x_{n+1} = A_n x_n + B_n u_n & (n = 0, \pm 1, \ldots) , \\ v_n = C_n x_n + D_n u_n & (n = 0, \pm 1, \ldots) , \end{cases}$$

where $(A_n)_{n=-\infty}^{\infty}$, $(B_n)_{n=-\infty}^{\infty}$, $(C_n)_{n=-\infty}^{\infty}$, $(D_n)_{n=-\infty}^{\infty}$ are bounded sequences of matrices of sizes $r'' \times r''$, $r'' \times r$, $r' \times r''$ and $r' \times r$, respectively, and such that the system

$$x_{n+1} = A_n x_n \qquad (n = 0, \pm 1 \ldots) \tag{4.2}$$

admits a dichotomy. Then the system (4.1) admits a well defined input-output operator $T : \ell_r^2(\mathbb{Z}) \to \ell_{r'}^2(\mathbb{Z})$ defined by $T((u_n)_{n=-\infty}^{\infty}) = (v_n)_{n=-\infty}^{\infty}$.

With the system (4.1) we associate the following system

$$\begin{cases} x_{n+1} = A_n x_n + (B_n \ I_{r''}) \begin{pmatrix} u_n \\ f_n \end{pmatrix} & (n = 0, \pm 1, \ldots)\,, \\ v_n = C_n x_n + (D_n \ 0) \begin{pmatrix} u_n \\ f_n \end{pmatrix} & (n = 0, \pm 1, \ldots)\,, \end{cases} \tag{4.3}$$

where the input is now $\begin{pmatrix} u_n \\ f_n \end{pmatrix}_{n=-\infty}^{\infty}$, with $(u_n)_{n=-\infty}^{\infty} \in \ell_r^2(\mathbb{Z})$ and $(f_n)_{n=-\infty}^{\infty} \in \ell_{r''}^2(\mathbb{Z})$. Denote by $\Theta : \ell_{r+r''}^2(\mathbb{Z}) \to \ell_{r'}^2(\mathbb{Z})$ the input output operator of (4.3). It is clear that

$$\Theta\left(\begin{pmatrix} u_n \\ 0 \end{pmatrix}_{n=-\infty}^{\infty} \right) = T\Big((u_n)_{n=-\infty}^{\infty} \Big)\,. \tag{4.4}$$

The following result enables us to transfer contractivity properties from T to Θ. In the sequel, we let $(J, J') = ((J_n)_{n=-\infty}^{\infty}, (J'_n)_{n=-\infty}^{\infty})$ be two bounded sequences of self-adjoint matrices of orders $r \times r$ and $r' \times r'$, respectively. Recall that T is a strict (J, J') contraction if there is a positive number δ such that

$$\sum_{n=-\infty}^{\infty} (u_n^* J_n u_n - v_n^* J'_n v_n) \geq \delta \sum_{n=-\infty}^{\infty} \|u_n\|^2\,, \tag{4.5}$$

whenever $(u_n)_{n=-\infty}^{\infty} \in \ell_r^2(\mathbb{Z})$, and $T((u_n)_{n=-\infty}^{\infty}) = (v_n)_{n=-\infty}^{\infty}$. We will also use the following notation. If α is a positive number, then the pair $\left(\begin{pmatrix} J & 0 \\ 0 & \alpha I \end{pmatrix}, J' \right)$ denotes

$$\left(\begin{pmatrix} J & 0 \\ 0 & \alpha I \end{pmatrix}, J' \right) = \left(\begin{pmatrix} J_n & 0 \\ 0 & \alpha I_{r''} \end{pmatrix}_{n=-\infty}^{\infty}, (J'_n)_{n=-\infty}^{\infty} \right)\,. \tag{4.6}$$

THEOREM 4.1. *If the input-output operator T of (4.1) is a (J, J') strict contraction, then there exists a positive number α such that the input-output operator Θ of (4.3) is a strict $\left(\begin{pmatrix} J & 0 \\ 0 & \alpha I \end{pmatrix}, J' \right)$ contraction.*

REMARK. Let us remark that even in the case when T is a strict contraction in the usual sense, namely $J_n = I_r$ and $J'_n = I_{r'}$, then $\begin{pmatrix} J_n & 0 \\ 0 & \alpha I_{r''} \end{pmatrix} = \begin{pmatrix} I_r & 0 \\ 0 & \alpha I_{r''} \end{pmatrix}$ is not the identity. Hence, Θ is a strict contraction in a metric different than the usual one.

PROOF. Let $u = (u_n)_{n=-\infty}^{\infty} \in \ell_r^2(\mathbb{Z})$, $f = (f_n)_{n=-\infty}^{\infty} \in \ell_{r''}^2(\mathbb{Z})$, and set

$$v = (v_n)_{n=-\infty}^{\infty} = \Theta \begin{pmatrix} u \\ f \end{pmatrix} , \tag{4.7}$$

and

$$\overline{v} = (\overline{v}_n)_{n=-\infty}^{\infty} = Tu = \Theta \begin{pmatrix} u \\ 0 \end{pmatrix} , \tag{4.8}$$

where we used (4.4). We also define a self-adjoint operator $J' = (\delta_{ij} J'_j)_{ij=-\infty}^{\infty}$ in $\ell_{r'}^2$. Since $(J'_n)_{n=-\infty}^{\infty}$ is a bounded sequence of self-adjoint matrices, J' is a bounded self-adjoint operator in $\ell_{r'}^2$. Then

$$\langle J'v, v\rangle = \sum_{n=-\infty}^{\infty} v_n^* J'_n v_n, \ \langle J'\overline{v}, \overline{v}\rangle = \sum_{n=-\infty}^{\infty} \overline{v}_n^* J'_n \overline{v}_n , \tag{4.9}$$

where $\langle\cdot,\cdot\rangle$ denotes the inner product in $\ell_{r'}^2$. Let us now remark that by (4.7)-(4.8) we have

$$\|v\| \le \|\Theta\|(\|u\| + \|f\|) , \quad \text{and} \quad \|\overline{v}\| \le \|\Theta\| \ \|u\| . \tag{4.10}$$

In addition, (4.7)-(4.8) lead to $v - \overline{v} = \Theta \begin{pmatrix} 0 \\ f \end{pmatrix}$, whence

$$\|v - \overline{v}\| \le \|\Theta\| \ \|f\| . \tag{4.11}$$

We now have the estimates

$$\begin{aligned} \langle J'v, v\rangle - \langle J'\overline{v}, \overline{v}\rangle &= \langle J'(v - \overline{v}), v\rangle + \langle J'\overline{v}, (v - \overline{v})\rangle \\ &\le \|J'\| \ \|v - \overline{v}\| \ \|v\| + \|J'\| \ \|\overline{v}\| \ \|v - \overline{v}\| \\ &\le \|J'\| \ \|v - \overline{v}\|(\|v\| + \|\overline{v}\|) . \end{aligned}$$

By (4.10)-(4.11), this leads to

$$\begin{aligned} \langle J'v, v\rangle - \langle J'\overline{v}, \overline{v}\rangle &\le \|J'\| \ \|\Theta\| \ \|f\| \ \|\Theta\|(2\|u\| + \|f\|) \\ &= 2\|J'\| \ \|\Theta\|^2 \|f\| \ \|u\| + \|J'\| \ \|\Theta\|^2 \|f\|^2 . \end{aligned} \tag{4.12}$$

Let us now recall that for all nonnegative numbers b, t and s, with $b > 0$, we have

$$0 \leq \left(\sqrt{b}t - \frac{s}{\sqrt{b}}\right)^2 = bt^2 - 2st + \frac{s^2}{b},$$

thus, $2st \leq bt^2 + \frac{s^2}{b}$. Now let δ be the positive number in inequality (4.5). Applying the previous inequality with $b = \frac{\delta}{2}$, $t = \|u\|$ and $s = \|J'\|\ \|\Theta\|^2\|f\|$ we obtain

$$2\|J'\|\ \|\Theta\|^2\|f\|\ \|u\| \leq \frac{\delta}{2}\|u\|^2 + \frac{2\|J'\|^2\|\Theta\|^4}{\delta}\|f\|^2 .$$

Combining this inequality with (4.12), it follows that

$$\langle J'v, v\rangle - \langle J'\overline{v}, \overline{v}\rangle \leq \frac{\delta}{2}\|u\|^2 + \left[\frac{2\|J'\|^2\|\Theta\|^4}{\delta} + \|J'\|\ \|\Theta\|^2\right]\|f\|^2 .$$

Put $\alpha = \frac{\delta}{2} + \frac{2\|J'\|^2\|\Theta\|^4}{\delta} + \|J'\|\ \|\Theta\|^2$. Then this inequality becomes

$$\langle J'v, v\rangle - \langle J'\overline{v}, \overline{v}\rangle \leq \frac{\delta}{2}\|u\|^2 + \left(\alpha - \frac{\delta}{2}\right)\|f\|^2 .$$

Thus,

$$\langle J'\overline{v}, \overline{v}\rangle - \langle J'v, v\rangle \geq -\left(\frac{\delta}{2}\|u\|^2 + \left(\alpha - \frac{\delta}{2}\right)\|f\|^2\right) .$$

We now rewrite the last inequality entrywise. Then we obtain

$$\sum_{n=-\infty}^{\infty} (\overline{v}_n^* J_n' \overline{v}_n - v_n^* J_n' v_n) \geq -\sum_{n=-\infty}^{\infty} \left(\frac{\delta}{2}\|u_n\|^2 + \left(\alpha - \frac{\delta}{2}\right)\|f_n\|^2\right) . \tag{4.13}$$

On the other hand, it follows from (4.8) and the strict (J, J') contractivity property (4.5) of T that

$$\sum_{n=-\infty}^{\infty} (u_n^* J_n u_n - \overline{v}_n^* J_n' \overline{v}_n) \geq \delta \sum_{n=-\infty}^{\infty} \|u_n\|^2 .$$

Adding this inequality and (4.13) we obtain

$$\sum_{n=-\infty}^{\infty} (u_n^* J_n u_n - v_n^* J_n' v_n) \geq \frac{\delta}{2} \sum_{n=-\infty}^{\infty} \|u_n\|^2 - \left(\alpha - \frac{\delta}{2}\right) \sum_{n=-\infty}^{\infty} \|f_n\|^2 .$$

Hence,

$$\sum_{n=-\infty}^{\infty} \left((u_n^*\ f_n^*) \begin{pmatrix} J_n & 0 \\ 0 & \alpha I_{r''} \end{pmatrix} \begin{pmatrix} u_n \\ f_n \end{pmatrix} - v_n^* J_n' v_n \right) \geq \frac{\delta}{2} \sum_{n=-\infty}^{\infty} (u_n\|^2 + \|f_n\|^2) .$$

In view of equality (4.7), this means that Θ is a strict $\left(\begin{pmatrix} J & 0 \\ 0 & \alpha I \end{pmatrix}, J'\right)$ contraction. □

Combining Theorem 4.1 with Theorem 3.4 we obtain the following result.

THEOREM 4.2. *If the input-output operator T of the system (4.1) is a strict (J, J') contraction, then there exists a bounded sequence $(M_n)_{n=-\infty}^{\infty}$ of $r'' \times r''$ self-adjoint matrices and a positive number γ such that*

$$\begin{pmatrix} M_n & 0 \\ 0 & J_n \end{pmatrix} - \begin{pmatrix} A_n^* & C_n^* \\ B_n^* & D_n^* \end{pmatrix} \begin{pmatrix} M_{n+1} & 0 \\ 0 & J_n' \end{pmatrix} \begin{pmatrix} A_n & B_n \\ C_n & D_n \end{pmatrix} \geq \gamma I . \tag{4.14}$$

PROOF. By Theorem 4.1, there exists a positive number α such that the input output operator Θ of (4.3) is a $\left(\begin{pmatrix} J & 0 \\ 0 & \alpha I_{r''} \end{pmatrix}, J'\right)$ strict contraction. Note also that the matrix $(B_n \ I_{r''})$ appearing in (4.3) admits the right inverse $B_n^{[-r]} = \begin{pmatrix} 0 \\ I_{r''} \end{pmatrix}$, and that $\sup_n \|B_n^{[-r]}\| < \infty$. Hence, we may apply Theorem 3.4 to Θ. It follows that there exists a bounded sequence $(M_n)_{n=-\infty}^{\infty}$ of $r'' \times r''$ self-adjoint matrices, and a positive number γ such that

$$\begin{pmatrix} M_n & 0 & 0 \\ 0 & J_n & 0 \\ 0 & 0 & \alpha I_{r''} \end{pmatrix} - \begin{pmatrix} A_n^* & C_n^* \\ B_n^* & D_n^* \\ I_{r''} & 0 \end{pmatrix} \begin{pmatrix} M_{n+1} & 0 \\ 0 & J_n' \end{pmatrix} \begin{pmatrix} A_n & B_n & I_{r''} \\ C_n & D_n & 0 \end{pmatrix} \geq \gamma I.$$

Multiplying by $\begin{pmatrix} I_{r''} & 0 \\ 0 & I_r \\ 0 & 0 \end{pmatrix}$ on the right, and $\begin{pmatrix} I_{r''} & 0 & 0 \\ 0 & I_r & 0 \end{pmatrix}$ on the left, we obtain (4.14). □

REMARK. As in the remark at the end of Section 3, note that if all the sequences $(A_n)_{n=-\infty}^{\infty}, (B_n)_{n=-\infty}^{\infty}, (C_n)_{n=-\infty}^{\infty}, (D_n)_{n=-\infty}^{\infty}$, $(J_n)_{n=-\infty}^{\infty}$ and $(J_n')_{n=-\infty}^{\infty}$ are periodic with period m, then in Theorem 3.4, hence also in Theorem 4.2, we can take $(M_n)_{n=-\infty}^{\infty}$ to be periodic with period m.

5. THE MAIN RESULTS

In this section, we prove Theorems 1.1 and 1.3. We begin with the latter.

PROOF OF THEOREM 1.3. Assume first that T is a strict (J, J') contraction. Then by Theorem 4.2 there exists a bounded sequence $(M_n)_{n=-\infty}^{\infty}$ of self-adjoint matrices of order $r'' \times r''$, and a positive number γ such that (4.14) holds. Moreover, by the remarks at the end of Sections 3 and 4, if all the sequences $(A_n)_{n=-\infty}^{\infty}$, $(B_n)_{n=-\infty}^{\infty}$, $(C_n)_{n=-\infty}^{\infty}$, $(D_n)_{n=-\infty}^{\infty}$, $(J_n)_{n=-\infty}^{\infty}$ and $(J_n')_{n=-\infty}^{\infty}$ are periodic with period m, then we can choose

$(M_n)_{n=-\infty}^{\infty}$ also periodic of period m. Here the stationary case corresponds to $m = 1$. Now, it follows from (4.14) that

$$\text{(5.1)} \qquad \begin{pmatrix} M_n - A_n^* M_{n+1} A_n - C_n^* J_n' C_n & -A_n^* M_{n+1} B_n - C_n^* J_n' D_n \\ -B_n^* M_{n+1} A_n - D_n^* J_n' C_n & J_n - B_n^* M_{n+1} B_n - D_n^* J_n' D_n \end{pmatrix} \geq \gamma I$$

for $n = 0, \pm 1, \ldots$. Since the left hand side of this inequality in a bounded sequence, we may apply Lemma 2.6 to the effect that (1.9) and (1.10) hold for some $\varepsilon > 0$.

Conversely, assume that (1.9)-(1.10) hold for some $\varepsilon > 0$ and some bounded sequence $(M_n)_{n=-\infty}^{\infty}$ of self-adjoint matrices of order $r'' \times r''$. Then, by Lemma 2.6, inequality (5.1) holds for some $\gamma > 0$, and therefore, inequality (4.14) also holds. Now let $(u_n)_{n=-\infty}^{\infty} \in \ell_r^2(\mathbb{Z})$, denote by $(x_n)_{n=-\infty}^{\infty} \in \ell_{r''}^2(\mathbb{Z})$ the unique solution to the first set of equations in (1.1), and let $(v_n)_{n=-\infty}^{\infty} \in \ell_{r'}^2(\mathbb{Z})$ be the sequence defined by the second set of equalities in (1.1). Then

$$\text{(5.2)} \qquad (v_n)_{n=-\infty}^{\infty} = T((u_n)_{n=-\infty}^{\infty}) \, .$$

Note that by (1.1) we have

$$\begin{pmatrix} x_{n+1} \\ v_n \end{pmatrix} = \begin{pmatrix} A_n & B_n \\ C_n & D_n \end{pmatrix} \begin{pmatrix} x_n \\ u_n \end{pmatrix} \qquad (n = 0, \pm 1, \ldots) \, .$$

Hence, if we multiply (4.14) by $\begin{pmatrix} x_n \\ u_n \end{pmatrix}$ on the right and $(x_n^* \; u_n^*)$ on the left, we obtain

$$x_n^* M_n x_n + u_n^* J_n u_n - x_{n+1}^* M_{n+1} x_{n+1} - v_n^* J_n' v_n \geq \gamma \|u_n\|^2 + \|x_n\|^2$$

for $n = 0, \pm 1, \ldots$. Since the vectors $(x_n)_{n=-\infty}^{\infty}$, $(u_n)_{n=-\infty}^{\infty}$ and $(v_n)_{n=-\infty}^{\infty}$ are square summable, and $(J_n)_{n=-\infty}^{\infty}$, $(J_n')_{n=-\infty}^{\infty}$ and $(M_n)_{n=-\infty}^{\infty}$, are bounded, we may sum the last inequality from $n = -\infty$ to $n = \infty$. We obtain

$$\sum_{n=-\infty}^{\infty} (u_n^* J_n u_n - v_n^* J_n' v_n) \geq \gamma \sum_{n=-\infty}^{\infty} \|u_n\|^2 \, ,$$

after disregarding $\|x_n\|^2$ on the right. The last inequality and (5.2) show that T is a strict (J, J') contraction. □

We can now prove Theorem 1.1.

PROOF OF THEOREM 1.1. Assume first that the system (1.4) admits a dichotomy of rank p and that $\|T\| < 1$. We apply Theorem 1.3 with $J_n = I_r$ and $J_n' = I_{r'}$ $(n = 0, \pm 1, \ldots)$. Since $\|T\| < 1$, T is a (J, J')-strict contraction, and hence

Theorem 1.3 shows that there exists a bounded sequence $(M_n)_{n=-\infty}^{\infty}$ of self-adjoint matrices such that the inequalities (1.5)-(1.6) hold for some $\varepsilon > 0$. Moreover, by Theorem 1.3 the sequence $(M_n)_{n=-\infty}^{\infty}$ can be chosen constant or periodic if the system (1.1) is such. Now note that inequalities (1.5) and (1.6) imply

$$M_n - A_n^* M_{n+1} A_n \geq \varepsilon I \qquad (n = 0, \pm 1, \ldots) . \tag{5.3}$$

Since the system (1.4) admits a dichotomy of rank p, Proposition 2.5 shows that the sequence $(M_n)_{n=-\infty}^{\infty}$ has constant inertia. This proves the implication I)→II), as well as the last sentence in the statement of the theorem.

Assume now that II) holds. From (1.5)-(1.6) we obtain $M_n - A_n^* M_{n+1} A_n \geq \varepsilon I$ $(n = 0, \pm 1, \ldots)$. Since the sequence of self-adjoint matrices $(M_n)_{n=-\infty}^{\infty}$ is bounded and of constant inertia, it follows from Theorem 2.4 that the system (1.4) admits a dichotomy. Consequently, the operators T is well defined. Thus, we may apply Theorem 1.3 to the inequalities (1.5)-(1.6) with $J_n = I_r$ and $J_n' = I_{r'}$ $(n = 0, \pm 1, \ldots)$. It follows that T is a $(J, J') = ((I_r)_{n=-\infty}^{\infty}, (I_{r'})_{n=-\infty}^{\infty})$ -strict contraction, and hence $\|T\| < 1$. Thus, II)→I).

Finally, assume that the system (1.4) admits a dichotomy of rank p, and that $(M_n)_{n=-\infty}^{\infty}$ is a bounded sequence of self-adjoint matrices such that (1.5) and (1.6) hold for some $\varepsilon > 0$. Again, (1.5) and (1.6) imply (5.3). Hence, Proposition 2.5 shows that $\text{In}(M_n) = (p, 0, r'' - p)$ $(n = 0, \pm 1, \ldots)$ and $\sup_n \|M_n^{-1}\| < \infty$. □

6. REFERENCES

[BG1] A. Ben-Artzi and I. Gohberg, Inertia theorems for nonstationary discrete systems and dichotomy, *Linear Algebra and its Applications* **120** (1989), 95-138.

[BG2] A. Ben-Artzi and I. Gohberg, Band matrices and dichotomy, *Operator Theory: Advances and Applications* **50** (1991), 137-170, Birkhäuser Verlag.

[BG3] A. Ben-Artzi and I. Gohberg, Inertia theorems for block weighted shifts and applications, *Operator Theory: Advances and Applications* **56** (1992), 120-152, Birkhäuser Verlag.

[BGK1] A. Ben-Artzi, I. Gohberg, and M.A. Kaashoek, Invertibility and dichotomy of singular difference equations, *Operator Theory: Advances and Applications* **48** (1990), 157-184, Birkhäuser Verlag.

[BGK2] A. Ben-Artzi, I. Gohberg, and M.A. Kaashoek, Exponentially dominated infinite block matrices of finite Kronecker rank, *Integral Equations and Operator Theory* **18** (1994), 30-77.

[CS] Ch.V. Coffman and J.J. Schäffer, Dichotomies for linear difference equations, *Math. Annalen* **172** (1967), 139-166.

[GKvS] I. Gohberg, M.A. Kaashoek and F. van Schagen, Non compact integral operators with semi separable kernels and their discrete analogues: inversion and Fredholm properties, *Integral Equations and Operator Theory* **7** (1984), 642-703.

[HI] A. Halanay and V. Ionescu, *Time-Varying Discrete Linear Systems*, Birkhäuser Verlag (1994).

[S1] J.J. Schäffer, A note on systems of linear difference equations, *Math. Annalen* **177** (1968), 23-30.

[S2] J.J. Schäffer, Linear difference equations: closedness of covariant sequences, *Math. Annalen* **187** (1968), 69-76.

[SX] C.E. de Souza and L. Xie, On the discrete-time bounded real lemma with applications in the characterization of static feedback H_∞ controllers, *Systems & Control Letters* **18** (1992), 61-71.

A. Ben-Artzi, I. Gohberg
School of Mathematical Sciences
Raymond and Beverly Sackler Faculty of Exact Sciences
Tel Aviv University
Tel Aviv 69978, Israel

M.A. Kaashoek
Faculteit Wiskunde en Informatica
Vrije Universiteit
1081 HV Amsterdam
The Netherlands

AMS subject classification: 93A25.

Operator Theory:
Advances and Applications, Vol. 80

REGULARITY OF FINITE TYPE CRITICAL POINTS FOR SELF-ADJOINT OPERATORS IN KREIN SPACE

P. Binding and B. Najman

1 Introduction and Notation

Following the work of Krein and Langer [11], the spectral function has become a basic tool in the study of self-adjoint operators in Krein spaces, cf. [2, §4.1], [5, §VIII 6], [12] and the references therein. The resulting spectral decompositions behave as in Hilbert space except near the (at most finite) set of critical points, where the spectral function is, at least initially, undefined. Our purpose is to study these critical points in situations that permit finite dimensional analysis. In particular, we shall give a number of finite dimensional tests for regularity of critical points.

To explain this notion, suppose first that S is a self-adjoint operator in a Pontryagin space K, and that its spectral function E has a single critical point at 0. Thus, by definition, the inner product $[\cdot,\cdot]$ in K is indefinite [definite] on the range of $E(\Omega)$ for any open interval Ω containing [not containing] 0. Using the notation N_j for the null space of S^j we see that $A_0 := \sum_{j=1}^{\infty} N_j$ is the algebraic eigenspace of S at 0 and it turns out (cf. [12, Theorem 5.7 et seq.],)that E can be extended to include an atom $E(\{0\})$ (an orthoprojector with range A_0) if and only if 0 is a regular critical point, i.e., the strong limits $\lim_{\lambda\searrow 0} E([\lambda,\mu])$ and $\lim_{\lambda\nearrow 0} E([\nu,\lambda])$ exist whenever $\nu < 0 < \mu$. In this case there is an orthogonal S-invariant decomposition

$$K = A_0[\dot{+}]B \tag{1}$$

where $S|_B$ is an injective self-adjoint operator without critical points. As we shall see, a decomposition (1) characterizes regularity in the present case.

It is well known (cf. [2, Corollary 2.3.15]) that the nonreal eigenvalues of S are normal, so our ability to make a spectral decomposition of K by S, with the resulting completeness and expansion theory, depends on testing regularity. A direct test (originating from the definition) is, however, quite difficult to perform, particularly since the $E(\cdot)$ generally have infinite rank. Instead, one could use the fact that 0 is regular critical point if and only if A_0 is nondegenerate (i.e. $A_0 \cap A_0^{\perp} = \{0\}$). Again, however, this is difficult to check if A_0 is infinite dimensional. We remark that this fact (which is due to H.Langer, see [13, Bem.

1.2.4]) has become part of the folklore of the subject. Since we use it in an essential way, we include a proof for the reader's convenience in Section 3.

Our main purpose is to give new, easier to verify, tests for regularity. We mention in particular the necessary and sufficient conditions in Theorems 3.2, 3.3 and 4.1. Our results use a decomposition of A_0 by Jordan chains, i.e. maximal linearly independent sequences $(x_1, \cdots, x_k)$ of links x_j satisfying $Ax_1 = 0, Ax_j := x_{j-1} \in N_{j-1}$, $j = 2, \cdots, k$. It is well known [5, Theorem IX.4.9] that the length k of such a chain is bounded, say by m, so in fact $A_0 = N_m$. The bound m can be estimated by the index of K, that is by the dimension of a maximal negative subspace of K.

After giving some preliminary results in Section 2, we present in Section 3 several necessary and sufficient conditions for regularity of 0, implicitly involving finitely many determinants of various sizes, all with entries of the form $[x_i, y_j]$ for links x_i and y_j. The important point is that these conditions are "finite dimensional" and that testing regularity reduces to calculating an easily computable number.

In Section 4 we introduce the following notion, denoting the range of S by R. A Jordan chain $(x_1, \cdots, x_k)$ is ECR (Extensible into the Closure of the Range) if $x_k \in \overline{R}$ (note that $x_k \notin R$ by maximality). It is shown in Theorem 4.1 that 0 is regular if and only if no chain is ECR. This result provides new insight into the difference between regular and singular critical points, cf. also [2, Theorem 4.3.1].

Section 5 concerns two developments. First, we allow K to be a general Krein space, but we require the quadratic form $[x, Sx]$ to be positive on a subspace of finite codimension. In this case it is possible to reformulate the problem in a related Pontryagin space and it is shown that the previously developed tests for regularity are still valid. Second, we "globalize" our local results: we show that the results for A_0 carry over to a global decomposition of K involving all the finite critical points of S. This decomposition has already been used to obtain a variational principle in [4], and indeed it was this application that led us to the present study.

Notation. We denote by $N(\cdot)$ the null space and by $R(\cdot)$ the range of an operator. In Sections 2-4, K is a Pontryagin space with finite negative index, and S is a self-adjoint operator in K with spectral function E and a critical point at 0. In addition to $N_j = N(S^j)$ with $A_0 = N_m$, we write $R_j = R(S^j)$, $R = R_1$ and $N = N_1$. We denote by Σ the span of any set of chains of length > 1 whose first links form a basis of $N \cap R$. We write $O = \{0\}$ and denote by C^- (resp. C^+) the following cones in K:

$$C^- = \{x \in K : [x, x] \leq 0\}\ ,\ C^+ = \{x \in K : [x, x] > 0\} \cup O$$

and $-C^+ = \{x \in K : [x, x] < 0\} \cup O$. For a subspace X of K, X^- denotes any (necessarily finite dimensional) maximal subspace of $X \cap C^-$. We abbreviate

finite dimensional to	fd
nondegenerate, nondegeneracy to	nd
linear span to	ls .

Acknowledgement. The authors are grateful to H. Langer for his various helpful comments and T. Azizov for contributing to the proof of Theorem 3.3.

2 Preliminary results

We start with a few properties of nd subspaces. The following lemma is a consequence of a basic decomposition theorem for Pontryagin spaces [5, Theorem IX.2.5].

Lemma 2.1 *If X is a fd subspace of A_0 and A_0 is nd, then there is a fd and nd subspace Y such that $X \subseteq Y \subseteq A_0$.*

Proof Let X_1 (resp. X_2) be a complementary subspace of $X_0 = X \cap X^\perp$ in X (resp. in $X^\perp$). As shown in [5, p.187], X_0 has a dual companion X_3 in $(X_1 + X_2)^\perp$ such that

$$\dim X_3 = \dim X_0 \; . \tag{2}$$

Moreover, [5, Theorem IX.2.5] gives $K = X_2[\dot{+}]Y$ with $Y = X_0 + X_1 + X_3 = X + X_3$. Thus Y is nd by [5, Corollary I.9.5] and from (2) we find $dim Y \leq dim X + dim X_0 \leq 2\, dim X$. □

Our next result is a development of [5, Lemma I.6.4].

Lemma 2.2 *Let X and Y be subspaces of K such that Y is nd and that there exists an $X^- \subseteq Y$. Then $Y^\perp \cap X \subseteq C^+$.*

Proof Assume $0 \neq z \in Y^\perp \cap X \cap C^-$ and define $L = ls\{X^-, z\}$. Since Y is nd, z does not belong to Y; hence $z \notin X^-$ so L strictly contains X^-. Let $y \in L$; then $y = \alpha x + \beta z$ for some $x \in X^-$, and $\alpha, \beta \in \mathbf{C}$. Moreover $z \perp x$ since $z \in Y^\perp$, so we find $[y,y] = |\alpha|^2[x,x] + |\beta|^2[z,z] \leq 0$. Thus L is nonpositive which contradicts maximality of X^-, so such z cannot exist. □

Corollary 2.3 *If X is a subspace of K then X is nd $\Longleftrightarrow$ a nd X^- exists.*

Proof $\Longrightarrow$ follows from [5, Theorem I.11.7 and Corollary I.11.2].
$\Longleftarrow$ By Lemma 2.2 with $Y = X^-$, $(X^-)^\perp \cap X$ is nd so $X = X^-[\dot{+}]((X^-)^\perp \cap X)$ is nd. □

The next result is elementary, but we need it for future reference; cf. also [5, §I.10].

Lemma 2.4 *Let X and Y be n-dimensional subspaces of K with bases $\{x_1, \cdots, x_n\}$ and $\{y_1, \cdots, y_n\}$ respectively. Then*

(i) *$X \cap Y^\perp = O$ if and only if $det M_{xy} \neq 0$ where $(M_{xy})_{ij} = [x_i, y_j]$.*

(ii) *X is nd $\Longleftrightarrow$ $det M_{xx} \neq 0$ (note that M_{xx} is the Gram matrix of $\{x_i\}$).*

Proof (i) If $x = \sum_{i=1}^{n} \alpha_i x_i$ then $x \perp Y$ if and only if $[\alpha_1 \cdots \alpha_n] M_{xy} = 0$. Thus $x \neq 0 \Longleftrightarrow det M_{xy} = 0$.
(ii) follows from (i) with $Y = X$. □

The final preparatory Lemma is also used in [1], but the proof given there is finite dimensional.

Lemma 2.5 *An S-invariant subspace X of K is nd $\Longleftrightarrow$ $X \cap X^\perp \cap N = O$.*

Proof Suppose $0 \neq x \in X \cap X^\perp$ where $x \notin N$, say $x \in N_{j+1} \backslash N_j$ with $j \geq 1$. Writing $y = S^j x$ we have $0 \neq y \in N \cap X$. Moreover, for any $z \in X$,

$$[y, z] = [x, S^j z] = 0 \text{ since } x \in X^\perp .$$

Thus $y \in X \cap X^\perp \cap N$. This proves $\Longleftarrow$, and $\Longrightarrow$ is trivial. □

Proposition 2.6 *Assume that for some integer j the following condition holds:*
(a_j) There exists a nd N_j^-.
Then $j = m$ and $A_0 = N_m$ is nd.

Proof From Corollary 2.3 it follows that N_j is nd. If $j < m$, we pick a chain $(x_1, \cdots, x_m)$ of maximal length. Then $x_1 \neq 0, x_1 \in R_{m-1} \subseteq R_j$, $x_1 \in N \subseteq N_j$, hence $x_1 \in N_j \cap N_j^\perp$, contradicting nondegeneracy of N_j. Since $N_j = N_m$ for $j > m$, it follows that $j = m$. □

The next Proposition shows that instead of searching for maximal subspaces of $N_j \cap C^-$, we can search the smaller cone $N \cap C^-$ provided that we also construct a subspace Σ as at the end of Section 1. Note that

$$A_0 = N + \Sigma \ , \ N + (\Sigma \cap N_j) = N_j.$$

Proposition 2.7 *Assume that for some integer j the following condition holds: (b_j) There exist N^- and Σ so that $Y = N^- + (\Sigma \cap N_j)$ is nd.*
Then (a_j) holds, so $j = m$ and A_0 is nd.

Proof Suppose N_j is degenerate, so by Lemma 2.5 there is a nonzero $y \in N \cap N_j^\perp$. Since $y \in N^\perp$, $[y, y] = 0$ whence $y \in N^- \subseteq Y$. Moreover $Y \subseteq N_j$ so $y \in N_j^\perp \subseteq Y^\perp$, contradicting the assumption that Y is nd. It follows that N_j is in fact nd, and the result follows from Corollary 2.3. □

There is some connection between Proposition 2.7 and the decomposition of [5, Theorem IX.4.3c)] but there is no discussion of nondegeneracy there, and our construction seems more direct than the one given on [5, p.195].

In the next section, we shall show that the conditions (a_j) and (b_j) of Proposition 2.6 and Proposition 2.7, which are sufficient for nd of A_0, are in fact necessary as well.

3 Regularity tests

We start with the promised equivalence between regularity at 0 and nd of A_0. We prove this by means of another equivalence in [12, Proposition II.5.6], related to the decomposition (1). Recall that S is a self-adjoint operator in the Pontryagin space K with a critical point 0.

Theorem 3.1 *Let A_0 be the algebraic eigenspace of S at 0. The following are equivalent:*
(i) 0 is a regular critical point
(ii) (1) holds where $S|_B$ is injective and self-adjoint
(iii) A_0 is nd

Proof (i)$\Longrightarrow$ (ii) Following Langer [12, p.37] we fix an interval $\Delta \subset \mathbf{R}$ containing 0 and no other critical point of S, and we consider $S_\Delta := S|_{K_\Delta}$ where $K_\Delta = E(\Delta)K$. Evidently 0 is a regular critical point for S_Δ, so by [12, Proposition II.5.6], $K_\Delta = A_0[\dot{+}]A$ where A_0 is also the algebraic eigenspace for S_Δ at 0 (cf. [12, p.37]) and A is another subspace. It follows that $K = A_0[\dot{+}][(I - E(\Delta))K + A]$.
(ii) $\Longrightarrow$(iii) follows from [5, Corollary I.9.5].
(iii) $\Longrightarrow$ (i) Since S is closed [5, Theorem VI.2.2], A_0 is closed. Thus by [5, Theorem IX.2.2], A_0 is orthocomplemented in K_Δ, say $K_\Delta = A_0[\dot{+}]A_1$. Let E_j be the spectral function of $S|_{A_j}, j = 0, 1$. Then for any open interval $\Omega \subset \Delta$, $E(\Omega)K_\Delta = A_0 + E_1(\Omega)K_\Delta$ if $0 \in \Omega$ and $E(\Omega)K_\Delta = E_1(\Omega)K_\Delta$ if $0 \notin \Omega$. Now S_Δ has a definitising polynomial $p(\lambda) = \lambda^{2k}$ for some integer k (cf.[12, pp. 12, 37]) so A_1 is a Hilbert space, whence E_1 is bounded. It follows that E is bounded and the result now follows from [12, Proposition II.5.6]. □

As mentioned in Section 1, (ii) and (iii) are generally infinite dimensional in nature and we turn now to fd tests. In view of Proposition 2.6 and Proposition 2.7 we supress subscripts and (noting $N_m = A_0$) write
(a) there exists a nd A_0^-
(b) there exist N^- and Σ so that $N^- + \Sigma$ is nd.

Theorem 3.2 *The conditions of Theorem 3.1 are equivalent to each of (a), (b),*
(c) A_0 admits an S-invariant orthogonal decomposition

$$A_0 = F_0[\dot{+}]H_0$$

where F_0 is fd and nd and $H_0 \subseteq N$ is a Hilbert space,
(d) for each Σ there is N^- so that $N^- + \Sigma$ is nd.

Proof (iii)$\Longrightarrow$ (c) Suppose A_0 is nd. Choosing N^- and Σ write $X = N^- + \Sigma$ and construct $(Y =)F_0$ as in Lemma 2.1. Since $SA_0 = S(N + \Sigma) = S\Sigma$, we have $SF_0 \subseteq SA_0 = S\Sigma \subseteq \Sigma \subseteq X \subseteq F_0$, so $T_0 = F_0^\perp \cap A_0$ is S-invariant. By definition of $\Sigma(\subseteq F_0)$, the space T_0 contains no Jordan chain of length ≥ 2, so

$$T_0 \subseteq N. \tag{3}$$

Noting that $N^- \subseteq X \subseteq F_0$, we can apply Lemma 2.2 to find $T_0 = F_0^\perp \cap N \subseteq C^+$, so (c) holds.
(c) $\Longrightarrow$ (d) Clearly A_0 is nd, so by Lemma 2.1 applied to $X = ls(\Sigma, F_0)$, we can assume without loss of generality that $\Sigma \subseteq F_0$. Applying [1, Theorem 2.6] to $S|_{F_0}$, we see that Σ is nd. Arguing as for (3), we have

$$T := \Sigma^\perp \cap A_0 \subseteq N\ . \tag{4}$$

Since $\Sigma \subseteq A_0$, it follows that $A_0 = \Sigma[\dot{+}]T$, whence T is nd, so there is an orthogonal decomposition

$$T = T_-[\dot{+}]T_+\ , \tag{5}$$

where $T_\pm \subseteq \pm C^+$ and $T_\pm \subseteq N$ by (4). Moreover (4) gives $N = (\Sigma \cap N)[\dot{+}]T$, so

$$N = (\Sigma \cap N)[\dot{+}]T_-[\dot{+}]T_+ \tag{6}$$

by (5). Since $T_- \subset \Sigma^\perp$ by (5), it follows that $T_- \cap \Sigma = O$. Thus if we define

$$N^- = T_-[\dot{+}](\Sigma \cap N) \tag{7}$$

then $N^- + \Sigma = T_-[\dot{+}]\Sigma$ is nd because T_- and Σ are both nd.
(d) $\Longrightarrow$ (b) is trivial, (b) $\Longrightarrow$ (a) comes from Proposition 2.7 and (a) $\Longrightarrow$ (iii) follows from Proposition 2.6. □

The equivalence of (c) in Theorem 3.2 and (iii) in Theorem 3.1 follows also from a theorem of Pontryagin, see [2, Theorem 2.2.26].

By Lemma 2.4(ii) conditions (c) and (d) involve a determinant of size $d := dim(N^- + \Sigma)$. It turns out that this can be replaced by a test involving smaller determinants whose sizes sum to d.

For a fixed choice of chains and $j \in \{2, \cdots, m\}$ we define L_j as the span of the chains of length j. Given an N^- we note that $N^- \cap R$ consists of those eigenvectors which extend to chains of length ≥ 2. We then choose a complementary subspace (of nonextensible eigenvectors) L_1 to $N^- \cap R$ in N^-. Then

$$\Sigma = \sum_{j=2}^{m} L_j\ , \quad N^- = (N^- \cap R) + L_1\ .$$

Theorem 3.3 *The conditions of Theorems 3.1, 3.2 are equivalent also to*

(e) there are nd L_j, $j = 1, \cdots, m$,

(f) L_j, $j = 2, \cdots, m$ are nd, and there is a nd choice of L_1.

Proof (c) $\Longrightarrow$(f) The subspaces $L_j, j = 2, \cdots, m$ are nd by [1, Theorem 2.6]. Again choose N^- as in (7). Then $N^- \cap R = \Sigma \cap N$ so T_- is an appropriate choice of L_1.
(f)$\Longrightarrow$ (e) is trivial.
(e) $\Longrightarrow$ (b) Since $N^- \cap R \subseteq \Sigma$, it suffices to prove that $L = \sum_{j=1}^{m} L_j$ is nd. If not then by Lemma 2.5 there is nonzero $x \in L \cap L^\perp \cap N$, say $x = \sum_{j=k}^{m} x_j$ where $k \in \{1, \cdots, m\}$, $x_j \in L_j$ and $x_k \neq 0$. From [1, Lemma 2.2 and Corollary 2.3], applied to $S|_L$ we conclude that if $j > k$ then $x_j \in L_j \cap N \subseteq R_{j-1} \subseteq R_k \subseteq N_k^\perp \subseteq L_k^\perp$. Since $L^\perp \subseteq L_k^\perp$ it follows that $x_k = x - \sum_{j=k+1}^{m} x_j \in L_k^\perp \cap L_k$, contradicting nd of L_k. □

Remark Condition (f) involves determinants of size $d_j = dim L_j$, but this can be reduced still further if we use the results of [1, §4]. For example, when $j > 1$, L_j is nd $\Longleftrightarrow$ $X \cap Y^\perp = O$ where $X = L_j \cap N, Y = L_j \cap R_{j-1}$, i.e. X and Y are the spans of the first and the last links, respectively, of the chains in L_j. Using Lemma 2.4 (i) this gives a test involving determinants of sizes d_j/j.

4 ECR chains

From Theorem 3.3(e) we see that a necessary condition for regularity of $\lambda = 0$ is that each L_j be nd. Moreover, even though the L_j are generally not orthogonal, by reducing to the finite dimensional case (applying Theorem 3.2 (c)) and then applying [1, Theorem 2.6] we see that it is also necessary that for any index set J the sum $\sum_{j\in J} L_j$ be nd. In particular if $k \in \{1, \cdots, m\}$ then the nd of $M_k = \sum_{j, j\neq k} L_j$ is a necessary condition. In the following example A_0 contains a single chain $(e_1, \cdots, e_m)$ and is degenerate, so nd of M_m (which is O in this case) is not a sufficient condition for regularity.

Example. Let $H = l_2(\mathbf{C})$ with orthogonal basis $e_1, e_2, \cdots$, and inner product $(\cdot, \cdot)$. Fix a positive integer m. For $i = 1, 2$ define the operator J_i on $ls\{e_i, \cdots, e_{m+1}\}$ by $J_i e_k = e_{m+1+i-k}$, the vector e by $e = \sum_{j=2}^{\infty} \frac{1}{j} e_{m+j} \in H$, and the rank one operator C by $Cx = (x, e)e_m$. Let $Z = ls\{e_j, j \geq m+2\}$ and let I_Z be the identity operator on Z. Further define $T = J_1 \oplus I_Z$ and $V = T + C + C^*$; T is a self-adjoint unitary operator and V is a bounded operator. A routine calculation gives $N(V) = O$; the compactness of C implies that V^{-1} is also bounded. Therefore H, endowed with the inner product $[x, y] = (x, Vy)$, becomes a Pontryagin space which we denote by K. Now define $E_1 = ls\{e_1\}$, $E_2 = ls\{e_2, \cdots, e_{m+1}\}$, the operator Q in Z by $Qe_{m+j} = \frac{1}{j} e_{m+j} (j \geq 2)$, and the operator W on $H = E_1 \oplus E_2 \oplus Z$ by $W = O \oplus J_2 \oplus Q$. Then $S = V^{-1}W$ is self-adjoint on K, $N = N(W) = E_1$ and $Se_j = e_{j-1}$ $(2 \leq j \leq m)$. We claim that $e_m \notin R$, implying that $e_1, \cdots, e_m$ is a Jordan chain. Indeed, if $S(\sum_{j=1}^{\infty} \alpha_j e_j) = e_m$ then $\sum_{j=1}^{\infty} \alpha_j We_j = Ve_m$ would imply $\alpha_{m+j} = 1$ for $j \geq 2$. It follows that $A_0 = L_m = E_1 + E_2$. Since $[e_1, e_j] = 0$, $1 \leq j \leq m$, A_0 is degenerate. As W is compact, so is S and consequently zero is a critical point which is not regular.

We note that in the above example $N = ls\{e_1\}$ so $[e_1, e_m] = 0$ gives $e_m \in \bar{R} = N^\perp$. This illustrates the definition of an ECR chain given in the Introduction. In the case $m = 1$, we see that e_1 belongs to $(N \cap \bar{R}) \setminus R$ cf. [2, Theorem 4.3.1] .

We now prove that the above connection between ECR chains and lack of regularity is general:

Theorem 4.1 *0 is a regular critical point if and only if the following holds:*
(g) no Jordan chain is ECR.

Proof (only if) If $\{x_1\}$ is an ECR chain then $x_1 \in N \cap \bar{R} = N \cap N^\perp$, so if $(y_1, \cdots, y_l)$ is any chain then $x_1 \perp y_1$ and, for $j > 1$, $[x_1, y_j] = [x_1, Sy_{j-1}] = [Sx_1, y_{j-1}] = 0$. Thus $0 \neq x_1 \in A_0 \cap A_0^\perp$ so the result follows from Theorem 3.1. If $(x_1, \cdots, x_k)$ is ECR for $k > 1$ and $(y_1, \cdots, y_k)$ is any chain of length k, then $[x_1, y_j] = [S^{k-1}x_k, y_j] = [x_k, S^{k-1}y_j] = [x_k, S^{k-j}y_1] = 0$ for $1 \leq j \leq k$ since $x_k \in \bar{R} = N^\perp$. Thus $x_1 \in L_k \cap L_k^\perp$, so the result follows from Theorem 3.3 (a)$\Longrightarrow$ (e).
(if) Since S is closed [5, Theorem VI.2.2], N is closed and thus admits a fundamental

decomposition $N_-[\dot{+}]N_0[\dot{+}]N_+$ for closed subspaces $N_\pm \subseteq \pm C^+$ and a neutral subspace N_0 [5, Theorem V.3.1]. Writing $X_1 = N_- + N_+$ and $X_n = X_1 + \sum_{l=2}^{n} L_l$ $(2 \leq n \leq m)$ we see that X_1 is nd, X_n is closed and $X_m = A_0$.

We shall prove by induction on n that if no Jordan chain is ECR then X_n is nd. This holds for X_1, so assume it holds for X_{n-1}. If the inductive step fails then X_{n-1} is nd and by Lemma 2.5 there is nonzero $x \in X_n \cap X_n^\perp \cap N$. Suppose x starts a chain of length $l < n$. By choosing a new L_l if necessary, we may assume $x \in L_l \subseteq X_{n-1}$ and, since $x \in X_n^\perp \subseteq X_{n-1}^\perp$, we contradict nd of X_{n-1}. Thus x must start a chain of length n, say with final link $y \in X_n \backslash X_{n-1}$ so

$$x = S^{n-1}y. \tag{8}$$

Now X_{n-1} is closed and nd, and hence is orthocomplemented [5, Theorem IX.2.2] and thus has an orthoprojector P [5, Theorem II.3.10] satisfying

$$R(P) = X_{n-1} \subseteq N_{n-1} .$$

It follows that $x_n = y - Py \in X_n \backslash X_{n-1}$, and we define $x_j = S^{n-j}x_n$ $(1 \leq j \leq n)$. Evidently $(x_1, \cdots, x_n)$ is a chain, and we complete the proof by showing that it is ECR, i.e., that $x_n \perp N$. Since $x_n \in X_{n-1}^\perp$ (by definition of P) $\subseteq X_1^\perp$, it suffices to show that $x_n \perp N \cap \sum_{l=2}^{n} L_l = N \cap R$. Let $(y_1, \cdots, y_k)$ be any chain of length $k > 1$. If $k < n$ then $y_1 \in X_{n-1}$ so $x_n \perp y_1$. Define

$$\alpha := [x_n, y_1] = [x_n, S^{k-1}y_k] = [S^{k-1}x_n, y_k] .$$

If $k > n$ then $S^{k-1}x_n \in S^{k-1}X_n = O$ so $\alpha = 0$. If $k = n$ then $S^{n-1}x_n = S^{n-1}y = x$ by (8), and since $x \in X_n^\perp$, we have

$$\alpha = [S^{n-1}x_n, y_n] = [x, y_n] = 0.$$

□

5 General Krein spaces

In this Section K is a Krein space, S a definitizable self-adjoint operator in K with spectral function E.

Definition A critical point λ is of <u>finite type</u> if $K_\Delta = E(\Delta)K$ is a Pontryagin space for some open interval Δ containing λ. The set of all critical points of S of finite type is denoted by $c_f(S)$.

Proposition 5.1 *Let* $0 \in c_f(S)$. *Then Theorems* 3.1 - 3.3 *and* 4.1 *remain valid.*

Proof Note that the algebraic eigenspace A_0 of S is also the algebraic eigenspace at 0 of the restriction $S_\Delta = S|_{K_\Delta}$. Applying the corresponding results to S_Δ, we obtain the stated result for S.

□

Our next result is a globalization of Proposition 5.1.

Proposition 5.2 *Assume that all finite critical points of S are of finite type. Then the following statements are equivalent:*

(i) All finite critical points are regular.

(ii) E extends to an orthoprojector valued measure on the Borel sets of $\mathbf{R}$.

(iii) For each $\lambda \in c_f(S)$ the corresponding algebraic eigenspace A_λ is nd.

Proof Since K admits an S-invariant decomposition

$$K = K_0[\dot{+}]K_1 \tag{9}$$

where K_0 is nd and $S|_{K_1}$ has real spectrum, we can additionally assume that S has real spectrum. Then (i) is equivalent to (ii) by [12, Proposition II.5.7 et seq.] and to (iii) by Theorem 3.1.

□

Remarks 1. For critical points which are not of finite type different tests are necessary. Note that if ∞ is a critical point, it is not of finite type. For corresponding regularity tests we refer to [6] or [8] and the references therein.
2. The space K_0 in (9) is not fd in general. To see this, it is sufficient to consider

$$K = \sum_{j=1}^{\infty} \oplus K_j \quad , \quad A = \sum_{j=1}^{\infty} \oplus A_j$$

where $A_j = \tilde{A}, K_j = \tilde{K}$ do not depend on j and $\tilde{K}$ is two dimensional indefinite inner product space, $\tilde{A}$ is self-adjoint in $\tilde{K}$ and has nonreal spectrum.
Then A is definitizable and $K_0 = K$ in (9).

If K is a Pontryagin space then the assumption of Proposition 5.2 holds trivially.

Another class of problems to which Proposition 5.2 applies is covered by the following definition. For further consequences, see [3], [9].
Definition. A self-adjoint operator T is a quasi-uniformly positive (qup) operator if there exists a subspace $W \subset D(T)$ of finite codimension in $D(T)$ such that

$$\inf\{[Tu, u] \, , u \in W \, , \, \|u\| = 1\} > 0.$$

A qup operator is definitizable, see [9, Proposition 1.1] .

Theorem 5.3 *Let S be a self-adjoint operator such that $S_\alpha := S - \alpha I$ is qup for some $\alpha \in$* $\mathbf{R}$. *Then*

a) all finite critical points of S are of finite type.

b) conditions (i)-(iii) of Proposition 5.2 are equivalent also to

(iv) K admits a S-invariant orthogonal decomposition $K = F[\dot{+}]G$ where F is fd and nd, G is a Krein space and $S|_G$ has real spectrum and no finite critical points.

Proof Let p be a definitizing polynomial of S_α and let $q(\mu) = p(\mu - \alpha)$. Then $q(S) \geq 0$. Thus S is definitizable; if E_α is the spectral function of S_α and E the spectral function of S, then $E_\alpha(\mu) = E(\mu + \alpha)$. Therefore it suffices to establish the result for $\alpha = 0$ and for notational simplicity we shall simply supress α.
a) Our proof is an elaboration of an argument on [7, p.40].
Let $\lambda > 0$ be a critical point and let $\Delta = (a, b)$ be an interval containing λ and no other critical point. Write $S_\Delta = S|_{K_\Delta}$. Let Γ be the rectangular contour with corners $a - \delta \pm i\delta$, $b + \delta \pm i\delta$ where $a > \delta > 0$. It follows that the resolvent $R_z = (S_\Delta - zI)^{-1}$ is continuous on Γ, so the operator

$$Q = \oint_\Gamma z^{1/2} R_z dz$$

is defined and continuous on K_Δ, since $\sigma(S_\Delta) \subseteq \bar{\Delta}$ [12, Theorem II.3.1].

Since Γ is symmetric about the real axis, $Q = Q^*$ and $Q^2 = S_\Delta$, see [10, Theorem VII 3.10]. Thus Q has a continuous inverse by the spectral mapping theorem [10, Theorem VII 3.11]. Hence it follows that

$$[Qx, Qx] = [Sx, x] \geq 0$$

for all $x \in W$, so the inner product is nonnegative on the subspace QW which is of finite codimension in K_Δ since Q is a homeomorphism. The proof is similar if $\lambda < 0$; $\lambda = 0$ is not a critical point of S by [9, Proposition 1.3].
b) (i)$\Longrightarrow$(iv). Since S is qup, the space K_0 in (9) can be taken fd. Therefore without loss of generality we can assume that the spectrum of S is real. Let $\lambda \in c_f(S)$. Then it is a regular critical point by b). It follows from Theorem 3.2 that there exists an S-invariant orthogonal decomposition $K = F_\lambda[\dot{+}]D_\lambda[\dot{+}]K_\lambda$ where F_λ is fd and nd, $D_\lambda \subseteq N(S - \lambda I)$ is a Hilbert space, K_λ is a Krein space and λ is not an eigenvalue of $S|_{K_\lambda}$. Since $D_\lambda[\dot{+}]K_\lambda =: G_\lambda$ is a S-invariant Krein space, since λ is not a critical point of $S|_{G_\lambda}$ and since the set $c_f(S)$ is finite, we see that $F := \sum_{\lambda \in c_f(S)} [\dot{+}] F_\lambda$ and $G := \bigcap_{\lambda \in c_f(S)} G_\lambda$ satisfy (iv).
(iv)$\Longrightarrow$(iii). Let $\lambda \in c_f(S)$, and consider the self-adjoint operators $T = S|_F$ and $U = (S - \lambda I)|_G$. Clearly λ is also an eigenvalue of T and $A_\lambda(S) = A_\lambda(T)[\dot{+}]N(U)$. Since $A_\lambda(T)$ is nd by standard finite dimensional theory (cf. [1]) and $N(U)$ is nd (it is even a Hilbert space) because U has no critical points, it follows that A_λ is nd. □

Remarks 1. A variation of the decomposition of T 5.3(iv) is used in Proposition 2.2 of [4]. In this variation $G = Y + Z$ is an $S-$invariant decomposition, Y is fd and nd, and $S_\beta|_Z$ is a positive operator for an appropriate real β.
2. It follows from Theorem 5.3 and Proposition 5.2 that the "local" condition (iii) from Proposition 5.2 can be replaced by any other local condition (d)-(g) from Theorems 3.2, 3.3 and 4.1; for example it can be replaced by
(v) For each $\lambda \in \sigma(S) \cap \mathbf{R}$ and fixed basis of Jordan chains in A_λ, there exists a nd choice of $L_1(\lambda)$ and all $L_j(\lambda)$, $j = 2, \cdots, m(\lambda)$ are nd.
Here $L_j(\lambda)$ and $m(\lambda)$ correspond to L_j and m in Theorem 3.3 (f).

References

[1] T. Ja. Azizov, P. A. Binding, J. Bognár, B. Najman, Nondegenerate subspaces of Jordan chains in indefinite spaces, Lin. Alg. Appl. (to appear).

[2] T. Ja. Azizov, I. S. Iohvidov, Linear Operators in Spaces with an Indefinite Metric, Wiley, 1989.

[3] P. A. Binding, A canonical form for self-adjoint pencils in Hilbert spaces, Integral Equations Operator Theory 12(1989),324-342.

[4] P. A. Binding, B. Najman, A variational principle in Krein space, Trans Amer. Math. Soc. 342(1994), 489-499.

[5] J. Bognár, Indefinite Inner Product Spaces, Springer Verlag, 1974.

[6] B. Ćurgus, On the regularity of the critical point infinity of definitizable operators, Integral Equations Operator Theory 8 (1985), 462-488.

[7] B. Ćurgus, H. Langer, A Krein space approach to symmetric ordinary differential operators with an indefinite weight function. J.Differential Equations 79 (1989), 31-61.

[8] B. Ćurgus, B. Najman, A Krein space approach to elliptic eigenvalue problems with indefinite weights, Differential Integral Equations, 7(1994), 1241-1252.

[9] B. Ćurgus, B. Najman, Quasi-uniformly positive operators in Krein spaces, Integral Equations Operator Theory, these Proceedings.

[10] N. Dunford, J. Schwartz, Linear Operators, Part I, Interscience (Wiley), 1963.

[11] M. G. Krein, H. Langer, On the spectral function of a selfadjoint operator in a space with indefinite metric, Dokl. Akad. Nauk SSSR 152 (1963), 39-42 (Russian).

[12] H. Langer, Spectral functions of definitizable operators in Krein spaces, Lecture Notes in Math. 948, Springer Verlag 1982, 1-46.

[13] H. Langer, Spektraltheorie linearer Operatoren in $J-R\ddot{a}umen$ und einige Anwendungen auf die Schar $L(\lambda) = \lambda^2 I + \lambda B + C$: Habilitationschrift, Dresden, 1965.

Paul Binding
Department of Mathematics and Statistics
University of Calgary
2500 University Drive N.W.
Calgary, Alberta, Canada T2N 1N4
e-mail: binding@acs.ucalgary.ca

Branko Najman
Department of Mathematics
University of Zagreb
Bijenička 30
41000 Zagreb, Croatia
e-mail: najman@cromath.math.hr

AMS Subject Classification: 47 B 50

Operator Theory:
Advances and Applications, Vol. 80

QUASI-UNIFORMLY POSITIVE OPERATORS IN KREIN SPACE

BRANKO ĆURGUS and BRANKO NAJMAN

Definitizable operators in Krein spaces have spectral properties similar to those of selfadjoint operators in Hilbert spaces. A sufficient condition for definitizability of a selfadjoint operator A with a nonempty resolvent set $\rho(A)$ in a Krein space $(\mathcal{H}, [\,\cdot\,|\,\cdot\,])$ is the finiteness of the number of negative squares of the form $[Ax|y]$ (see [10, p. 11]).

In this note we consider a more restrictive class of operators which we call *quasi-uniformly positive.* A closed symmetric form s is called *quasi-uniformly positive* if its isotropic part $\mathcal{N}_s$ is finite dimensional and the space $(\mathcal{D}(s), s(\,\cdot\,,\,\cdot\,))$ is a direct sum of a Pontryagin space with a finite number $\pi(s)$ of negative squares and $\mathcal{N}_s$. The number $\kappa(s) := \dim \mathcal{N}_s + \pi(s)$ is the number of nonpositive squares of s; it is called the *negativity index* of s. A selfadjoint operator A in a Krein space $(\mathcal{H}, [\,\cdot\,|\,\cdot\,])$ is *quasi-uniformly positive* if the form $a(x, y) = [Ax|y]$ defined on $\mathcal{D}(A)$ is closable and its closure $\overline{a}$ is quasi-uniformly positive. The number $\kappa(A) := \kappa(\overline{a})$ is the *negativity index* of A. Such operators often appear in applications, see [3, 4, 5] and Section 3 of this note.

It turns out that this class of operators is stable under relatively compact perturbations, see Corollaries 1.2 and 2.3. The perturbations as well as the operators are usually defined as forms, so the above definition is natural.

Most of the results in this note are known. In particular the perturbation results from Section 2 are consequences of the results of [7]. We have found it useful to state the results in the framework of quadratic forms and quasi-uniformly positive operators since the proofs and the statements are simpler but still sufficiently general for several important applications.

As an illustration of these results we consider the operator associated with the Klein-Gordon equation

$$\left[\left(\frac{\partial}{\partial t} - ieq\right)^2 - \sum_j \left(\frac{\partial}{\partial x_j} - ieA_j\right)^2 + m^2\right] u = 0.$$

Setting

$$u_1 = u, \quad u_2 = \left(-i\frac{\partial}{\partial t} - eq\right) u$$

we get a system of equations for (u_1, u_2). The associated operator is quasi-uniformly positive in a Krein space suggested by the physical interpretation of the equation. The obtained results are essentially known, see [8, 11].

In the first two sections of this note $(\mathcal{H}, [\cdot|\cdot])$ is a Krein space, $(\mathcal{H}, (\cdot|\cdot))$ is a Hilbert space and J is the corresponding fundamental symmetry.

1 Quasi-uniformly positive operators

In this section we prove that quasi-uniformly positive operators in a Krein space are definitizable.

PROPOSITION 1.1 *A quasi-uniformly positive operator A in the Krein space $(\mathcal{H}, [\cdot|\cdot])$ is definitizable.*

PROOF [1] Since $S = JA$ is quasi-uniformly positive in $(\mathcal{H}, (\cdot|\cdot))$, there exists a selfadjoint operator F_1 of finite rank such that $S + F_1$ is uniformly positive. Since $\mathcal{D}(S) = \mathcal{D}(A)$ is dense in $\mathcal{H}$, perturbing F_1 we see that there exists a selfadjoint operator F such that $\mathcal{R}(JF) \subset \mathcal{D}(S) = \mathcal{D}(A)$ and such that $H := S + F$ is uniformly positive. The operator JH is uniformly positive in the Krein space $\mathcal{H}$. Since 0 and all nonreal numbers are in the resolvent set of JH, the resolvent identity yields

$$(JH - z)^{-1} = zH^{-1/2}(H^{1/2}JH^{1/2} - z)^{-1}H^{-1/2}J + (JH)^{-1}, \quad z \neq \overline{z}.$$

Therefore

$$\sup_{\eta \in \mathbb{R}} \|(JH - i\eta)^{-1}\| < \infty. \tag{1}$$

From the resolvent identity and $\mathcal{R}(JF) \subset \mathcal{D}(JH)$, for arbitrary real numbers η, η_0 we get

$$\begin{aligned}(JH - i\eta)^{-1}JF &= (JH - i\eta)^{-1}(JH - i\eta_0)^{-1}(JH - i\eta_0)JF = \\ &-i(\eta - \eta_0)^{-1}\{(JH - i\eta)^{-1} - (JH - i\eta_0)^{-1}\}(JH - i\eta_0)JF.\end{aligned}$$

Now (1) implies

$$\lim_{\eta \to \pm\infty} \|(JH - i\eta)^{-1}JF\| = 0.$$

Therefore, for sufficiently large $|\eta|$ the operator $I + (JH - i\eta)^{-1}JF$ has bounded inverse. Since

$$A - i\eta = JH - i\eta - JF = (JH - i\eta)(I + (JH - i\eta)^{-1}JF), \tag{2}$$

it follows that $i\eta \in \rho(A)$ for sufficiently large $|\eta|$. Consequently [10, (c) p. 11] implies that A is definitizable. □

In the next proposition we use the concept of relative compactness for operators. For its definition and properties see [9].

PROPOSITION 1.2 *The class of quasi-uniformly positive operators in a Krein space is closed with respect to relatively compact additive perturbations.*

---

[1] The authors are grateful to Prof. Peter Jonas for providing this proof which is significantly shorter than the original one.

PROOF Let A be a quasi-uniformly positive operator in the Krein space $(\mathcal{H}, [\cdot|\cdot])$ and let V be a symmetric operator in $(\mathcal{H}, [\cdot|\cdot])$ which is relatively compact with respect to A. For every $\lambda \in \rho(JA) \cap \rho(A)$ the identity

$$\begin{aligned} &JV(JA - \lambda I)^{-1} - JV(A - \lambda I)^{-1}J \\ &= \lambda JV(A - \lambda I)^{-1}(J - I)(JA - \lambda I)^{-1} \\ &= \lambda JV(JA - \lambda I)^{-1}(J - I)(A - \lambda I)^{-1}J \end{aligned}$$

holds. Therefore the operator V is A-compact if and only if the operator JV is JA-compact. Since the operator JA is quasi-uniformly positive in the Hilbert space $(\mathcal{H}, (\cdot|\cdot))$, it follows from [9, Theorem IV.5.35] that the operator $JA + JV$ is quasi-uniformly positive in the Hilbert space. Consequently $A + V$ is quasi-uniformly positive in the Krein space. □

PROPOSITION 1.3 *Let A be a quasi-uniformly positive operator in the Krein space $(\mathcal{H}, [\cdot|\cdot])$ and let 0 be in the spectrum of A. Then 0 is an isolated eigenvalue of A of finite multiplicity. In particular, 0 is not a singular critical point of a quasi-uniformly positive operator.*

PROOF Let H be the operator introduced in the proof of Proposition 1.1. Then 0 is in the resolvent set of JH and $A - JH$ is an operator of finite rank. The proposition follows from the Weinstein-Aronszajn formulas, see [9, IV, §6]. □

PROPOSITION 1.4 *Let S be a quasi-uniformly positive operator in $(\mathcal{H}, (\cdot|\cdot))$ with discrete spectrum. Then the spectrum of JS is also discrete.*

PROOF Let H be the operator introduced in the proof of Proposition 1.1. From the Weinstein-Aronszajn formulas it follows that the spectrum of H is also discrete. Therefore $H^{-1} = (JH)^{-1}J$ is a compact operator. The resolvent identity implies that the resolvent of JH is compact. It follows from the equality (2) that the resolvent of JS is also compact. □

The converse of Proposition 1.4 is not true. As we show in the example below, there exists a uniformly positive operator S in $(\mathcal{H}, (\cdot|\cdot))$ with nonempty continuous spectrum such that the spectrum of JS is discrete.

EXAMPLE Consider the Hilbert space ℓ^2. Let e_n, $n = 1, 2, \ldots$ be the standard orthonormal basis and $(\cdot|\cdot)$ the standard scalar product in ℓ^2. Let $\mathcal{H}_k$, $k = 1, 2, \ldots$ be a subspace of ℓ^2 spanned by e_{2k-1}, e_{2k}. Then $\ell^2 = \bigoplus_{k=1}^{\infty} \mathcal{H}_k$. Let J be a fundamental symmetry on ℓ^2 such that the matrix representation of the restriction of J on $\mathcal{H}_k$ is $\begin{pmatrix} 1 & 0 \\ 0 & -1 \end{pmatrix}$. This and all the other matrix representations in $\mathcal{H}_k$ are with respect to the basis $\{e_{2k-1}, e_{2k}\}$. Let S be a uniformly positive operator in the Hilbert space $(\ell^2, (\cdot|\cdot))$ such that the matrix representation of the restriction of S on $\mathcal{H}_k$, is $\begin{pmatrix} k & -(k-1) \\ -(k-1) & k \end{pmatrix}$. Clearly 1 is an eigenvalue of S of infinite multiplicity, i.e. the spectrum of S is not discrete. The operator JS is uniformly positive in the Krein space $(\ell^2, (J\cdot|\cdot))$. The eigenvalues of JS are $\pm\sqrt{2k-1}$ and the linear span of the corresponding eigenvectors is dense in ℓ^2. Therefore, the spectrum of JS is discrete.

However, if ∞ is not a singular critical point of JS, then the following proposition holds.

PROPOSITION 1.5 *Let A be a quasi-uniformly positive operator in the Krein space $(\mathcal{H},[\cdot|\cdot])$. Assume that ∞ is not a singular critical point of A. Then A has discrete spectrum if and only if JA has discrete spectrum.*

PROOF We only have to prove that the discreteness of the spectrum of A implies the discreteness of the spectrum of JA. By [4, Proposition 2.3] there exists a Riesz basis consisting of eigenvectors and associated eigenvectors of A. This implies that A has compact resolvent. From the identity

$$\begin{aligned}(A-\lambda I)^{-1}-(JA-\lambda I)^{-1} &= \\ &(A-\lambda I)^{-1}(I-J)JA(JA-\lambda I)^{-1}\end{aligned}$$

it follows that JA also has compact resolvent. □

The discreteness of the spectra of A and JA does not imply the nonsingularity of ∞. This can be seen from the following example.

EXAMPLE In the notation of the previous example, let A be an operator in ℓ^2 such that the matrix representation of the restriction of A on $\mathcal{H}_k$ is $\begin{pmatrix} k^2 & -k(k-1) \\ k(k-1) & -k^2 \end{pmatrix}$ in $\mathcal{H}_k$. The matrix representation of the restriction of JA in $\mathcal{H}_k$ is $\begin{pmatrix} k^2 & -k(k-1) \\ -k(k-1) & k^2 \end{pmatrix}$. The operator JA is uniformly positive in $(\ell^2,(\cdot|\cdot))$ and it has discrete spectrum. Therefore, the operator A is uniformly positive in the Krein space $(\ell^2,(J\cdot|\cdot))$ and its spectrum is also discrete. Since the cosine of the angle between the eigenvectors of A in $\mathcal{H}_k$ converges to 1, the point ∞ is a singular critical point of A.

Quasi-uniformly positive operators have important spectral properties. We list them for reader's convenience.

Let E be the spectral function of the quasi-uniformly positive operator A (see [10]). Let $\lambda \in \sigma(A)\cap\mathbb{R}$. Then λ is of *positive* type (*negative* type, respectively) if there exists an open interval Δ containing λ such that $(E(\Delta)\mathcal{H},[\cdot|\cdot])$ $((E(\Delta)\mathcal{H},-[\cdot|\cdot])$, respectively) is a Hilbert space. Further λ is a *critical point* if $[\cdot|\cdot]$ is indefinite on $E(\Delta)\mathcal{H}$ for every open interval Δ containing λ. The set of all spectral points of A of positive type (negative type, respectively) is denoted by $\sigma_+(A)$ ($\sigma_-(A)$, resp.). The set of all critical points of A is denoted by $c(A)$.

A critical point λ is said to be of *finite negative* (*positive*, respectively) *index* $\kappa_-(\lambda)$ ($\kappa_+(\lambda)$, *respectively*) if $(E(\Delta)\mathcal{H},[\cdot|\cdot])$ is a Pontryagin space with a finite number $\kappa_-(\lambda)$ ($\kappa_+(\lambda)$, respectively) of negative (positive, respectively) squares for all sufficiently small open intervals Δ containing λ.

A critical point is of *finite index* if it is of finite positive or finite negative index. In the terminology of [1] such a point is said to be of finite type. Every critical point of finite index is an eigenvalue.

Recall that the negativity index $\kappa(A)$ of the quasi-uniformly positive operator A equals the total multiplicity of the nonpositive eigenvalues of the selfadjoint operator JA in the Hilbert space $(\mathcal{H},(\cdot|\cdot))$.

If 0 is an eigenvalue of A then by Proposition 1.3 it is an isolated eigenvalue of finite algebraic multiplicity. From the canonical form of a Hermitian operator in a finite dimensional Krein space [6, Theorem 3.3] it follows that in the corresponding algebraic eigenspace there exists a basis consisting of mutually orthogonal Jordan chains $\{x_{i1},...,x_{in_i}\}$, $i=1,...,p$ with the property that $[x_{i1}|x_{in_i}]\neq 0$. We denote $\varepsilon_i = \text{sgn}\,[x_{i1}|x_{in_i}]$.

Note that while the Jordan chains are not unique, the number p of the chains, their lengths $n_i, i=1,...,p$ and the signs $\varepsilon_i, i=1,...,p$ are invariants. We say that $\{p;n_1,...,n_p;\varepsilon_1,...,\varepsilon_p\}$ is *the Jordan chain data of A at* 0.

The following proposition follows from the results in [10] and [3, Section 1.3].

PROPOSITION 1.6 *Let A be a quasi-uniformly positive operator in the Krein space $(\mathcal{H},[\cdot|\cdot])$ with the negativity index $\kappa(A)$.*

(a) The set of nonreal eigenvalues of A with positive imaginary parts consists of finitely many eigenvalues with finite total algebraic multiplicity κ_a.

(b) The sets $\sigma_+(A)\cap\mathbb{R}_-$ and $\sigma_-(A)\cap\mathbb{R}_+$ consist of finitely many isolated eigenvalues of finite total (geometric) multiplicities κ_b^- and κ_b^+.

(c) All finite critical points of A are of finite index; the set $c(A)\cap\mathbb{R}_-$ ($c(A)\cap\mathbb{R}_+$, respectively) consists of negative (positive, resp.) critical points of finite positive (negative, resp.) index. If 0 is a critical point than it is a critical point of finite both positive and negative index. Moreover, in that case $\kappa_-(0)+\kappa_+(0)$ equals the algebraic multiplicity of the eigenvalue 0.

(d) Let $\{p;n_1,...,n_p;\varepsilon_1,...,\varepsilon_p\}$ be the Jordan chain data of A at 0. Let $n^-(0)$ denote the number of indices i with the property $(-1)^{n_i}\varepsilon_i=-1$, and $n^+(0)$ the number of indices i with the property $\varepsilon_i=-1$. Then

$$\kappa_a+\kappa_b^+ +\kappa_b^- + \sum_{\lambda\in c(A)\cap[0,\infty)}\kappa_-(\lambda)+\sum_{\lambda\in c(A)\cap(-\infty,0)}\kappa_+(\lambda)+n^-(0)=\kappa(A), \tag{3}$$

and

$$\kappa_a+\kappa_b^+ +\kappa_b^- + \sum_{\lambda\in c(A)\cap(0,\infty)}\kappa_-(\lambda)+\sum_{\lambda\in c(A)\cap(-\infty,0]}\kappa_+(\lambda)+n^+(0)=\kappa(A). \tag{4}$$

(e) Every Jordan chain of A is of finite length. There are finitely many linearly independent Jordan chains of length ≥ 2; the sum of the lengths of these Jordan chains does not exceed $3\,\kappa(A)$.

The proof of part (d) uses the canonical form of the Hermitian operators JP and JAP in the finite dimensional Krein space $(P\mathcal{H},[\cdot|\cdot])$, where P is the orthogonal projection onto the algebraic eigenspace of the eigenvalue 0. The formulas (3) and (4) explain how the nonpositive squares of the form a are "used". The estimate in (e) is very crude. Note that $\kappa(A)$ is the maximal codimension of a subspace of $\mathcal{D}(A)$ on which a is uniformly positive definite. Therefore, parts (d) and (e) can be used to estimate the respective spectral quantities.

2 Quasi-uniformly positive forms

In this section we consider sesquilinear forms a and v in the Hilbert space $(\mathcal{H}, (\cdot|\cdot))$ satisfying

(A) The form a is closed and uniformly positive.

(B) The form v is relatively a-bounded with the a-bound $\Gamma < 1$.

This means (see [9, page 319]) that $\mathcal{D}(v) \supseteq \mathcal{D}(a)$ and that for all $\gamma > \Gamma$ there exists $C \geq 0$ such that

$$|v(x,x)| \leq \gamma a(x,x) + C\|x\|^2, \quad x \in \mathcal{D}(a). \tag{5}$$

Let B be the positive operator associated with the form a in the Hilbert space $(\mathcal{H}, (\cdot|\cdot))$, see [9, Theorem VI.2.1]. Then $\mathcal{D}(B^{1/2}) = \mathcal{D}(a)$ by [9, Theorem VI.2.23]. It follows from [9, Lemma VI.3.1] that there exists a bounded selfadjoint operator D on $\mathcal{H}$ such that

$$v(x,y) = (DB^{1/2}x|B^{1/2}y), \quad x, y \in \mathcal{D}(a). \tag{6}$$

By [9, Theorem VI.3.9] the form $a_1 = a + v$ is closed, symmetric and bounded from below. Let B_1 be the selfadjoint operator associated with the form a_1 in the Hilbert space $(\mathcal{H}, (\cdot|\cdot))$. Then

$$\mathcal{D}(|B_1|^{1/2}) = \mathcal{D}(B^{1/2}). \tag{7}$$

Let $A = JB$ and $A_1 = JB_1$. The operator B is uniformly positive and $0 \in \rho(B)$. Consequently, $0 \in \rho(A)$ and A is definitizable in the Krein space $(\mathcal{H}, [\cdot|\cdot])$.

PROPOSITION 2.1 *Assume that the selfadjoint operator A_1 is definitizable in the Krein space $(\mathcal{H}, [\cdot|\cdot])$. Then ∞ is not a singular critical point of A_1 if and only if it is not a singular critical point of A.*

PROOF This follows from (7) and [2, Corollary 3.6]. □

It remains to find sufficient conditions to establish the definitizabilty of the operator A_1.

In the next proposition we need the notion of relative compactness of quadratic forms. We refer to [12, page 369]. It is equivalent to the compactness of the operator D in (6).

PROPOSITION 2.2 1. *If there exists $\gamma < 1$ such that the relation* (5) *holds with $C = 0$, then $a_1 = a + v$ is a uniformly positive form. Therefore the operator A_1 is uniformly positive in the Krein space $(\mathcal{H}, [\cdot|\cdot])$.*
2. *If the form v is a-compact, then the form $a_1 = a + v$ is quasi-uniformly positive in the Hilbert space $(\mathcal{H}, (\cdot|\cdot))$. Therefore A_1 is a definitizable operator in the Krein space $(\mathcal{H}, [\cdot|\cdot])$..*

PROOF 1. The form $a + v$ is uniformly positive. Hence B is uniformly positive.
2. Since v is a-bounded with the a-bound < 1, the form $a + v$ and therefore also the operator B, is bounded from below. By [12, page 369] the operators B and B_1 have the same essential spectrum. Therefore B is quasi-uniformly positive in $(\mathcal{H}, (\cdot|\cdot))$ and A_1 is quasi-uniformly positive in the Krein space $(\mathcal{H}, [\cdot|\cdot])$. By Proposition 1.1 the operator A_1 is definitizable. □

COROLLARY 2.3 *Let s be a quadratic form in a Hilbert space* $(\mathcal{H}, (\cdot|\cdot))$. *The following statements are equivalent:*

(i) *s is a quasi-uniformly positive form.*

(ii) *s is a relatively form-compact symmetric perturbation of a uniformly positive form in* $(\mathcal{H}, (\cdot|\cdot))$.

PROOF The implication (i) $\Rightarrow$ (ii) follows from the corresponding statement about operators. The converse implication is the statement 2 of Proposition 2.2. □

In the next corollary we summarize the results of this section.

COROLLARY 2.4 *If any of the two assumptions of the* Proposition 2.2 *is satisfied, then* ∞ *is not a singular critical point of* A_1 *if and only if it is not a singular critical point of* A.

PROPOSITION 2.5 *If the form v is a-compact, then the essential spectra of* A *and* A_1 *coincide. Additionally,* A_1 *has compact resolvent if and only if* A *has a compact resolvent.*

PROOF From (6) and the definition of B_1 it follows that for all $x \in \mathcal{D}(B_1)$ and for all $y \in \mathcal{D}(B^{1/2})$ we have

$$((B_1 - \lambda J)x|y) = ((I + D - \lambda B^{-1/2}JB^{-1/2})B^{1/2}x|B^{1/2}y). \tag{8}$$

The operator $Q = B^{-1/2}JB^{-1/2}$ is a bounded selfadjoint operator in the Hilbert space $\mathcal{H}$. From (8) we have

$$(B_1 - \lambda J)x = B^{1/2}(I + D - \lambda Q)B^{1/2}x, \quad x \in \mathcal{D}(B_1). \tag{9}$$

For $\lambda \in \rho(A_1)\backslash\mathbb{R}$ its conjugate $\overline{\lambda}$ is also in $\rho(A_1)$. Therefore the range of the operator $I + D - \overline{\lambda}Q$ contains $\mathcal{D}(B^{1/2})$ and consequently its adjoint $I + D - \lambda Q$ is injective. Since the operator $I - \lambda Q$ is bounded and boundedly invertible it follows from the Fredholm alternative that the injective operator $I + D - \lambda Q$ has a bounded inverse. Inverting (9) we get

$$(B_1 - \lambda J)^{-1} = B^{-1/2}(I + D - \lambda Q)^{-1}B^{-1/2}. \tag{10}$$

We also note that $\lambda \in \rho(B)$ and

$$(B - \lambda J)^{-1} = B^{-1/2}(I - \lambda Q)^{-1}B^{-1/2}. \tag{11}$$

It follows from (10) and (11) that

$$(B_1 - \lambda J)^{-1} - (B - \lambda J)^{-1} = B^{-1/2}[(I + D - \lambda Q)^{-1} - (I - \lambda Q)^{-1}]B^{-1/2} =$$

$$= -B^{-1/2}(I - \lambda Q)^{-1}D(I + D - \lambda Q)^{-1}B^{-1/2}.$$

Thus the operator $(A_1 - \lambda I)^{-1} - (A - \lambda I)^{-1}$ is compact. By [9, Theorem IV.5.35] the operators $(A - \lambda I)^{-1}$ and $(A_1 - \lambda I)^{-1}$ have the same essential spectrum. As a consequence the operators A and A_1 have the same essential spectrum. □

3 Klein-Gordon equation

Let $\mathcal{G}$ be a Hilbert space with a scalar product $(\cdot|\cdot)$, H a positive selfadjoint operator in $\mathcal{G}$ such that $H \geq m^2 I > 0$. For $-1 \leq \alpha \leq 1$, let $\mathcal{G}_\alpha$ be the Hilbert space completion of $(\mathcal{D}(H^\alpha), (H^\alpha \cdot |H^\alpha \cdot))$. Denote by $\|\cdot\|_\alpha$ the norm of this Hilbert space.

If $\alpha \leq 0$ the space $\mathcal{G}_\alpha$ coincides with $\mathcal{D}(H^\alpha)$. The operator H can be extended to an isometry between $\mathcal{G}_\alpha$ and $\mathcal{G}_{\alpha-1}$.

Denote by $\mathcal{H}$ the Hilbert space $\mathcal{G}_{1/4} \oplus \mathcal{G}_{-1/4}$ and by $\langle\cdot|\cdot\rangle$ its natural scalar product. If $x \in \mathcal{G}_{1/4}$ then $|(x|y)| \leq \|x\|_{1/4}\|y\|_{-1/4}$ $(y \in \mathcal{G})$. Therefore the scalar product $(\cdot|\cdot)$ can be extended by continuity from $\mathcal{G}_{1/4} \times \mathcal{G}$ to $\mathcal{G}_{1/4} \times \mathcal{G}_{-1/4}$ and similarly from $\mathcal{G} \times \mathcal{G}_{1/4}$ to $\mathcal{G}_{-1/4} \times \mathcal{G}_{1/4}$. Define an indefinite scalar product on $\mathcal{H}$ by

$$[x|y] = (x_1|y_2) + (x_2|y_1), \quad x = (x_1, x_2),\ y = (y_1, y_2) \in \mathcal{H}.$$

The space $\mathcal{H}$ with the indefinite scalar product $[\cdot|\cdot]$ is a Krein space. The fundamental symmetry is

$$\mathbf{J} = \begin{bmatrix} 0 & H^{-1/2} \\ H^{1/2} & 0 \end{bmatrix}$$

Define the operator $\mathbf{A}$ in $\mathcal{H}$ on $\mathcal{D}(\mathbf{A}) = \mathcal{G}_{3/4} \oplus \mathcal{G}_{1/4}$ by

$$\mathbf{A} = \begin{bmatrix} 0 & I \\ H & 0 \end{bmatrix}.$$

The operator $\mathbf{A}$ is a selfadjoint operator in $(\mathcal{H}, [\cdot|\cdot])$. Since

$$[\mathbf{A}x|x] = (Hx_1|x_1) + (x_2|x_2), \quad x = (x_1, x_2) \in \mathcal{D}(\mathbf{A}), \tag{12}$$

the operator $\mathbf{A}$ is uniformly positive in $(\mathcal{H}, [\cdot|\cdot])$. The form $[\mathbf{A}x|y]$, $x, y \in \mathcal{D}(\mathbf{A})$ is closable. Let $\mathbf{a}$ be its closure. Let $\mathbf{B}$ be the uniformly positive operator associated with the form $\mathbf{a}$ in the Hilbert space $(\mathcal{H}, \langle\cdot|\cdot\rangle)$. It follows from (12) that the domain of $\mathbf{a}$ is $\mathcal{D}(\mathbf{a}) = \mathcal{H}_1(\mathbf{A}) = \mathcal{G}_{1/2} \oplus \mathcal{G}$ and that

$$\mathbf{a}(x, y) = \langle \mathbf{P}x|\mathbf{P}y\rangle, \quad x, y \in \mathcal{G}_{1/2} \oplus \mathcal{G}$$

with

$$\mathbf{P} = \mathbf{B}^{1/2} = \begin{bmatrix} H^{1/4} & 0 \\ 0 & H^{1/4} \end{bmatrix}.$$

The following lemma follows from the fact that the operators $\mathbf{A}$ and $\mathbf{J}$ commute.

LEMMA 3.1 *Infinity is not a singular critical point of* $\mathbf{A}$.

Let V be a $H^{1/2}$-bounded symmetric operator in $\mathcal{G}$. We define the form

$$\mathbf{v}(x, y) = (Vx_1|y_2) + (x_2|Vy_1), \quad x = (x_1, x_2), y = (y_1, y_2) \in \mathcal{G}_{1/2} \oplus \mathcal{G}.$$

LEMMA 3.2 *Let V be a $H^{1/2}$-bounded symmetric operator with the relative bound β_0. Then the form $\mathbf{v}$ is $\mathbf{a}$-bounded in $\mathcal{H}$ with the relative $\mathbf{a}$-bound $\leq \sqrt{\beta_0}$.*

PROOF Let $\beta > \beta_0$. Then there exists $C > 0$ such that

$$\|Vx_1\|^2 \leq \beta \|H^{1/2}x_1\|^2 + C\|x_1\|^2.$$

Noting that $\mathbf{v}(x,x) = 2\mathrm{Re}\,(Vx_1|x_2)$ it follows that

$$|\mathbf{v}(x,x)| \leq 2\|Vx_1\|\,\|x_2\| \leq \sqrt{\beta}\|x_2\|^2 + \frac{1}{\sqrt{\beta}}\|Vx_1\|^2.$$

Since H is uniformly positive, $\|x_1\|^2$ can be replaced by $\|H^{1/4}x_1\|^2$. Therefore

$$|\mathbf{v}(x,x)| \leq \sqrt{\beta}\mathbf{a}(x,x) + \frac{C}{\sqrt{\beta}}\langle x|x\rangle.$$

□

COROLLARY 3.3 *If the $H^{1/2}$-bound of V is < 1 then the form $\mathbf{a}+\mathbf{v}$ defined on $\mathcal{G}_{1/2} \oplus \mathcal{G}$ is closed, symmetric and bounded from below.*

PROOF This follows from [9, Theorem VI.3.9]. □

In the rest of this section we assume that the operator V is $H^{1/2}$-bounded with the relative bound < 1.

Let $\mathbf{B}_1$ be the selfadjoint operator associated with $\mathbf{a} + \mathbf{v}$ in the Hilbert space $(\mathcal{H}, \langle\cdot|\cdot\rangle)$ and let $\mathbf{A}_1 = \mathbf{J}\mathbf{B}_1$. The operator $\mathbf{A}_1$ is selfadjoint in the Krein space $(\mathcal{H}, [\cdot|\cdot])$. From Proposition 2.1 we conclude:

PROPOSITION 3.4 *If the selfadjoint operator $\mathbf{A}_1$ is definitizable then ∞ is not its singular critical point.*

It follows from the symmetry of V that it can be extended to a bounded operator from $\mathcal{G}_\alpha$ to $\mathcal{G}_{\alpha-1/2}$ for $0 \leq \alpha \leq 1/2$. A calculation shows that

$$\mathbf{v}(x,y) = \langle \mathbf{D}\mathbf{P}x|\mathbf{P}y\rangle, \quad x,y \in \mathcal{G}_{1/2} \oplus \mathcal{G}$$

with

$$\mathbf{D} = \begin{bmatrix} 0 & H^{-3/4}VH^{-1/4} \\ H^{1/4}VH^{-1/4} & 0 \end{bmatrix}.$$

The operator $\mathbf{D}$ is bounded in $\mathcal{H}$ and

$$\|\mathbf{D}\| = \|VH^{-1/2}\|. \tag{13}$$

If the operator V is $H^{1/2}$-compact than it is $H^{1/2}$-bounded with the relative bound 0. Moreover $H^{1/4}VH^{-1/4}$ is a compact operator from $\mathcal{G}_{1/4}$ into $\mathcal{G}_{-1/4}$ and $H^{-3/4}VH^{-1/4}$ is a compact operator from $\mathcal{G}_{-1/4}$ into $\mathcal{G}_{1/4}$. Consequently $\mathbf{D}$ is a compact operator in $\mathcal{H}$. From (13), Lemma 3.1, Corollary 2.4, Propositions 2.2 and 2.5 we conclude:

THEOREM 3.5 *Let V be a symmetric $H^{1/2}$-bounded operator with the relative bound < 1. Let $\mathbf{A}_1$ be the selfadjoint operator in the Krein space $(\mathcal{H}, [\cdot|\cdot])$ defined above.*
1. *Assume that $\|VH^{-1/2}\| < 1$. Then $\mathbf{A}_1$ is a uniformly positive operator which is similar to a selfadjoint operator in the Hilbert space $(\mathcal{H}, \langle\cdot|\cdot\rangle)$.*
2. *Assume that $VH^{-1/2}$ is compact. Then $\mathbf{A}_1$ is a definitizable operator and ∞ is not its singular critical point. The essential spectrum of $\mathbf{A}_1$ equals the essential spectrum of $\mathbf{A}$ and this is the set of all λ such that λ^2 is in the essential spectrum of H.*

References

[1] Binding. P., Najman, B.: Regularity of finite type critical points for self-adjoint operators in Krein space. Preprint.

[2] Ćurgus, B.: On the regularity of the critical point infinity of definitizable operators. Integral Equations Operator Theory 8 (1985), 462-488.

[3] Ćurgus, B., Langer, H.: A Krein space approach to symmetric ordinary differential operators with an indefinite weight function. J. Differential Equations 79 (1989), 31-61.

[4] Ćurgus, B., Najman, B.: A Krein space approach to elliptic eigenvalue problems with indefinite weights. Differential and Integral Equations 7 (1994), 1241-1252. .

[5] Ćurgus, B., Najman, B.: Quadratic eigenvalue problems. Mathematische Nachrichten, to appear.

[6] Gohberg, I., Lancaster, P., Rodman, L.: Matrices and Indefinite Scalar Products. Birkhäuser, Basel, 1983.

[7] Jonas, P.: On a problem of the perturbation theory of selfadjoint operators in Krein spaces. J. Operator Theory 25 (1991), 183-211.

[8] Jonas, P.: On the spectral theory of operators associated with perturbed Klein-Gordon and wave type equations. Preprint Karl-Weierstrass-Institut für Mathematik, Berlin, 1990.

[9] Kato, T.: Perturbation Theory for Linear Operators. 2nd ed. Springer-Verlag, Berlin, 1976.

[10] Langer, H.: Spectral function of definitizable operators in Krein spaces. Functional Analysis, Proceedings, Dubrovnik 1981. Lecture Notes in Mathematics 948, Springer-Verlag, Berlin, 1982, 1-46.

[11] Langer, H., Najman, B.: A Krein space approach to the Klein-Gordon equation. Unpublished manuscript.

[12] Reed, M., Simon, B.: Methods of Modern Mathematical Physics, vol. IV: Analysis of Operators. Academic Press, New York, 1978.

B. Ćurgus
Department of Mathematics,
Western Washington University,
Bellingham, WA 98225, USA
curgus@cc.wwu.edu

B. Najman
Department of Mathematics,
University of Zagreb,
Bijenička 30, 41000 Zagreb, Croatia
najman@cromath.math.hr

Math Review 1991 Mathematics Subject Classification 46B50 45C20

Operator Theory:
Advances and Applications, Vol. 80

Functional-Differential and Functional Equations with Rescaling.

Gregory Derfel[1]

A brief survey of the present state of functional- differential equations with rescaling is given. Various applications of equations with rescaling in probability, spectral theory of Schrödinger operator, subdivision processes and wavelets are discussed, as well.

1. Introduction

Functional-differential equations provide a mathemaical model for a physical system in which the rate of change of the system may depend upon its past history: that is, the future state of the system depends not only upon the present state, but also on part of its past history. A special case of such an equation is a differential-difference equation

$$x'(t) = f(t, x(t), x(t-\tau)) \tag{1}$$

where τ is a nonnegative constant.

Functional-differential equations have been discussed in the literature since the eighteenth century, by the Bernoullis, Laplace and Condorcet. However, only during the last forty years the subject has been, and is continuing to be, investigated at a very rapid pace. The impetus has mainly been due to the developments in the theory of control, mathematical biology, medicine and mathematical economics. The first pioneer books on the subject are monographs by Mishkis [Mish], Bellman and Cooke [BK], El'sgol'tz [El]. For recent overview see Hale [H]. The theory of functional-differential equations is closely connected with the theory of difference equations. A beautiful exposition of functional equations is given in [Ku], [PelSh]; for difference equations and chaos see [ShMR].

A very natural and important class of functional differential equations is the class of functional differential equations with linearly transformed arguments:

$$\sum_{j=0}^{\ell}\sum_{k=0}^{m} a_{jk}y^{(k)}(\alpha_j t + \beta_j) = 0, \quad -\infty < t < \infty \tag{2}$$

[1]Research supported in part by a grant of the Israel Ministry of Science and by a grant of Israel Academy of Science and Humanities

$$a_{jk} \in \mathbb{C}\,;\ \alpha_j(\neq 0,1),\ \beta_j \in \mathbb{R}$$

Such equations have direct applications in physics of quasi-crystals, actuarial theory (ruin problems), dynamical systems and their transitions to chaos and approximation theory. These applications are discussed in 2-6.

2. Asymptotic behavior of the solutions.

Functional-differential equations with linearly transformed arguments arise in numerous scientific application and the question about the asymptotics of their solutions is of great importance in these problems. Thus, asymptotics of the solutions of the equation

$$y'(t) = ay(\alpha t),\ 0 < \alpha < 1 \tag{3}$$

have been investigated by K. Mahler [Mah] and N.G. de Brujn [Br] in connection with the so-called "partition problem" in number theory.

Equation

$$y'(t) = ay(\alpha t) + by(t), \tag{4}$$

where α may be < 1 or > 1 arises in the oscillation theory [FMOT], in astrophysics [Amb] and in ruin problems [Gav]. An outstanding analysis of equation (4) is given in the paper by T. Kato and J. B. McLeod [KM], [K].

We mention here only one result concerning the general equation

$$y^{(m)}(t) = \sum_{j=0}^{\ell} \sum_{k=0}^{m-1} a_{jk} y^{(k)}(\alpha_j t + \beta_j). \tag{5}$$

Let us denote

$$\alpha = \min|\alpha_j|;\quad A = \max|\alpha_j|,\quad B = \max|\beta_j|.$$

Theorem 1. **[Der1]**. *If $\alpha > 1$ then every solution of equation (5), which satisfies the estimate*

$$|y^{(k)}(t)| \leq cexp\{-\gamma \ln^2(1+|t|)\},\ k = 0,1,\ldots,m-1 \tag{6}$$

where $c > 0$ and

$$\gamma > \bar{\gamma}_1 = m^2 \ln A/(2\ln^2 \alpha)$$

is a function supported on the interval $[-B/(\alpha-1), B/(\alpha-1)]$.

The following theorem shows that Theorem 1 cannot be improved essentially.

Theorem 2. **[Der 2]** *If $\alpha > 1$ and $\beta_j = 0$ $(j = 0,\ldots,\ell)$ then equation (5) has a solution which satisfies estimate (6) for $c > 0$ and for any $\gamma < \bar{\gamma}_1 = m/(2\ln A)$, but which is not compactly supported.*

Results of the same type can be established also for the equation (5) in the case when $A = \max|\alpha_j| < 1$ [Der1, Der2, V1, V2,W,WCS], and also for the equation

$$y(t) = \sum_{j=0}^{\ell}\sum_{k=1}^{m} a_{jk}y^{(k)}(\lambda_j t + \mu_j). \tag{7}$$

(with an isolated term without differentiation) both in case $\lambda = \min|\lambda_j| > 1$ and $\Lambda = \max|\lambda_j| < 1$ [Der1].

The results mentioned above may be considered as the results about the classes of the uniqueness of the solutions of the equations (5) and (7) respectively.

For functional-differential equations

$$y(t) = \sum_{j=0}^{\ell}\sum_{k=0}^{m} a_{jk}y^{(k)}(\lambda_j t + \mu_j).$$

similar to (7), but with the inner sum from $k = 0$ and at least one of $a_{jo0} \neq 0$ the result similar to Theorem 1, but with the estimate

$$|y(t) \leq c(1 + |t|)^{\gamma} \tag{8}$$

is valid. Here $\gamma > 0$ if $L = \sum_{j=0}^{\ell} |a_{j0}|/\lambda_j$ is a small value and $\gamma < 0$ if L is large enough, e.g. the class of uniqueness in that case is a class of functions of power growth (or power decay). [Der3].

Intensive investigation (both analytic and numerical) of so-called generalized pantograph equation

$$y'(t) = Ay(t) + By(at) + Cy'(qt)$$

where $q \in (0,1), A, B, C$ are $d \times d$ complex matrices has been fulfilled by A. Iserles, M. Buhmann [I], [BI1], [BI2].

3.Compactly supported solutions. Wavelets and subdivision processes.

One of the specific features of functional-differential equations with several transformations of arguments is the possibility of the existence of compactly supported solutions for such equations. The first example of a solution with compact support for such equations was discovered by V.L. Rvachov and V.A.. Rvachov in [Rv1]. These authors considered the equation

$$y'(t) = 2y(2t + 1) - 2y(2t - 1) \tag{9}$$

and its compactly supported solutions. They named that solution *up-function* and gave numerous applications of that function (and other similar functions - Fup_n, Ξ_n, etc.) in approximation theory [Rv1], [Rv2], [Rv3].

The dilation equation

$$f(t)=\sum_{j=0}^{m} a_j f(2t-k) \tag{10}$$

and its compactly supported solutions were intensively studied in connection with subdivision schemes, and wavelets in [Str], [MP], [DGL], [DL], [Berg1], [Berg2], [Da], [DaLa], [DDL].

Sufficient conditions of non-existence of compactly supported solutions for general equation (2) was given in [Der1]: If

$$\sum_{j=0}^{\ell} a_{j0} a_j^{-n} \neq 0, \quad n=1,2,\ldots \tag{11}$$

for any $n \in N$, then equation (2) has no compactly supported solutions. It means that equation (2) has compactly supported solutions in exceptional cases only. (It is worth to mention that necessary condition of the existencce of compactly supported solutions for dilation equation (10): $\sum_{j=0}^{\ell} a_j = 2^n$ [DaLa] is a special case of (11)).

However, it has been proved in [Der6] that every equation

$$z'(t)=\sum_{j=0}^{\ell} q_j \frac{\alpha_j^2}{2\beta_j}\left[z(\alpha_j t+\beta_j)-z(\alpha_j t-\beta_j)\right], \tag{12}$$

where $q_j \geq 0, \sum_{j=0}^{\ell} q_j = 1$ $\alpha_j > 1$, $\beta_j > 0$ has a nontrivial, compactly supported solution. (It should be mentioned that almost all special functions by V.A. Rvachov and V.L. Rvachov (up, Fup_n, Ξ_n, etc.) may be deduced from (12), when coefficients q_j, α_j, β_j are chosen in a special way.) It also has been proved in [Der1] (see Theorem 1, section 2) that every rapidly decaying solution of equation (5) is compactly supported. It is important to determine conditions of existence and nonexistence of compactly supported solutions for wide classes of functional-differential equations, having in mind various applications in approximation theory.

4. Spectral methods in the theory of functional-differrential equations, applications to quasi-crystals and localization theory

The question about phase transaction "conductor-insulator" is of great importance in Anderson localization theory and in the quasi-crystal theory. From the mathematical point of view that question may be reduced to the problem of the description of the spectrum of Schrödinger operators

$$y''(x)+v(x)y(x)=\lambda y(x) \tag{13}$$

or

$$c_{n+1}+c_{n-1}+v(n)x_r=\lambda c_n \tag{14}$$

with bounded but non-periodic potential [Sim], [B1], [B2].

On the other hand in [Der4], [DM1], [DM2], [DM3], [DM4], [Der5] we have investigated the T. Kato problem [KM], [K] on the existence of bounded (almost periodic) solutions of functional-differential equations with "compressed" and "stretched" arguments. We have considered the model equations

$$\lambda y(t) = y(qt) + y(t/q) + \sigma[y(t+1) + y(t-1)] \tag{15}$$

$$y''(t) = \sum_{j=0}^{\ell} a_j y(\alpha_j t) + \lambda y(t) \tag{16}$$

where $\alpha_j(\neq 1)$ are multiplicatively commensurable values, that is, $\alpha_j = q^{r_j}$, $q > 1$, r_j-rational numbers and

$$y''(t) = y(\alpha_1 t) + y(t/\alpha_1) + y(\alpha_2 t) + y(t/\alpha_2) + \lambda y(t) \tag{17}$$

where α_1, α_2 ($\neq 1$) are multiplicatively uncommensurable, and it turns out that the question about bounded solutions of eq. (15), (16), (17), also may be reduced to the problem about purely point spectrum of Schorödinger difference equation

$$c_{n+1} + c_{n-1} + 2\sigma \cos(2\pi q^n \omega) c_n = \lambda c_n, \ \omega \in [1, q). \tag{18}$$

of "almost-Mathieu" type, Jacobi difference equation

$$\sum_{\substack{j=-\ell \\ j\neq 0}}^{\ell} a_j c_{n-j} + (\lambda + \omega^2 q^{2n}) c_n = 0, \ \omega \in [1, q). \tag{19}$$

or two-dimensional difference Schrödinger equation

$$-\Delta c_{m,n} - e^{\beta m + \gamma n} \omega^2 c_{m,n} = \lambda c_{m,n}, \ \omega \in [1, q).$$

(where $\beta = \ln \alpha_1$, $\lambda = \ln \alpha_2$) respectively. For example the fact that the spectrum of the Jacobi equation (19) in the half-plane $Re\lambda < -K$

$$K = M \max\{q^{2^\ell}, 2/(1 - 1/q^2)\}, M = \sum_{\substack{j=-\ell \\ j\neq 0}}^{\ell} |a_j|$$

is purely point, real valued and unbounded, implies the following:

Theorem 3. [DM2],[DM3]

(1) *If $\lambda < -K$, then equation (16) has a nontrivial bounded solution.*

(2) *If $\lambda < -K$, any bounded solution is almost periodic.*

(3) *If $\lambda > K$, the equation (15) has no solution bounded on the whole axis.*

It should be pointed out that the problem of description of spectrum for difference Schrödinger operator with bounded, but not periodic potential is still far from its consummation. For example, S.M. Molchanov's conjecture that the "almost-Mathieu" equation (16) for almost all ω has purely point, dence spectrum when $\lambda \in [-2\sigma - 2, 2\sigma - 2]$, has not yet been proved, nor has it been refuted.

5. Probabilistic methods in functional-differential equations theory

Let us consider the point $x \in \mathbb{R}$, which is iterated randomly by two linear maps $L_1 x = \alpha_1 x + \beta_1$, $L_2 x = \alpha_2 x + \beta_2$ with probabilities p_1 and p_2, respectively, $p_1 + p_2 = 1$. The problem may be considered as a random walk along two lines on the plane. The limiting behavior of a walking particle according to the position of the lines was studied in [Maks], [Grin], [Berg1], [Berg2], [Der6]. Denoting by ζ independent, identically distributed random matrices which accept the values $\begin{pmatrix} \alpha_1, & \beta_1 \\ 0, & 1 \end{pmatrix}$, $(i = 1, 2)$, with probabilities p_1 and p_2, respectively, $\begin{pmatrix} \alpha, & \beta \\ 0, & 1 \end{pmatrix} x \overset{def}{=} \alpha x + \beta$ and x_n-the nth random iteration of x, one can readily express x_n in the terms of random matrix product [FurKif]

$$x_n = \zeta_n(\ldots(\zeta_2(\zeta_1(x)))\ldots). \tag{20}$$

A key role in the study of the limiting behavior of a walking particle is played by the fact, that corresponding stationary distribution function $y(t) = \lim_{n\to\infty} P\{x_n \le t\}$ satisfies functional equation

$$y(t) = p_1 y\left(\frac{t-\beta_1}{\alpha_1}\right) + p_2 y\left(\frac{t-\beta_2}{\alpha_2}\right). \tag{21}$$

All above considerations are applicable to the more general case of a random walk along n lines and the corresponding equation is [Der6]

$$y(t) = \sum_{j=0}^{\ell} p_j y\left(\frac{t-\beta}{\alpha_j}\right), \quad p_j \ge 0, \ \sum p_j = 1. \tag{22}$$

The functional-differential equation

$$y'(t) + y(t) = \sum_{j=0}^{\ell} p_j y(\alpha_j t), \quad p_j \ge 0, \ \sum p_j = 1. \tag{23}$$

similar to (22) arises in acturial theory [Gav]. Let us consider the point (fortune of a gambler), which moves with constant velocity 1 to the left and sometimes jumps from point x to points $\alpha_j x$ (some $\alpha_j > 1$ and others < 1) with probabilities p_j, respectively (gambler plays a series of games). Suppose that the probability of non-jump during the time dt is

$1 - dt$ and the probability of jump during the time dt is dt. Let $y(x)$ denote the probability that a gambler starting with initial fortune x will eventually be ruined (his fortune drops to zero). Then $y(x)$ satisfies equation (23).

Equations (22) and (23) may be considered as special cases of equation

$$y(t) = \int\int_{\substack{-\infty < \lambda < \infty \\ 0 < \mu < \infty}} y\left(\frac{t-\lambda}{\mu}\right) F(d\lambda, d\mu), \tag{24}$$

where $F(d\lambda, d\mu)$ is a probabilitstic measure defined in upper half-plain [Der6] ((22), (23) may be deduced from (24) when $F(d\lambda, d\mu)$ is chosen in a special way). Necessary and sufficient conditions of the existence of continuous, bounded nontrivial solutions of (24) are established in [Der6].

Theorem 4. **[Der6]** *If*

$$I = \int_{R^2_+} \ln \mu F(d\lambda, d\mu) < 0,$$

then equation (24) possesses a continuous, bounded, nontrivial ($\neq c$) solution and if $I > 0$, then equation (24) does not possess such a solution.

Some interesting applications of probabilistic methods to functional-differential equations of type (12) were developed in [KI1], [KI2], [MIK].

It was proved by Grincevicius [Grin] that limiting distribution of a particle walking along the lines is continuous, but it can be absolutely continuous or singular continuous. An open question is when a probability distribution function $y(t)$ is absolutely continuous and when it is singular continuous.

6. Invariant measures and chaos

Iterations of a continuous map of the interval into itself serve as simplest examples for a dynamical system with chaotic behavior. Let us consider piecewise linear map $L_\lambda(x) = \lambda g(x),\ 0 < \lambda < 2$,

$$g(x) = \begin{cases} x & 0 \le x \le 1 \\ 2 - x & 1 \le x \le 2 \end{cases}$$

Chaotic behavior of successive iterates of an initial point x under the influence of L_λ, can be described either in terms of a deterministic matrix product of type (20) or by means of absolutely continuous invariant measure. The density $y(x)$ of that measure satisfies so-called Perron-Frobenius equation

$$\begin{cases} \lambda y(x) = y(x/\lambda) + y(2 - x/\lambda) & 2\lambda - \lambda^2 < x < \lambda, \\ y(x) = 0 & \text{otherwise,} \end{cases}$$

which was studied by B. Derrida, A. Gervois, V. Pomeau [DGP]. To determine invariant measures of general piecewise linear maps and clarify the relations between that problem and the problem described above in the section 5 is an important open problem.

7. Oscillations

Another branch of the research of functional-differential equations is the study of oscillations; that is, when one solution or all solutions of an equation have infinitely many zeros. This question is widely studied for first order equations

$$x'(t) = \sum_{i=1}^{k} p_i(t)x(g_i(t))$$

(see, for example, [LLZ, Tho, GLS, FJ, KF]. An intimate connection between asymptotic behavior and oscillation was established for the equations with linearly transformed argument in [MFB]. Even for a simple-looking equation $x'(t) = -x(t/k)$, $k > 1$, the distribution of the zeros of the solutions is unknown. The conjectures of [MFB] about the zeros of the solutions are still unresolved.

Bibliography

[AMb] V. A. Ambartsumian, On the fluctuation of brightness in galaxy, Sov. Math. Doklady **44** (1944), 223-226.

[B1] J. Bellisard, Almost periodicity in solid state physics and C^*-algebras, In: The Harald Bohr Centenary (C. Berg, F. Fuglede eds.), Royal Danish Acad. Sci., MfM **42:3**, 1989, 35-75.

[B2] J. Bellisard, Gap labelling theorems for Schrödinger Operators, In: Number Theory and Physics, 1990, 1-22.

[Berg1] M. Berger, Random affine iterated systems: smooth curve generation (to appear)

[Berg2] M. Berger, Wavelets as attractors of random dynamical systems (to appear).

[BI1] M.D. Buhmann, A. Iserles, On the dynamics of the discretized neutral equation, IMA. J. Num. Anal. **12** (1992), 339-363.

[BI2] M.D. Buhmann, A. Iserles, Numerical analysis of functional equations with a variable delay, In: Numerical Analysis, 1991 (D.F. Griffits and G.A. Watson eds.) Longman 1992, 17-33.

[BK] R. Bellman, K.Cooke, Differential-Difference Equations, Academic Press, New-York, 1963.

[**Br**] N.G. de Brujin, The difference-differential equation $F'(x) = e^{\alpha x+\beta}F(x-1)$. Nederl. Acad. Wetesh. Proc. Ser. A, **15** (1953), 449-464.

[**Da**] I. Daubechies, Orthogonal basis of compactly supported wavelets. Comm. Pure Applied Math., **41** (1988), 909-996.

[**DaLa**] I. Daubechies, J. Lagarias, Two-scale difference equations I and II,
SIAM J. Math Anal. **22** (1991), 1388-1410.
SIAM J. Math Anal. **23** (1992), 1031-1079.

[**DDL**] G. Derfel, N. Dyn, D. Levin, Generalized refinement equations and subdivision processes (to appear in J. Approx. Theory).

[**Der1**] G. Derfel, Asymptotic properties of the solutions of functional differential equations with linearly transformed arguments, Ph. D. Thesis, 1977, Tbilisi.

[**Der2**] G. Derfel, On the asymptotics of the solutions of some linear functional differential equations. Reports of I. Vekua Institute of Applied Math, Tbilisi, **12-13**, (1978), 21-23.

[**Der3**] G. Derfel, Functional differential equations with linearly transformed arguments and their applications. In: Differential Equations. Equadiff-91 Vol.1, (C.Perello, C.Simo, Y.Sola-Morales eds.), World Scientific, 1992, 421-424.

[**Der4**] G. Derfel, About spectrum of difference Schrödinger equation and about behaviour of solutions of functional equations with linear transformations of arguments In: Dynamical Systems and Differential Difference Equations, Inst. of Math of Ukrainian Ac. of Sci., Press, Kiev, 1986, 14-20,

[**Der5**] G. Derfel, T. Kato problem for functional-differential equations and difference Schrödinger operator. Operator Theory: Advances and applications **46** (1990), Birkhäuser Verlag, Basel, 319-321.

[**Der6**] G. A. Derfel, Probabilistic methods for a class of functional-differential equation, Ukrainian Math. **41**, (1989), 1137-1141.

[**DGL**] N. Dyn, J. Gregory, D. Levin, A 4-point interpolatory subdivision scheme for curve design. Computer Aided Geometric Design, **4** (1987), 257-268.

[**DGP**] B. Derrida, A. Gervois, Y. Pomeau, Iteration of endomorphism on the real axis and representation of numbers, Ann. Inst. Henri Poincare, Sect. A. **29**, (1978), 305-356.

[**DL**] N. Dyn, D. Levin, Smooth interpolations by bisectional algorithm, In: Approximation Theory (C.K. Chui, L.L. Shumaker and Y.D. Wards, eds.) Academic Press, New York, 1986, 335-337.

[**DM1**] G. Derfel and S.A. Molchanov, Probability methods and spectral methods in the theory of functional-differential equations, Uspekhi Math. Nauk **42**, (1987), 126.

[**DM2**] G. Derfel and S.A. Molchanov, About T. Kato problem for functional-differential equations with commensurable and uncommensurable transformations of argument. Russian Math. Surv. **44** (1989) 181.

[**DM3**] G. Derfel and S.A. Molchanov, On T. Kato problem on bounded solutions of functional-differential equations, Functional Analysis and its Applications, **24**, (1990,) 67-69.

[**DM4**] G. Derfel and S.A. Molchanov, Spectral methods in the theory of functional-differential equations, Mathematical Notes of the Ac. Sci. USSR, **47** (1990), 42-51.

[**El**] L. El'sgol'tz, Introduction to the Theory of Differential Equations with Deviating Argument, Nauka, Moscow, 1964.

[**FJ**] A. Feldstein, L. Jackiewicz, Unstable neutral functional differential equations, Canadian Math. Bull., **33**, (1990), 428-433.

[**FMOT**] L. Fox, D. Mayers, J. Ockendon, A. Tayler, On a functional-differential equation, J. Inst. Math. Appl. **8** (1971), 271-307.

[**Fr**] P. Frederickson, Dirichlet series solutions for certain functional differential equations, Lect. Notes Math. **243** (1971), 249-254.

[**FurKif**] H. Furstenberg and Y. Kifer, Random matrix products and measures on projective space, Israel J. of Math. **46** (1983), 12-32.

[**Gav**] D.P. Gaver, An absorption probability problem, J. Math. Anal. Appl. **9** (1964), 384-393.

[**GLS**] E. Grove, G. Ladas, J. Schinas, Sufficient conditions for the oscillation of delay and neutral delay equations Canad. Math. Bull. **31** (1988), 459-466.

[**Grin**] A.K. Grincevicius, On the continuity of the distribution of a sum of dependent variables connected with independent walks on lines, Theory of Probability and Appl. **19**, (1974) 163-168.

[**H**] J. Hale, Theory of Functional-Differential Equations, Springer-Verlag,New-York/Berlin, 1977.

[**I**] A. Iserles, On the generalized pantograph functional-differential equation, Euro. J. of Applied Math., **4** (1993), 339-363.

[**K**] T. Kato, Asymptotic behavour of solutions of the functional-differential equation $y'(x) = ay(\lambda x) + by(x)$, In: Delay and Functional-Differential Equations and their Apllications, Acad. Press, New-York. 1972, 197-217.

[**KF**] Y. Kuang, A. Feldstein, Monotonic and oscilatory solutions of a linear neutral delay equation with infinite lag, SIAM J. Math. Anal. **21**, (1990), 1633-1641.

[**KI1**] K. Kabaya, M. Iri, Sum of uniformly distributed random variables and a family of nonanalytic C^∞-functions, Japan J. Appl. Math. **4** (1987), 1-22.

[**KI2**] K. Kabaya, M. Iri, On operators defining a family of nonanalytic C^∞-functions. Japan J. Appl. Math. **5** (1988), 333-365.

[**KM**] T. Kato and J.B. Mcleod, The functional-differential equation $y'(x) = = ay(\lambda x) + by(x)$, Bull. Amer. Math. Soc. **77** (1971), 891-937.

[Ku] M. Kuczma, Functional Equations in a Single Variable, Polish Sci. Publ., Warszawa, 1968.

[LLZ] Ladde, Laksmikantham and Zhang, Oscillation Theory of Differential Equations with Deviating Arguments, Marcel Dekker, New York and Basel, 1987.

[Mah] K. Mahler, On special functional equation, J. London Math. Soc. **15** (1940), 115-123.

[Maks] V.M. Maksimov, A generalized Bernoulli scheme and its limit distribution. Theory of Probability and Appl. **18**, (1973), 547-556.

[MFB] G. Morris, A. Feldstein, E. Bowen, The Phragmen-Lindelöf principle and a class of functional differential equations, In: Ordinary Differential Equations (L. Weiss ed.), Academic Press, New-York, 1972, 513-540.

[MIK] S. Moriguti, M. Iri, K. Kabaya, On asymptotic properties of the eigenfunctions of a linear operator, Japan J. Appl. Math. **7**. (1990), 203-229.

[Mish] A.D. Mishkis, Linear Equations with Retarded Argument, GNTI, Moscow, 1951.

[MP] Ch. Michelli, A. Pinkus, Descartes systems for corner cutting (to appear).

[PelSh] G.P. Peluh, A.N. Sharkowsky, Introduction to the Theory of Functional Equations, Naukova Dumka, Kiev, 1974.

[Rv1] V.L. Rvachov V.A. Rvachov, On a function with compact support, Dokl. Acad. Nauk Ukranian SSR, Ser. A, (1971), 705-707.

[Rv2] V.L. Rvachov V.A. Rvachov, Non-classical Methods of Approximation Theory in Boundary-value Problems, Naukova Dumka, Kiev, 1979.

[Rv3] V.L. Rvachov V.A. Rvachov, Compactly supported solutions of functional-differential equations and their applications, Russian Math. Surv. **45** (1990), 87-120.

[ShMR] A. Sharkovsky, Ju, Maistrenko, E. Romanenko, Difference Equations and their Applications, Naukova Dumka, Kiev, 1986.

[Sim] B. Simon, Almost periodic Schrödinger operators: a review, Advances in Applied Math. **3** (1982), 463-390.

[Str] G. Strang, Wavelets and dilation equations: a brief introduction, SIAM Review **31** (1989), 614-627.

[Tho] A. Thomaras, Oscillation of an equation relevant to an industrial problem, Bull., Austral. Math. Soc.**12** (1975), 425-431.

[V1] F. Vogl, Das Wachstum ganzer Lözungen gewisser linearer Functional-Differentialgleichungen Monats. Math. **86** (1978), 239-250.

[V2] F. Vogl, Uber ein System linearer Functional-Differentialgleichungen, ZAMM, **60**, (1980), 7-17.

[**W**] Y. Wiener, Distributional and entire solutions of linear functional-differential equations, International Y. Math & Math. Sci. **5** (1982), 729-736.

[**WKS**] Y. Wiener, K.L.Cooke, S.M. Shah, Coexistence of analytic and distributional solutions for linear differential equations. J. Math Anal. Appl. **159** (1991), 271-289

Department of Mathematics and Computer Sciences
Ben-Gurion University of the Negev, P.O.B 653
Beer-Sheva 84105, Israel

MSC: 34Kxx, 39B22, 47B39

Operator Theory:
Advances and Applications, Vol. 80

ON THE SIGNATURES OF SELFADJOINT PENCILS

AAD DIJKSMA AURELIAN GHEONDEA*

For bounded selfadjoint operators F and G on a Hilbert space with $F\overline{\mathcal{R}(G)} \subseteq \mathcal{R}(G)$, we give formulae and estimates of the positive/negative signatures $\kappa^{\pm}(\lambda G - F)$ of the pencil $\lambda G - F$. For example, we prove that for all real λ,

$$\kappa^{\pm}(\lambda G - F) \leq \min\{\kappa^{\pm}((\lambda - 0)G - F), \kappa^{\pm}((\lambda + 0)G - F)\}.$$

These results are reformulated in terms of the pencil $\lambda I - A$ where A is a selfadjoint operator in a Kreĭn space and then proved by means of the spectral theory of definitizable operators.

1. Introduction

Let A be a bounded selfadjoint operator on a Hilbert space $\mathcal{H}$ with inner product $(\cdot,\cdot)$ and denote by $\kappa^{\pm}(A)$ the supremum of $\dim \mathcal{L}$ where $\mathcal{L}$ runs through the set of all finite dimensional subspaces of $\mathcal{H}$ such that

$$\pm(Ax, x) > 0, \quad x \in \mathcal{L} \setminus \{0\}.$$

If no such $\mathcal{L}$ exists we set $\kappa^{\pm}(A) = 0$. The numbers $\kappa^{+}(A)$ and $\kappa^{-}(A)$ are called the positive and negative signatures (or Hermitian indices) of A with respect to the inner product $(\cdot,\cdot)$. They coincide with the number of points in the spectrum of A, counted with their multiplicities, which lie in the open intervals $(0,\infty)$ and $(-\infty, 0)$, respectively; see N. I. Akhiezer and I. M. Glazman [2], Section 82. For any real λ the limits

$$\kappa^{\pm}((\lambda + 0)I - A) = \lim_{\mu \downarrow \lambda} \kappa^{\pm}(\mu I - A), \ \kappa^{\pm}((\lambda - 0)I - A) = \lim_{\mu \uparrow \lambda} \kappa^{\pm}(\mu I - A)$$

exist (possibly equal to infinity) and

$$\kappa^{+}(\lambda I - A) = \kappa^{+}((\lambda - 0)I - A) \leq \kappa^{+}((\lambda + 0)I - A), \tag{1.1}$$

$$\kappa^{-}(\lambda I - A) = \kappa^{-}((\lambda + 0)I - A) \leq \kappa^{-}((\lambda - 0)I - A). \tag{1.2}$$

*The second author was supported by a grant from the Netherlands Organization for Scientific Research, N. W. O.

This statement can be proved as Satz 1 in [2], p.229, by means of the left continuous spectral function $E(t)$ with respect to which A can be decomposed as the integral

$$A = \int_{\mathbf{R}} t\, dE(t).$$

It can be shown that for all $\mu \in \mathbf{R}$,

$$\kappa^-(\mu I - A) = \dim E(\mu, \infty)\mathcal{H},$$

and hence

$$\kappa^-((\lambda - 0)I - A) = \dim E[\lambda, \infty)\mathcal{H},$$

$$\kappa^-((\lambda + 0)I - A) = \dim E(\lambda, \infty)\mathcal{H} = \dim E[\lambda, \infty)\mathcal{H} - \dim \ker(\lambda I - A).$$

This proves (1.2) for $\kappa^-(\lambda I - A)$. Formula (1.1) for $\kappa^+(\lambda I - A)$ can be obtained from (1.2) by using the relation $\kappa^+(A) = \kappa^-(-A)$. This discussion serves to show how spectral theory plays a role in the study of the numbers $\kappa^\pm(\lambda I - A)$.

In this paper we study $\kappa^\pm(\lambda I - A)$ in the case where the inner product $(\cdot, \cdot)$ is indefinite that is, where A is a selfadjoint operator in the Kreĭn space $\mathcal{H}$. If for some real λ the number $\kappa^-(\lambda I - A)$ or the number $\kappa^+(\lambda I - A)$A is finite, then, as a consequence of a theorem of L. S. Pontryagin, the operator A is necessarily definitizable (see [22]). This allows us to use the spectral theory of definitizable operators which has been initiated and developed by M. G. Kreĭn and H. Langer [15], H. Langer [21], [22] (see also P. Jonas [12]). Using the properties of the spectral function $E(t)$ associated with the definitizable operator A, one can prove a formula to compute the number $\kappa^-(\lambda I - A)$ (cf. [9]). We recall this formula in Lemma 3.1. ¿From this formula other formulae for $\kappa^\pm(\lambda I - A)$ can be deduced. Special cases occur when $\mathcal{H}$ is a finite dimensional or a Pontryagin space. If λ is an eigenvalue of A and its root subspace is degenerate we give an estimate of $\kappa^\pm(\lambda I - A)$ (see Theorem 4.6). The upper semicontinuity of the mappings $A \mapsto \kappa^\pm(A)$ in conjunction with the spectral theory of definitizable operators implies that for all real λ the following inequalities hold (see Theorem 3.4):

$$\kappa^\pm(\lambda I - A) \leq \min\{\kappa^\pm((\lambda + 0)I - A), \kappa^\pm((\lambda - 0)I - A)\}. \tag{1.3}$$

We show in Section 6 that if F and G are selfadjoint operators in a Hilbert space and $F\overline{\mathcal{R}(G)} \subset \mathcal{R}(G)$ then $\kappa^\pm(\lambda G - F)$ can be calculated in terms of $\kappa^\pm(\lambda I - A)$ where A is a certain selfadjoint operator in a Kreĭn space and that, on account of (1.3), for all real numbers λ the following inequalities hold:

$$\kappa^\pm(\lambda G - F) \leq \min\{\kappa^\pm((\lambda + 0)G - F), \kappa^\pm((\lambda - 0)G - F)\}. \tag{1.4}$$

To construct A in the general case we use the technique of induced Kreĭn spaces. In the special case where G is boundedly invertible the construction is as follows: We write G in the polar decomposition

$$G = J|G| = |G|^{\frac{1}{2}} J |G|^{\frac{1}{2}},$$

where J is a symmetry operator on $\mathcal{H}$, that is, $J = J^{-1} = J^*$, and set

$$A = J|G|^{-\frac{1}{2}} F |G|^{-\frac{1}{2}}.$$

Then $\kappa^\pm(\lambda G - F) = \kappa^\pm(\lambda I - A)$, where the term on the righthand side is to be interpreted as the positive/negative signature of the operator $\lambda I - A$ with respect to the inner product

$[x, y] = (Jx, y)$, $x, y \in \mathcal{H}$. Note that $\mathcal{H}$ equipped with $[\cdot, \cdot]$ is a Kreĭn space and that A is selfadjoint with respect to this inner product, that is, $[Ax, y] = [x, Ay]$, $x, y \in \mathcal{H}$.

The definition of $\kappa^{\pm}(\lambda I - A)$ as a number in $\mathbf{N}$ or the symbol ∞ is related with the assumption that $\mathcal{K}$ is separable. In the nonseparable case the inequality (1.3) is not so relevant because our definition of $\kappa^{-}(\lambda I - A)$ does not distinguish between the different infinite numbers that may appear. However we did not assume that the space $\mathcal{K}$ is separable because the proof does not require it. But even in the nonseparable case the inequality (1.3) has a meaning. In the larger framework of transfinite cardinal numbers the inequality (1.3) is closely related with the multiplicity theory of spectral measures. It is beyond our present aim to deal with this, although we consider the connection to be interesting.

The interest in these results is twofold. On the one hand, we were led to the search of criteria insuring the finiteness of the number $\kappa^{-}(\lambda I - A)$ for some nonnegative λ and the properties of the function

$$\mathbf{R} \ni \lambda \mapsto \kappa^{-}(\lambda I - A) \in \mathbf{N} \cup \{\infty\}, \tag{1.5}$$

by the recent investigations on quasi–contractions. These were initiated by the second author in [9] in connection with the generalization of the geometric theory of contractions in Kreĭn spaces of M. G. Kreĭn and Yu. S. Shmulyan [16], and the spectral theory of these operators in M. G. Kreĭn and Yu. S. Shmulyan [17]. On the other hand, these results shed some light on the pencil $\lambda G - F$ which was studied from a closely related point of view by P. A. Binding and K. Seddighi [5], P. Lancaster and Q. Ye [19], and P. Lancaster, A. Shkalikov, and Q. Ye [20] in connection with boundary value problems with the spectral parameter in the boundary conditions. We make some comments on this relation in the final section.

We thank H. Langer for providing the reference [20] and for his encouragement. We also thank P. Jonas for his comments on an earlier version of this paper and for providing us with a second proof of Theorem 3.4.

2. The Signatures of Selfadjoint Operators in Kreĭn Spaces

2.1 Kreĭn Spaces. A *Kreĭn space* $(\mathcal{K}, [\cdot, \cdot])$, or $\mathcal{K}$ for short, is a complex linear space $\mathcal{K}$ equipped with a Hermitian sesquilinear form $[\cdot, \cdot]$, which admits a decomposition of the form

$$\mathcal{K} = \mathcal{K}^{+}[+]\mathcal{K}^{-},$$

where $\mathcal{K}^{+}$ and $\mathcal{K}^{-}$ are linear manifolds in $\mathcal{K}$ such that $(\mathcal{K}^{\pm}, \pm[\cdot, \cdot])$ are Hilbert spaces and $[\mathcal{K}^{+}, \mathcal{K}^{-}] = \{0\}$. This kind of decomposition is called a *fundamental decompositon* of the Kreĭn space $\mathcal{K}$ and, if $J^{\pm} : \mathcal{K} \to \mathcal{K}$ are the projections from $\mathcal{K}$ onto the summands $\mathcal{K}^{\pm}$, then $J = J^{+} - J^{-}$ is called the corresponding *fundamental symmetry.* The form $(x, y)_J = [Jx, y]$, $x, y \in \mathcal{K}$, defines a positive definite inner product on $\mathcal{K}$, which turns $\mathcal{K}$ into a Hilbert space. The norm associated with this inner product depends on the fundamental symmetry, but the norms obtained from different fundamental symmetries are equivalent and hence define the same topology on $\mathcal{K}$. All topological notions on $\mathcal{K}$ refer to this common norm toplogy. In this paper a *subspace* of $\mathcal{K}$ is by definition a closed linear manifold in $\mathcal{K}$, and if $\mathcal{A}$ and $\mathcal{B}$ are subspaces in $\mathcal{K}$ we use the symbol $\mathcal{A}[+]\mathcal{B}$ for the sum space $\mathcal{A} + \mathcal{B}$, if $\mathcal{A} \cap \mathcal{B} = \{0\}$, $\mathcal{A} \perp \mathcal{B}$ with respect to the inner product $[\cdot, \cdot]$, and $\mathcal{A} + \mathcal{B}$ is closed.
The dimensions of the spaces $\mathcal{K}^{\pm}$ of a fundamental decomposition of the Kreĭn space $\mathcal{K}$ are the same for each decomposition and the numbers $\kappa^{\pm}(\mathcal{K}) = \dim \mathcal{K}^{\pm}$ are called the *positive*

and *negative signature* of the Kreĭn space $\mathcal{K}$. The space $\mathcal{K}$ is called a *Pontryagin space* if $\kappa(\mathcal{K}) = \min\{\kappa^+(\mathcal{K}), \kappa^-(\mathcal{K})\} < \infty$; the number $\kappa(\mathcal{K})$ is called the *rank of indefiniteness* of the space $\mathcal{K}$.

We denote by $\mathcal{L}(\mathcal{K}_1, \mathcal{K}_2)$ the set of all bounded linear operators from the Kreĭn space $\mathcal{K}_1$ to the Kreĭn space $\mathcal{K}_2$; we write $\mathcal{L}(\mathcal{K}_1)$ for $\mathcal{L}(\mathcal{K}_1, \mathcal{K}_1)$. If $T \in \mathcal{L}(\mathcal{K}_1, \mathcal{K}_2)$ then $T^\sharp \in \mathcal{L}(\mathcal{K}_2, \mathcal{K}_1)$ stands for the *adjoint* of T with respect to the indefinite innner products $[\cdot,\cdot]_i$ on $\mathcal{K}_i$, i.e.,

$$[T^\sharp x, y]_1 = [x, Ty]_2, \quad x \in \mathcal{K}_2, \quad y \in \mathcal{K}_1.$$

If J_i is any fundamental symmetry on $\mathcal{K}_i$, then $T^\sharp = J_1 T^* J_2$, where T^* stands for the adjoint of T with respect to the Hilbert spaces $(\mathcal{K}_i, (\cdot,\cdot)_{J_i})$. Note that a fundamental symmetry J on a Kreĭn space $\mathcal{K}$ belongs to $\mathcal{L}(\mathcal{K})$ and $J = J^{-1} = J^\sharp = J^*$. If A is an operator on $\mathcal{K}$ we denote by $\rho(A)$, $\sigma(A)$, $\sigma_c(A)$, and $\sigma_p(A)$ the resolvent set, the spectrum, the continuous and the point spectrum of A, respectively. For more information about operator theory in Kreĭn or Pontryagin spaces we refer to the monographs [1], [3], [4], and [11].

2.2 The Signatures of Selfadjoint Operators. Let $\mathcal{K}$ be a Kreĭn space and $A \in \mathcal{L}(\mathcal{K})$ be a selfadjoint operator, i.e., $A = A^\sharp$ and we set $[x,x]_A = [Ax, x]$, for $x \in \mathcal{K}$. We denote by $\kappa^\pm(A)$ the supremum of $\dim \mathcal{L}$ where $\mathcal{L}$ runs through the set of all finite dimensional subspaces of $\mathcal{K}$ such that

$$\pm[x,x]_A > 0, \quad x \in \mathcal{L} \setminus \{0\}.$$

If no such $\mathcal{L}$ exists we set $\kappa^\pm(A) = 0$. The numbers $\kappa^+(A)$ and $\kappa^-(A)$ are called the *positive* and *negative signatures* (or Hermitian indices) of A with respect to the inner product $[\cdot,\cdot]$. These numbers are either a nonnegative integer or ∞. The number $\kappa^\pm(A)$ can also be characterized as the greatest number of positive/negative eigenvalues, counted with their multiplicities, of all Hermitian matrices of the form $([x_i, x_j]_A)_{i,j=n}^n$, where the points $x_1, x_2, \cdots, x_n$ and the number n vary over $\mathcal{K}$ and $\mathbb{N}$, respectively.

2.3 The Signatures of an Operator Block–Matrix. In this subsection $\mathcal{H}_1$ and $\mathcal{H}_2$ are Hilbert spaces. Let $A \in \mathcal{L}(\mathcal{H}_1)$, $B \in \mathcal{L}(\mathcal{H}_2, \mathcal{H}_1)$, $C \in \mathcal{L}(\mathcal{H}_1, \mathcal{H}_2)$, and $D \in \mathcal{L}(\mathcal{H}_2)$. The *Frobenius-Schur factorizations* are the following identities:

$$\begin{pmatrix} A & B \\ C & D \end{pmatrix} = \begin{pmatrix} I & 0 \\ CA^{-1} & I \end{pmatrix} \begin{pmatrix} A & 0 \\ 0 & D - CA^{-1}B \end{pmatrix} \begin{pmatrix} I & A^{-1}B \\ 0 & I \end{pmatrix}, \tag{2.1}$$

where it is assumed that the operator A has a bounded inverse, and

$$\begin{pmatrix} A & B \\ C & D \end{pmatrix} = \begin{pmatrix} I & BD^{-1} \\ 0 & I \end{pmatrix} \begin{pmatrix} A - BD^{-1}C & 0 \\ 0 & D \end{pmatrix} \begin{pmatrix} I & 0 \\ D^{-1}C & I \end{pmatrix}, \tag{2.2}$$

where it is assumed that the operator D has a bounded inverse. These factorizations play a role in the calculation of the signatures of selfadjoint operators represented in block–matrix form (e.g., see [6]).

Here we recall a fact established in [6] for the case of finite dimensional Hilbert spaces, but which can be carried over to infinite dimensional Hilbert spaces with essentially the same proof.

LEMMA 2.1 *Let $A \in \mathcal{L}(\mathcal{H}_1)$ and $D \in \mathcal{L}(\mathcal{H}_2)$ be selfadjoint, let $B \in \mathcal{L}(\mathcal{H}_1, \mathcal{H}_2)$, and let H be the selfadjoint operator block–matrix*

$$H = \begin{pmatrix} A & B^* \\ B & D \end{pmatrix}.$$

If A has a closed range and $B_1 = B|\mathcal{R}(A)$ and $B_2 = B|\ker(A)$ then

$$\kappa^{\pm}(H) = \kappa^{\pm}(A) + \operatorname{rank} B_2 + \kappa^{\pm}(P_{\ker B_2^*}(D - B_1 A^{-1} B_1^*)|\ker B_2^*),$$

where A^{-1} is calculated on $\mathcal{R}(A)$. If $A = 0$ we set $B_1 A^{-1} B_1^* = 0$.

2.4 The Upper Semicontinuity of the Signatures. We prove that the signatures of selfadjoint operators are upper semicontinuous with respect to the uniform topology. In the following $\mathcal{K}$ is a Kreĭn space. We fix a fundamental symmetry and denote the corresponding norm by $\|\cdot\|$. First we prove a lemma.

LEMMA 2.2 *If A is a selfadjoint operator in $\mathcal{L}(\mathcal{K})$ and $\mathcal{L}$ is a finite dimensional subspace which is negative definite with respect to the inner product $[\cdot,\cdot]_A$ then there exists $\delta > 0$ such that for all selfadjoint operators $B \in \mathcal{L}(\mathcal{K})$ with $\|A - B\| < \delta$ the space $\mathcal{L}$ is negative definite with respect to the inner product $[\cdot,\cdot]_B$, and hence $\dim(\mathcal{L}) \leq \kappa^-(B)$.*

Proof. Let $\mathcal{L}$ be a finite dimensional subspace in $\mathcal{K}$ such that

$$[Ax, x] < 0, \quad x \in \mathcal{L} \setminus \{0\}. \tag{2.3}$$

For any selfadjoint operator $B \in \mathcal{L}(\mathcal{K})$ we have

$$|[Ax, x] - [Bx, x]| \leq \|A - B\| \cdot \|x\|^2, \quad x \in \mathcal{K}. \tag{2.4}$$

Since $\mathcal{L}$ is finite dimensional the set $\{x \in \mathcal{L} \mid \|x\| = 1\}$ is compact. The map $x \mapsto -[x,x]_A$ is continuous and hence has a minimum on this set

$$\min\{-[Ax, x] \mid \|x\| = 1, x \in \mathcal{L}\} = \delta.$$

By (2.3),$\delta > 0$, and from (2.4) it follows that for any selfadjoint operator $B \in \mathcal{K}$ with $\|A - B\| < \delta$ the space $\mathcal{L}$ is negative definite for the inner product $[\cdot,\cdot]_B$. ∎

THEOREM 2.3 *For any selfadjoint operator $A \in \mathcal{L}(\mathcal{K})$ the following inequalities hold:*

$$\kappa^{\pm}(A) \leq \liminf_{B^\sharp = B \to A} \kappa^{\pm}(B). \tag{2.5}$$

Proof. We prove the inequality (2.5) only for κ^-. First we assume that $\kappa^-(A)$ is finite and we choose $\mathcal{L}$ of dimension $\kappa^-(A)$ such that (2.3) holds. Using Lemma 2.2 we obtain that for $B = B^\sharp$ close to A, $\kappa^-(A) = \dim(\mathcal{L}) \leq \kappa^-(B)$. This implies (2.5).

If $\kappa^-(A)$ is infinite then there exists a sequence of subspaces $\mathcal{L}_n \in \mathcal{K}$ which are negative definite with respect to the inner product $[\cdot,\cdot]_A$ and such that $\dim(\mathcal{L}_n) = n$. Then applying again Lemma 2.2 we see that

$$\liminf_{B \to A} \kappa^-(B) \geq n, \quad n \in \mathbf{N},$$

and hence the inequality (2.5) also holds. ∎

The following lemma is known in perturbation theory, but we give an elementary proof based on Frobenius–Schur factorizations.

LEMMA 2.4 *Let $A \in \mathcal{L}(\mathcal{K})$ be a boundedly invertible selfadjoint operator. Then there exists a $\delta > 0$ such that for all $B = B^\sharp \in \mathcal{L}(\mathcal{K})$ with $\|B - A\| < \delta$ we have $\kappa^\pm(B) = \kappa^\pm(A)$.*

Proof. Without restricting the generality we can assume that $\mathcal{K}$ is a Hilbert space (simply replace A by JA where J is a fundamental symmetry on $\mathcal{K}$). Since A is boundedly invertible $\mathcal{K}$ can be decomposed as

$$\mathcal{K} = \mathcal{L} \oplus \mathcal{L}^\perp, \tag{2.6}$$

with respect to which A has a diagonal form

$$A = \begin{pmatrix} A_1 & 0 \\ 0 & A_2 \end{pmatrix}$$

such that $A_1 \geq 2\delta I$ and $A_2 \leq -2\delta I$ for some $\delta > 0$. Let $B \in \mathcal{L}(\mathcal{K})$ be selfadjoint and write its block matrix with respect to (2.6) as

$$B = \begin{pmatrix} B_{11} & B_{12} \\ B_{12}^* & B_{22} \end{pmatrix}.$$

If $\|B - A\| < \delta$ then $B_{11} \geq \delta I$ and $B_{22} \leq -\delta I$. In particular, B_{11} and B_{22} are boundedly invertible, and applying a Frobenius–Schur factorization we obtain that B is congruent (via boundedly invertible operators) with

$$B' = \begin{pmatrix} B_{11} - B_{12}B_{22}^{-1}B_{12}^* & 0 \\ 0 & B_{22} \end{pmatrix}.$$

Since B_{22}^{-1} is negative and B_{11} is positive, $B_{11} - B_{12}B_{22}^{-1}B_{12}^*$ is positive and hence $\kappa^-(B) = \kappa^-(B') = \dim(\mathcal{L}) = \kappa^-(A)$. Applying the dual Frobenius–Schur factorization we obtain that for $\|B - A\| < \delta$, $\kappa^+(B) = \kappa^+(A)$. ∎

3. The Kreĭn Space Environment

3.1 Definitizable Operators. A selfadjoint operator $A \in \mathcal{L}(\mathcal{K})$, where $\mathcal{K}$ is a Kreĭn space, is called *definitizable* if there exists a nontrivial polynomial p such that

$$[p(A)x, x] \geq 0, \; x \in \mathcal{K}.$$

If A is definitizable then $\sigma_0(A) = \sigma(A) \backslash \mathbf{R}$ is a finite set and there exists a finite set $c(A) \subset \mathbf{R}$, called the set of *critical points* of A, such that on $\mathcal{R}_A$, the Boolean algebra generated by all intervals Δ of $\mathbf{R}$ with $\partial\Delta \cap c(A) = \emptyset$, there exists a mapping

$$E: \mathcal{R}_A \to \mathcal{L}(\mathcal{K})$$

(called the *spectral function* of A) with the following properties:

(i) $E(\Delta_1 \cap \Delta_2) = E(\Delta_1)E(\Delta_2), \quad \Delta_1, \Delta_2 \in \mathcal{R}_A.$

(ii) $E(\Delta_1 \cup \Delta_2) = E(\Delta_1) + E(\Delta_2) - E(\Delta_1 \cap \Delta_2), \quad \Delta_1, \Delta_2 \in \mathcal{R}_A.$

(iii) $\sigma(A|E(\Delta)\mathcal{K}) \subseteq \overline{\Delta}, \quad \Delta \in \mathcal{R}_A.$

(iv) $E(\Delta)$ is positive if $p|\Delta > 0$ and $E(\Delta)$ is negative if $p|\Delta < 0$, $\Delta \in \mathcal{R}_A$.

(v) If $\Delta \in \mathcal{R}_A$ and $\Delta \cap c(A) = \emptyset$ then $E|\Delta$ is a spectral measure and

$$AE(\Delta) = \int_\Delta t \, dE(t),$$

where the integral is convergent in the strong operator topology.

(vi) $E(\mathbf{R}) = I - E(\sigma_0(A); A)$.

(vii) For any $\Delta \in \mathcal{R}_A$, $E(\Delta)$ belongs to the bicommutant of the resolvents $(A - \lambda I)^{-1}$, $\lambda \in \rho(A)$.

For a detailed investigation of definitizable operators see [15], [21], [22], and also [12].

Whenever σ is a spectral set of the operator A we denote by $E(\sigma; A)$ the corresponding spectral projection obtained by the Riesz–Dunford functional calculus. If λ is an isolated point in the spectrum of A we set $E(\lambda; A) = E(\{\lambda\}; A)$.

In the sequel we will intensively use a formula for the computation of the signatures of a definitizable operator which can be found in [9]. The assumption that the selfadjoint operator is definitizable is not restrictive because, as a consequence of a theorem of Pontryagin on the existence of invariant maximal semidefinite subspaces, each selfadjoint operator A such that $\kappa^-(A) < \infty$ is definitizable (see [15], [21], [22]).

LEMMA 3.1 *Let $A \in \mathcal{L}(\mathcal{K})$ be a selfadjoint definitizable operator and denote by $E(t)$ its spectral function. Then, for $\varepsilon > 0$ and such that $c(A) \cap [-\varepsilon, \varepsilon] \subseteq \{0\}$, we have*

$$\begin{aligned} \kappa^\pm(A) &= \sum_{\Im\lambda>0} \operatorname{rank} E(\lambda; A) + \kappa^\pm(A|E(-\varepsilon,\varepsilon)\mathcal{K}) \\ &\quad + \kappa^\pm(E[\varepsilon, +\infty)) + \kappa^\mp(E(-\infty, -\varepsilon]). \end{aligned} \tag{3.1}$$

Proof. We only prove (3.1) for $\kappa^-(A)$. Since $\sigma_0(A) = \sigma(A) \setminus \mathbf{R}$ is a finite set, it is a spectral set of A. It is well–known that if

$$\sigma_0^+(A) = \{\lambda \in \sigma_0(A) | \ \Im\lambda > 0\}, \quad \sigma_0^-(A) = \{\lambda \in \sigma_0(A) | \ \Im\lambda < 0\},$$

then we have

$$\sigma_0^+(A) = \{\bar{\lambda} | \ \lambda \in \sigma_0^-(A)\},$$

and the spectral subspaces $E(\sigma_0^+(A); A)\mathcal{K}$ and $E(\sigma_0^-(A); A)\mathcal{K}$ are unitary equivalent (as Hilbert spaces). Identifying both spaces with the same Hilbert space $\mathcal{H}$ we have

$$E(\sigma_0(A); A)\mathcal{K} = \mathcal{H} \oplus \mathcal{H},$$

where the fundamental symmetry is given by $J(x \oplus y) = y \oplus x$, $x, y \in \mathcal{H}$. With respect to this decomposition

$$A|E(\sigma_0(A); A)\mathcal{K} = \begin{bmatrix} B & 0 \\ 0 & B^* \end{bmatrix},$$

where $B \in \mathcal{L}(\mathcal{H})$ is invertible. This shows that

$$\begin{aligned}\kappa^{-}(A|E(\sigma_0(A);A)\mathcal{K}) &= \operatorname{rank} B = \dim \mathcal{H} \\ = \dim E(\sigma_0^{+}(A);A)\mathcal{K} &= \sum_{\Im\lambda>0} \operatorname{rank} E(\lambda;A).\end{aligned} \tag{3.2}$$

Let now $\varepsilon > 0$ be sufficiently small such that $c(A) \cap [-\varepsilon, \varepsilon] \subset \{0\}$. Since

$$\sigma(A|E[\varepsilon,+\infty)\mathcal{K}) \subseteq [\varepsilon,+\infty),$$

it follows by the Riesz–Dunford functional calculus that there exists an operator $R \in \mathcal{L}(E[\varepsilon,+\infty)\mathcal{K})$, $R = R^{\sharp}$ and $\sigma(R) \subset [\varepsilon^{1/2},+\infty)$ (in particular R is invertible) such that

$$A|E[\varepsilon,+\infty)\mathcal{K} = R^2.$$

This means

$$[Ax,x] = [Rx,Rx], \ x \in E[\varepsilon,+\infty)\mathcal{K}.$$

Since R is invertible we obtain

$$\kappa^{-}(A|E[\varepsilon,+\infty)\mathcal{K}) = \kappa^{-}(E[\varepsilon,+\infty)). \tag{3.3}$$

In the same way we obtain

$$\kappa^{-}(A|E(-\infty,-\varepsilon]\mathcal{K}) = \kappa^{+}(E(-\infty,-\varepsilon]). \tag{3.4}$$

The formula (3.1) follows now from (2.4)-(2.6). ∎

Corollary 3.2 *If 0 is not a singular critical point of A, or, equivalently, if the strong operator limits*

$$\lim_{\mu\downarrow 0} E(\mu,\infty) = E(0,\infty), \quad \lim_{\mu\uparrow 0} E(-\infty,\mu) = E(-\infty,0), \tag{3.5}$$

exist, then the formula (3.1) can be written as

$$\begin{aligned}\kappa^{\pm}(A) \;=\; & \sum_{\Im\lambda>0} \operatorname{rank} E(\lambda;A) + \kappa^{\pm}(A|E(\{0\})\mathcal{K}) \\ & +\kappa^{\pm}(E(0,+\infty)) + \kappa^{\mp}(E(-\infty,0)),\end{aligned} \tag{3.6}$$

where $E(\{0\})$ is the regular root subspace of A corresponding to 0.

Proof. We only prove (3.6) for $\kappa^{-}(A)$. Assume that 0 is not a singular critical point of A then for $\varepsilon > 0$ and sufficiently small the spectral subspaces $E(-\varepsilon,0)\mathcal{K}$ and $E(0,\varepsilon)\mathcal{K}$ are definite. We claim that

$$\kappa^{-}(A|E(-\varepsilon,\varepsilon)\mathcal{K}) = \kappa^{-}(A|E(\{0\})\mathcal{K}) + \kappa^{+}(E(-\varepsilon,0)) + \kappa^{-}(E(0,\varepsilon)). \tag{3.7}$$

Let $\mathcal{D}$ be the linear span of the elements of the spaces $E(\delta,\varepsilon)\mathcal{K}$ with $0 < \delta < \varepsilon$. Then $\mathcal{D}$ is dense in $E(0,\varepsilon]\mathcal{K}$ and for any $x \in \mathcal{D}$ there exists a $\delta \in (0,\varepsilon)$ with the property

$$Ax = \int_{\delta}^{\varepsilon} t\,dE(t)x,$$

where the integral is convergent in the strong operator topology. Then for $t \in (\delta, \varepsilon)$ $[Ax, x]$ and $[E(t)x, x]$ have the same sign. Using a Pontryagin Lemma type argument, we conclude that

$$\kappa^-(A|E(0,\varepsilon)\mathcal{K}) = \kappa^-(E(0,\varepsilon)).$$

Similarly one can prove that

$$\kappa^-(A|E(-\varepsilon,0)\mathcal{K}) = \kappa^+(E(-\varepsilon,0)).$$

These two equalities prove the formula (3.7). Inserting the formula (3.7) in (3.1) we obtain the formula (3.6). ∎

In the following we will use the formula (3.1) for $\lambda I - A$ instead of A and for the readers' convenience we make the corresponding transcription.

COROLLARY 3.3 *Let A be a definitizable operator in a Kreĭn space $\mathcal{K}$, denote by $E(t)$ its spectral function, and let λ be a real number. Then*

$$\begin{aligned}\kappa^{\pm}(\lambda I - A) &= \sum_{\Im\lambda>0} \operatorname{rank} E(\lambda; A) + \kappa^{\pm}((\lambda I - A)|E(\lambda-\varepsilon, \lambda+\varepsilon)\mathcal{K}) \\ &\quad + \kappa^{\mp}(E[\lambda+\varepsilon, +\infty)) + \kappa^{\pm}(E(-\infty, \lambda-\varepsilon]).\end{aligned} \tag{3.8}$$

Proof. We only prove (3.8) for κ^-. We use the formula (3.1) replacing A by $\lambda I - A$ and notice that the interval $(-\infty, -\varepsilon)$ corresponds to the interval $(\lambda+\varepsilon, \infty)$, and the interval $(\varepsilon, +\infty)$ corresponds to the interval $(-\infty, \lambda-\varepsilon)$. The contribution of the non-real part of the spectrum of A can be described with the same formula, since the transformation $\mu \mapsto \lambda - \mu$ (which is the composition of a symmetry with respect to the imaginary axis with a translation) maps $\sigma(A) \setminus \mathbf{R}$ into $\sigma(\lambda I - A) \setminus \mathbf{R}$. ∎

THEOREM 3.4 *Let $A \in \mathcal{L}(\mathcal{K})$ be selfadjoint. Then for all real λ the four one-sided limits $\kappa^{\pm}((\lambda \pm 0)I - A)$ exist and the following inequalities hold*

$$\kappa^{\pm}(\lambda I - A) \leq \min\{\kappa^{\pm}((\lambda-0)I - A), \kappa^{\pm}((\lambda+0)I - A)\}. \tag{3.9}$$

Proof. We only prove the statement for κ^-. First we note that if $\kappa^-(\lambda I - A)$ is infinite for all real λ then the statement is trivial. If for some real μ we have $\kappa^-(\mu I - A) < \infty$ then, according to [21], by a theorem of Pontryagin on the existence of invariant maximal nonpositive subspaces, it follows that A is definitizable. Let $\lambda \in \mathbf{R}$ be fixed. We distinguish three cases:

(a) λ is isolated from the right with respect to the spectrum of A. Then by Lemma 2.4 the function of $\lambda \mapsto \kappa^-(\lambda I - A)$ is constant on the interval $(\lambda, \lambda+\varepsilon)$ for some $\varepsilon > 0$. Hence the limit $\kappa^-((\lambda+0)I - A)$ exists.

(b) There exists a decreasing sequence of points of positive type which converges to λ. Then for all $n \in \mathbf{N}$ there exists an $\varepsilon > 0$ such that for all $\mu \in (\lambda, \lambda+\varepsilon)$ we have $\kappa^+(E(\mu,\infty) \geq n$ and hence from Corollary 3.3 we obtain $\kappa^-((\lambda+0)I - A) = \infty$.

(c) There exists a decreasing sequence of points of negative type which converges to λ. Then for any $\mu > \lambda$ there exists an $\varepsilon > 0$ such that $\lambda < \mu - \varepsilon$ and $\kappa^-(E(-\infty, \mu-\varepsilon)) = \infty$. Hence, again by Corollary 3.3, we conclude $\kappa^-((\lambda+0)I - A) = \infty$.

Since A is definitizable these are all the possible cases, hence we have proved that the limit $\kappa^-((\lambda+0)I-A)$ always exists. Similarly one proves that the limit $\kappa^-((\lambda-0)I-A)$ always exists. Since

$$\liminf_{\mu\to\lambda} \kappa^-(\mu I - A) = \min\{\kappa^\pm((\lambda-0)I-A), \kappa^\pm((\lambda+0)I-A)\},$$

the inequality (3.9) follows from Theorem 2.3. ∎

3.2 The case of an unbounded operator. Let A be an unbounded operator in the Kreĭn space $\mathcal{K}$, $A: \mathcal{D}(A) \to \mathcal{K}$, with domain $\mathcal{D}(A)$ dense in $\mathcal{K}$. The operator A is *selfadjoint* if $A = A^\sharp$, where the operator $A^\sharp: \mathcal{D}(A^\sharp) \to \mathcal{K}$ is defined by

$$\mathcal{D}(A^\sharp) = \{y \in \mathcal{K} \mid \mathcal{D}(A) \ni x \mapsto [x,y] \text{ is bounded }\},$$

$$[Ax, y] = [x, A^\sharp y], \quad x \in \mathcal{D}(A), y \in \mathcal{D}(A^\sharp).$$

By definition, A is *definitizable*, if it has a nonvoid resolvent set and if there exists a nontrivial polynomial p such that

$$[p(A)x, x] \geq 0, \quad x \in \mathcal{D}(A^n),$$

where n is the degree of p. The assumption on the resolvent set of A is essential since it guarantees that $\sigma(A) \setminus \mathbf{R}$ is a finite set of points and that there exists a spectral function $E(t)$ with similar properties as in the bounded case. On the real axis, the point ∞ can be a critical point, regular as well as singular. For any bounded interval $\Delta \in \mathcal{R}_A$ we have that $E(\Delta)\mathcal{K} \subseteq \mathcal{D}(A)$. For a detailed investigation and proofs see [21], [22], and [12].

We define $\kappa^\pm(A)$ as the positive/negative signature of the indefinite inner product space $(\mathcal{D}(A), [\cdot,\cdot]_A)$, where, as before,

$$[x,y]_A = [Ax, y], \quad x, y \in \mathcal{D}(A).$$

The formula in (3.1) can be carried over now as

$$\begin{aligned}\kappa^\pm(A) &= \sum_{\Im\lambda>0} \operatorname{rank} E(\lambda; A) + \kappa^\pm(A|E(-\varepsilon,\varepsilon)\mathcal{K}) \\ &\quad + \lim_{n\to\infty} \kappa^\pm(E[\varepsilon, n)) + \lim_{n\to\infty} \kappa^\mp(E(-n,-\varepsilon]),\end{aligned} \tag{3.10}$$

where $\varepsilon > 0$ is such that $c(A) \cap (-\varepsilon, \varepsilon) \subseteq \{0\}$.

Formula (3.10) implies that Theorem 3.4 also holds for unbounded selfadjoint operators.

THEOREM 3.5 *Let A be an unbounded selfadjoint operator in a Kreĭn space $(\mathcal{K}, [.,.])$ with $\rho(A) \neq \emptyset$. For any real λ we have*

$$\kappa^\pm(\lambda I - A) \leq \min\{\kappa^\pm((\lambda+0)I-A), \kappa^\pm((\lambda-0)I-A)\}. \tag{3.11}$$

Proof. Either $\kappa^\pm(\lambda I - A)$ is infinite for all λ and then (3.11) holds trivially, or $\kappa^\pm(\lambda I - A)$ is finite for some λ in which case A is definitizable. Using the spectral function of A one can then reduce the problem to the bounded case by choosing a compact interval Δ sufficiently large. ∎

3.3 Some criteria of finiteness of the signatures. In the proof of Theorem 3.4 we showed that in many cases the left limit as well as the right limit of $\kappa^-(\mu I - A)$ at λ is infinite and thus that in these cases the inequality (3.9) is trivial. Using Corollary 3.3 we obtain some criteria concerning the finiteness of the numbers $\kappa^\pm(\lambda I - A)$ where A is a selfadjoint operator in a Kreĭn space $\mathcal{K}$, with $\rho(A) \neq \emptyset$ if A is unbounded.

LEMMA 3.6 *If* $\kappa^+(\lambda I - A) < \infty$ *(or* $\kappa^-(\lambda I - A) < \infty$*) for some* $\lambda \in \mathbf{R}$ *then* rank $E(\sigma_0(A); A) < \infty$, *i.e., the nonreal spectrum of A consists of a finite number of points of finite algebraic multiplicity, and for* $\varepsilon > 0$ *and sufficiently small* $\mathcal{K}$ *can be written as*

$$\mathcal{K} = E(\lambda - \varepsilon, \lambda + \varepsilon)\mathcal{K}[+]E(-\infty, \lambda - \varepsilon]\mathcal{K}[+]E[\lambda + \varepsilon, \infty)\mathcal{K},$$

where the last two spaces are Pontryagin spaces with $\kappa^-(E(-\infty, \lambda - \varepsilon]) < \infty$ *and* $\kappa^+(E[\lambda + \varepsilon, \infty)) < \infty$ *(or* $\kappa^+(E(-\infty, \lambda - \varepsilon]) < \infty$ *and* $\kappa^-(E[\lambda + \varepsilon, \infty)) < \infty$*, respectively).*

THEOREM 3.7 *If for two real numbers* λ_1 *and* λ_2 *with* $\lambda_1 < \lambda_2$ *we have* $\kappa^-(\lambda_1 I - A) < \infty$ *and* $\kappa^-(\lambda_2 I - A) < \infty$ *then for all* $\lambda \in (\lambda_1, \lambda_2)$ *we have*

$$\kappa^-(\lambda I - A) < \kappa^-(\lambda_1 I - A) + \kappa^-(\lambda_2 I - A) - \sum_{\Im\lambda > 0} \operatorname{rank} E(\lambda; A) < \infty$$

and

$$\operatorname{rank} E(\lambda_1, \lambda_2) < \kappa^-(\lambda_1 I - A) + \kappa^-(\lambda_2 I - A) - 2\sum_{\Im\lambda > 0} \operatorname{rank} E(\lambda; A) < \infty.$$

Theorem 3.7 also holds if we replace κ^- by κ^+.

Proof of Theorem 3.7. By Lemma 3.6, $S = \sum_{\Im\lambda>0} \operatorname{rank} E(\lambda; A) < \infty$. If Δ is a compact subinterval of (λ_1, λ_2) then we have for small $\varepsilon > 0$,

$$\kappa^+(E(\Delta)) \leq \kappa^+(E(\lambda_1 + \varepsilon, \infty)) \leq \kappa^-(\lambda_1 I - A) - S,$$

$$\kappa^-(E(\Delta)) \leq \kappa^-(E(-\infty, \lambda_2 - \varepsilon)) \leq \kappa^-(\lambda_2 I - A) - S,$$

and hence

$$\operatorname{rank} E(\Delta) < \kappa^-(\lambda_1 I - A) + \kappa^-(\lambda_2 I - A) - 2S.$$

As Δ is arbitrary, it follows that the projection $E(\lambda_1, \lambda_2)$ exists and has finite rank with bound given as in the statement.

Let $\lambda \in (\lambda_1, \lambda_2)$. Then $E(\{\lambda\}) = \mathcal{S}_\lambda$ is a regular finite dimensional subspace which is invariant under $\lambda I - A$. By Theorem 3.4 with $\mathcal{K} = \mathcal{S}_\lambda$ we have for small $\varepsilon > 0$,

$$\kappa^-(\lambda I - A)|\mathcal{S}_\lambda) \leq \kappa^-(((\lambda + \varepsilon)I - A)|\mathcal{S}_\lambda) = \kappa^-(\mathcal{S}_\lambda).$$

Hence

$$\begin{aligned}
\kappa^-(\lambda I - A) &= \kappa^-((\lambda I - A)|\mathcal{S}_\lambda) + \kappa^-(E(-\infty, \lambda)) + \kappa^+(E(\lambda, \infty)) + S \\
&\leq \kappa^-(\mathcal{S}_\lambda) + \kappa^-(E(-\infty, \lambda)) + \kappa^+(E(\lambda_1, \infty)) + S \\
&= \kappa^-(E(-\infty, \lambda + \varepsilon]) + \kappa^+(E(\lambda_1, \infty)) + S \\
&\leq \kappa^-(E(-\infty, \lambda_2)) + \kappa^+(E(\lambda_1, \infty)) + S \\
&\leq \kappa^-(\lambda_2 I - A) + \kappa^-(\lambda_1 I - A) - S,
\end{aligned}$$

which proves the first inequality in the statement. ∎

COROLLARY 3.8 *If A is also bounded in* $\mathcal{K}$*, the following assertions are equivalent:*

(i) $\kappa^-(\lambda I - A) < \infty$ *for all* $\lambda \in \mathbf{R}$.

(ii) $\kappa^+(\lambda I - A) < \infty$ *for all* $\lambda \in \mathbf{R}$.

(iii) $\mathcal{K}$ *is finite dimensional.*

To see that (i) or (ii) imply (iii) simply take two real numbers λ_1 and λ_2 such that $\sigma(A) \subset (\lambda_1 - \varepsilon, \lambda_2 + \varepsilon)$ and apply Lemma 3.6.

COROLLARY 3.9 *If A is unbounded and definitizable, the following assertions are equivalent:*

(i) $\kappa^-(\lambda I - A) < \infty$ *for all* $\lambda \in \mathbf{R}$.

(ii) $\mathcal{K}$ *is separable, A has discrete spectrum, all spectral points are normal eigenvalues, and there exists $M \in \mathbf{R}$ such that for all $\mu > \lambda > M$ $E(\lambda, \mu)$ is negative and $E(-\mu, -\lambda)$ is positive.*

The following corollary will be used in the final section. Its proof is also an immediate application of Corollary 3.3.

COROLLARY 3.10 *If A is bounded and $\rho \in \mathbf{R}$, then the following assertions are equivalent:*

(i) $\kappa^+(\beta I - A) < \infty$ *for all* $\beta < \rho$ *and* $\kappa^-(\alpha I - A) < \infty$ *for all* $\alpha > \rho$.

(ii) $\mathcal{K}$ *is a Pontryagin space with* $\kappa^-(\mathcal{K}) < \infty$ *and* $A - \rho I$ *is a compact operator.*

Corollaries 3.9 and 3.10 remain true if we interchange everywhere the positive and negative signatures.

4. The Pontryagin Space Case

4.1 The Finite Dimensional Case. In this subsection we give explicit formulae for the computation of the numbers $\kappa^\pm(\lambda I - A)$ when A is a selfadjoint operator in a finite dimensional Kreĭn space and λ is a real number. We first recall an important result in the spectral theory of selfadjoint operators in finite dimensional Kreĭn spaces to be found in I. Gohberg, P. Lancaster, and L. Rodman [10], Theorem I.3.3.

A *sip matrix* (standard involutory permutation matrix) J is by definition a matrix of the form

$$J = \begin{pmatrix} 0 & 0 & \cdots & 0 & 1 \\ 0 & 0 & \cdots & 1 & 0 \\ \vdots & & . \cdot \dot{} & & \vdots \\ 0 & 1 & \cdots & 0 & 0 \\ 1 & 0 & \cdots & 0 & 0 \end{pmatrix}. \tag{4.1}$$

Clearly a sip matrix J and $-J$ are always symmetries .

Let A be a selfadjoint operator in the finite dimensional Kreĭn space $\mathcal{K}$. From the spectral theory we know that the spectrum $\sigma(A)$ of A can be written as the union of two disjoint sets, the first set consists of pairs of nonreal eigenvalues $\{\alpha_j, \bar{\alpha}_j\}_{j=1}^r$, with $\Im \alpha_j > 0$ for all $j \in \{1, 2, \ldots, r\}$, and the second set $\{\lambda_j\}_{j=1}^n$ consists of distinct real eigenvalues.

Moreover, there exist a fundamental symmetry J and an orthonormal basis of $\mathcal{K}$ with respect to which A is represented as the direct sum of Jordan blocks $A_j^{(p)}$

$$A = \bigoplus_{j=1}^{r} \bigoplus_{p=1}^{l_j} \left(H_j^{(p)} \oplus \overline{H}_j^{(p)} \right) \oplus \bigoplus_{j=1}^{n} \bigoplus_{p=1}^{m_j} A_j^{(p)}, \tag{4.2}$$

and J has the representation

$$J = \bigoplus_{j=1}^{r} \bigoplus_{p=1}^{l_j} \widetilde{J}_j^{(p)} \oplus \bigoplus_{j=1}^{n} \bigoplus_{p=1}^{m_j} \varepsilon_j^{(p)} J_j^{(p)}, \tag{4.3}$$

where

(i) $\{H_j^{(p)}\}_{p=1}^{l_j}$ are the Jordan blocks, respectively of size $t_j^{(p)}$, corresponding to α_j, for all $j \in \{1, 2, \ldots, r\}$,

(ii) $\{\overline{H}_j^{(p)}\}_{p=1}^{l_j}$ are the Jordan blocks, respectively of the same size $t_j^{(p)}$, corresponding to $\bar{\alpha}_j$, for all $j \in \{1, 2, \ldots, r\}$,

(iii) $\{A_j^{(p)}\}_{p=1}^{m_j}$ are the Jordan blocks, respectively of size $s_j^{(p)}$ corresponding to λ_j, for all $j \in \{1, 2, \ldots, n\}$,

(iv) $\{\widetilde{J}_j^{(p)}\}_{p=1}^{l_j}$ are sip matrices, respectively of size $2t_j^{(p)}$,

(v) $\{J_j^{(p)}\}_{p=1}^{m_j}$ are sip matrices, respectively of size $s_j^{(p)}$.

(vi) The numbers $\varepsilon_j^{(p)}$ are signs ± 1.

The set of signs $\{\varepsilon_j^{(p)}\}$ is uniquely determined by A up to a permutation corresponding to a permutation of the Jordan blocks.

In the sequel $[r]$ stands for the integral part of the real number r.

LEMMA 4.1 *Asume that the selfadjoint operator A in the finite dimensional Kreĭn space has the afore mentioned canonical form and let the real eigenvalues of A be ordered such that $\lambda_n < \lambda_{n-1} < \ldots < \lambda_1$. Let μ be an arbitrary real number.*

(a) *If $\lambda_1 < \mu$ then*

$$\kappa^-(\mu I - A) = \sum_{j=1}^{r} \sum_{p=1}^{l_j} t_j^{(p)} + \sum_{k=1}^{n} \sum_{p=1}^{m_k} \left[\frac{2s_k^{(p)} + 1 - \varepsilon_k^{(p)}}{4} \right]. \tag{4.4}$$

(b) *If $\lambda_{i+1} < \mu < \lambda_i$ then*

$$\kappa^-(\mu I - A) = \sum_{j=1}^{r} \sum_{p=1}^{l_j} t_j^{(p)} + \sum_{k=1}^{i} \sum_{p=1}^{m_k} \left[\frac{2s_k^{(p)} + 1 + \varepsilon_k^{(p)}}{4} \right]$$
$$+ \sum_{k=i+1}^{n} \sum_{p=1}^{m_k} \left[\frac{s_k^{(p)} + 1 - \varepsilon_k^{(p)}}{4} \right]. \tag{4.5}$$

(c) *If* $\mu = \lambda_i$ *then*

$$\kappa^-(\mu I - A) = \sum_{j=1}^{r}\sum_{p=1}^{l_j} t_j^{(p)} + \sum_{k=1}^{i-1}\sum_{p=1}^{m_k}\left[\frac{2s_k^{(p)} + 1 + \varepsilon_k^{(p)}}{4}\right]$$
$$+ \sum_{k=i+1}^{n}\sum_{p=1}^{m_k}\left[\frac{s_k^{(p)} + 1 - \varepsilon_k^{(p)}}{4}\right] + \sum_{p=1}^{m_i}\left[\frac{s_i^{(p)} - 1 + \varepsilon_i^{(p)}}{4}\right]. \tag{4.6}$$

(d) *If* $\mu < \lambda_n$ *then*

$$\kappa^-(\mu I - A) = \sum_{j=1}^{r}\sum_{p=1}^{l_j} t_j^{(p)} + \sum_{k=1}^{n}\sum_{p=1}^{m_k}\left[\frac{s_k^{(p)} + 1 + \varepsilon_k^{(p)}}{4}\right]. \tag{4.7}$$

REMARK 4.2 The formulae for κ^+ can be obtained from those for κ^- by replacing $\varepsilon_k^{(p)}$ by $-\varepsilon_k^{(p)}$.

Proof of Lemma 4.1. We prove the formula for the item (c). Let $\mu = \lambda_i$. Since the Kreĭn space $\mathcal{K}$ is finite dimensional, we can apply the formula in Corollary 3.3 combined with the formula in Corollary 3.2. Thus, we have

$$\kappa^-(\lambda_i I - A) = \sum_{j=1}^{r} \operatorname{rank} E(\alpha_j; A) + \sum_{k=i+1}^{n} \kappa^-(E(\lambda_k))$$
$$+ \kappa^-((\lambda_i I - A)|E(\lambda_i)\mathcal{K}) + \sum_{k=1}^{i-1} \kappa^+(E(\lambda_k)). \tag{4.8}$$

To calculate this number we first calculate the signatures of a sip matrix. If J is a sip matrix of size s and $\varepsilon = \pm 1$, then

$$\kappa^+(\varepsilon J) = \left[\frac{2s + 1 + \varepsilon}{4}\right], \quad \kappa^-(\varepsilon J) = \left[\frac{2s + 1 - \varepsilon}{4}\right]. \tag{4.9}$$

This can be easily seen, e.g., from Lemma 2.1. From the canonical form (4.3) it follows that for each $1 \le k \le n$

$$\kappa^+(E(\lambda_k)) = \sum_{p=1}^{m_k}\left[\frac{2s_k^{(p)} + 1 + \varepsilon_k^{(p)}}{4}\right], \tag{4.10}$$

and

$$\kappa^-(E(\lambda_k)) = \sum_{p=1}^{m_k}\left[\frac{2s_k^{(p)} + 1 - \varepsilon_k^{(p)}}{4}\right]. \tag{4.11}$$

Also, from the canonical form (4.2) it is clear that for any $j \in \{1, 2, \ldots, r\}$,

$$\operatorname{rank} E(\alpha_j) = \sum_{p=1}^{l_j} t_j^{(p)}. \tag{4.12}$$

To compute the remaining term in the lefthand side of (4.8) we note that for any Jordan block $A_i^{(p)}$ of A corresponding to the eigenvalue λ_i we have

$$J_i^{(p)}(\lambda_i - A_i^{(p)}) = \begin{pmatrix} 0 & 0 & \cdots & 0 \\ 0 & 0 & \cdots & -1 \\ \vdots & & \iddots & \\ 0 & -1 & \cdots & 0 \end{pmatrix},$$

hence, using again (4.9) we have

$$\kappa^-(\lambda_i - A_i^{(p)}) = \left[\frac{s_i^{(p)} - 1 + \varepsilon_i^{(p)}}{4}\right]. \tag{4.13}$$

Inserting now the formulae (4.10), (4.11), and (4.13) in (4.8) we get the formula in item (c). The other formulae in items (a), (b), and (d) can be obtained in a similar way. ∎

4.2 The Pontryagin Space Case. In this subsection we calculate or give estimates of $\kappa^-(\lambda I - A)$ for the case that λ is real and A is a bounded selfadjoint operator in a Pontryagin space $\mathcal{K}$. As the analysis on cases performed in the proof of Theorem 3.4 shows, the calculation of the number $\kappa^-(\lambda I - A)$ is of interest only in case λ is an eigenvalue of A, hence we consider only this case .

We consider first the case where the root subspace corresponding to λ is nondegenerate.

LEMMA 4.3 *Let λ be an eigenvalue of A and assume that $\mathcal{S}_\lambda$ is nondegenerate. Then the Pontryagin space $\mathcal{K}$ can be decomposed into an orthogonal direct sum of regular subspaces,*

$$\mathcal{K} = \mathcal{K}_1[+]\mathcal{K}_2[+]\mathcal{K}_3, \tag{4.14}$$

where $\mathcal{K}_1 = \mathcal{S}_\lambda^\perp$, $\mathcal{K}_2 \subseteq \ker(\lambda I - A)$, $\mathcal{K}_3$ is finite dimensional, $\mathcal{K}_1$,$\mathcal{K}_2$, and $\mathcal{K}_3$ reduce A and

$$\mathcal{S}_\lambda = \mathcal{K}_2[+]\mathcal{K}_3.$$

Moreover, with respect to the decomposition (4.14), the operator A is represented by the operator block-matrix

$$A = \begin{pmatrix} A_1 & 0 & 0 \\ 0 & \lambda I & 0 \\ 0 & 0 & A_3 \end{pmatrix}, \tag{4.15}$$

such that $\lambda \notin \sigma_p(A_1)$. In addition, the space $\mathcal{K}_2$ can be chosen maximal with these properties, in which case the finite rank operator A_3 has no semisimple eigenvalues.

Proof. Since the root subspace $\mathcal{S}_\lambda$ is regular, $\mathcal{K}_1$ is a regular subspace of $\mathcal{K}$ and it reduces the operator A. Clearly, denoting by A_1 the compression of A to the subspace $\mathcal{K}_1$, we have $\lambda \notin \sigma_p(A_1)$. Since the space $\mathcal{K}$ is a Pontryagin space, the kernel of $\lambda I - A$ can be decomposed into an orthogonal and direct sum of a regular subspace $\mathcal{K}_2$ and its isotropic subspace

$$\ker(\lambda I - A) = \mathcal{K}_2[+](\ker(\lambda I - A))^0.$$

Define $\mathcal{K}_3 = \mathcal{S}_\lambda \cap \mathcal{K}_2^\perp$. It is also a regular subspace and it reduces A. As a consequence of Theorem 7.7.2 in I. S. Iokhvidov, M. G. Kreĭn, and H. Langer [11] (this theorem is stated for isometric operators, but it holds also for symmetric operators) the subspace $\mathcal{K}_3$ is finite dimensional.

The maximality of $\mathcal{K}_2$ follows from the fact that it has no regular extensions in ker A and the last assertion follows from the fact that all eigenvectors of A corresponding to the semisimple eigenvalues are included in $\mathcal{K}_2$. ∎

COROLLARY 4.4 *Let λ be an eigenvalue of the operator A with decomposition (4.15). Then*

$$\kappa^-(\lambda I - A) = \kappa^-(\lambda I - A_3) + \kappa^+(E(\lambda, +\infty)) + \kappa^-(E(-\infty, \lambda)).$$

Proof. Use Corollary 3.2 and Lemma 4.14. ∎

Since $\mathcal{K}_3$ is finite dimensional, $\kappa^-(\lambda I - A_3)$ in Corollary 4.4 can be calculated by means of the formula in Lemma 4.1. Hence the case of the nondegenerate root subspace is completely solved.

The following result is well-known in the spectral theory of definitizable operators (see [21]).

LEMMA 4.5 *Let λ be an eigenvalue of A and denote by $\mathcal{N} = \mathcal{S}_\lambda^0$ the isotropic subspace of the root subspace $\mathcal{S}_\lambda$. Let J be a fundamental symmetry of $\mathcal{K}$. Then the subspace $\hat{\mathcal{K}} = (\mathcal{N} \oplus J\mathcal{N})^\perp$ is regular and, with respect to the decomposition*

$$\mathcal{K} = \mathcal{N} \oplus \hat{\mathcal{K}} \oplus J\mathcal{N}, \tag{4.16}$$

A is represented in the block-matrix form

$$A = \begin{pmatrix} A_{11} & A_{12} & A_{13} \\ 0 & A_{22} & A_{23} \\ 0 & 0 & A_{33} \end{pmatrix}, \tag{4.17}$$

such that A_{22} is the compression of A to the subspace $\hat{\mathcal{K}}$, it is definitizable selfadjoint, $\sigma(A_{22}) \subseteq \sigma(A)$, the spectral function of A_{22} is the compression of E to $\hat{\mathcal{K}}$, and

$$\mathcal{S}_\lambda = \hat{\mathcal{S}}_\lambda[+]\mathcal{N},$$

where $\hat{\mathcal{S}}_\lambda$ denotes the root subspace of λ corresponding to A_{22}, in particular $\hat{\mathcal{S}}_\lambda$ is a regular subspace.

The main result of this section is the following inequality.

THEOREM 4.6 *If λ is an eigenvalue of A, then in the notation of Lemma 4.5, we have*

$$\kappa^-(\lambda I - A) \leq \kappa^-(\lambda I - A_{22}) + \dim \mathcal{N}. \tag{4.18}$$

Proof. To simplify the notation, without restricting the generality, we assume that $\lambda = 0$ and replace A by $-A$.

With respect to the decomposition (4.16) the fundamental symmetry J is represented by

$$J = \begin{pmatrix} 0 & 0 & J \\ 0 & \hat{J} & 0 \\ J & 0 & 0 \end{pmatrix}, \tag{4.19}$$

where $\hat{J}$ is fundamental symmetry on $\hat{\mathcal{K}}$ and the entries denoted by J are compressions of J to $\mathcal{N}$ and respectively $J\mathcal{N}$. Then

$$JA = \begin{pmatrix} 0 & 0 & JA_{33} \\ 0 & \hat{J}A_{22} & \hat{J}A_{23} \\ JA_{11} & JA_{12} & JA_{13} \end{pmatrix}.$$

This block–matrix operator is rearranged with respect to the decomposition of the Hilbert space $(\mathcal{K}, (\cdot,\cdot)_J)$

$$\mathcal{K} = \hat{\mathcal{K}} \oplus \mathcal{N} \oplus J\mathcal{N}, \tag{4.20}$$

as

$$JA = \begin{pmatrix} \hat{J}A_{22} & 0 & \hat{J}A_{23} \\ 0 & 0 & JA_{33} \\ JA_{12} & JA_{11} & JA_{13} \end{pmatrix}.$$

Set $H = JA$. Then H is selfadjoint in the Hilbert space $(\mathcal{K}, (\cdot,\cdot)_J)$ and, to simplify the notation we rewrite it as

$$H = \begin{pmatrix} B & 0 & C^* \\ 0 & 0 & D^* \\ C & D & E \end{pmatrix}, \tag{4.21}$$

where B and E are selfadjoint. In the remaining part of this proof the signatures are calculated with respect to the positive definite inner product $(\cdot,\cdot)_J$. Identifying the corresponding entries in (4.17) and (4.21) we see that the inequality (4.18) is equivalent to the inequality

$$\kappa^-(H) \leq \kappa^-(B) + \dim \mathcal{N}. \tag{4.22}$$

We divide the proof of (4.22) into two steps.

(1) We first assume that the range of B is closed. Then we can apply Lemma 2.1 to H in (4.21) and get

$$\begin{aligned} \kappa^-(H) \quad &= \kappa^-(B) + \operatorname{rank}(C_2) \\ &\quad + \kappa^-(\begin{pmatrix} 0 & D^*|\ker C_2^* \\ P_{\ker C_2^*} D & P_{\ker C_2^*}(E - C_1 B^{-1} C_1^*)|\ker C_2^* \end{pmatrix}) \\ &= \kappa^-(B) + \operatorname{rank}(C_2) + \operatorname{rank}(P_{\ker C_2^*} D) \\ &\quad + \kappa^-\left(P_{\ker(D^ast|\ker C_2^*)}(E - C_1 B^{-1} C_1^*)|\ker(D^*|\ker C_2^*)\right) \\ &\leq \kappa^-(B) + \operatorname{rank}(C_2^*) + \operatorname{rank}(D|\ker C_2^*) + \dim(\ker D|\ker C_2^*) \\ &= \kappa^-(B) + \operatorname{rank}(C_2^*) + \dim(\ker(C_2^*) \\ &= \kappa^-(B) + \dim \mathcal{N}, \end{aligned} \tag{4.23}$$

where $C_1 = C|\mathcal{R}(B)$ and $C_2 = C|\ker B$.

(2) In general, the range of B is not necessarily closed, but we always have that $\kappa^-(B) \leq \kappa^-(H)$. Hence if $\kappa^-(B) = \infty$, then the inequality (4.22) trivially holds. Therefore we may assume $\kappa^-(B) < \infty$. Then

$$\kappa^-(H) \leq \kappa^-(B) + 2\dim(\mathcal{N}) < \infty,$$

and hence we have for $\varepsilon > 0$ and sufficiently small,

$$\kappa^-(B+\varepsilon I) \le \kappa^-(B), \quad \kappa^-(H+\varepsilon I) = \kappa^-(H).$$

Denote by P the orthogonal projection of $\mathcal{K}$ onto $\overline{\mathcal{R}(B)}$. For $\varepsilon > 0$ and sufficiently small the operator $B+\varepsilon P$ has closed range, and therefore, by **(1)**,

$$\begin{aligned}\kappa^-(H+\varepsilon P) &\le \kappa^-(B+\varepsilon P) + \dim\mathcal{N} \\ &= \kappa^-(B) + \dim\mathcal{N}.\end{aligned} \tag{4.24}$$

Since $H \le H+\varepsilon P \le H+\varepsilon I$,

$$\kappa^-(H) \ge \kappa^-(H+\varepsilon P) \ge \kappa^-(H+\varepsilon I) = \kappa^-(H).$$

Hence, from (4.24), we obtain the inequality (4.22). ∎

5. Applications to Selfadjoint Pencils

5.1 Reduction to the Kreĭn Space Setting. Let $(\mathcal{H},(\cdot,\cdot))$ be a Hilbert space and let $G \in \mathcal{L}(\mathcal{H})$ be selfadjoint. Following [7] we associate with G two unitarily equivalent Kreĭn spaces $\mathcal{H}_G$ and $\mathcal{K}_G$. The space $\mathcal{H}_G$ is defined as the closure of the range of G equipped with the inner product $(x,y)_{\mathcal{H}_G} = (Jx,y)$, where J is the selfadjoint partial isometry which appears in the polar decomposition of G. The space $\mathcal{K}_G$ is defined via the spectral function $E(t)$ of G: Decompose the range of G as

$$\mathcal{R}(G) = GE(-\infty,0)\mathcal{H} \oplus GE(0,\infty)\mathcal{H}$$

and complete the linear manifolds $GE(-\infty,0)\mathcal{H}$ and $GE(0,\infty)\mathcal{H}$ with respect to the negative and positive inner product $(x,y)_{\mathcal{K}_G} = (Gx,y)$, respectively. The spaces $\mathcal{K}_G^-$ and $\mathcal{K}_G^+$ so obtained form a fundamental decomposition for the space $\mathcal{K}_G$. The spaces $\mathcal{H}_G$ and $\mathcal{K}_G$ are unitarily equivalent. The unitary operator is an extension of $|G|^{\frac{1}{2}}: \mathcal{R}(G) \subseteq \mathcal{K}_G \to \mathcal{H}_G$. Indeed, the relation

$$(x,y)_{\mathcal{K}_G} = (Gx,y)_{\mathcal{H}} = (J|G|^{\frac{1}{2}}x, |G|^{\frac{1}{2}}y), \quad x,y \in \mathcal{R}(G),$$

shows that $|G|^{\frac{1}{2}}$ is isometric. Moreover, the range of $|G|^{\frac{1}{2}}$ is dense in $\mathcal{H}_G$ and $|G|^{\frac{1}{2}}$ is bounded, as the strong topology on $\mathcal{H}_G$ is given via the Hilbert space norm $\|x\|_{\mathcal{H}}$ and the strong topology on $\mathcal{K}_G$ is given by the seminorm $\||G|^{\frac{1}{2}}x\|_{\mathcal{H}}$. For more details and the generalization to the case where $\mathcal{H}$ is a Kreĭn space we refer to [7]. Let $F \in \mathcal{L}(\mathcal{H})$ be another selfadjoint operator and assume that $F\overline{\mathcal{R}(G)} \subseteq \overline{\mathcal{R}(G)}$. Then, with respect to the decomposition $\mathcal{H} = \overline{\mathcal{R}(G)} \oplus \ker G$, the operator F has the representation

$$G = \begin{pmatrix} G_{11} & 0 \\ 0 & 0 \end{pmatrix}, \quad F = \begin{pmatrix} F_{11} & 0 \\ 0 & F_{22} \end{pmatrix}$$

and hence

$$\kappa^{\pm}(\lambda G - F) = \kappa^{\pm}(\lambda G_{11} - F_{11}) + \kappa^{\mp}(F_{22}). \tag{5.1}$$

This shows that without restricting the generality we can assume G is one-to-one and $\mathcal{R}(F) \subseteq \overline{\mathcal{R}(G)}$. Then there exists a unique operator $B \in \mathcal{L}(\mathcal{H})$ such that $F = GB$. Note that $\mathcal{R}(B) \subseteq \overline{\mathcal{R}(G)}$. From the relation

$$\begin{aligned}(Bx, y)_{\mathcal{K}_G} &= (GBx, y) = (Fx, y) = (x, Fy)\\ &= (Gx, By) = (x, y)_{\mathcal{K}_G}, \quad x, y \in \mathcal{R}(G),\end{aligned}$$

and the indefinite version of the Kreĭn Lemma (see [8]) it follows that B has a unique extension to a bounded selfadjoint operator in $\mathcal{K}_G$ which we also denote by B.

The operator $A \in \mathcal{L}(\mathcal{H}_G)$ is defined such that the diagram in Figure 1 below commutes, that is, such that A is equivalent to B under the unitary operator defined by $|G|^{\frac{1}{2}}$.

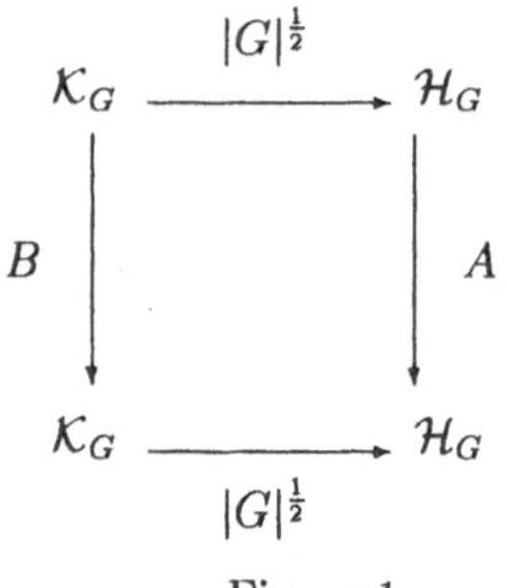

Figure 1.

The operator A is selfadjoint and is the unique extension of the densely defined operator

$$J|G|^{-\frac{1}{2}}F|G|^{-\frac{1}{2}} : \mathcal{R}(|G|^{\frac{1}{2}}) \subseteq \mathcal{H}_G \to \mathcal{H}_G.$$

LEMMA 5.1 *Let $\mathcal{H}$ be a Hilbert space and let $F, G \in \mathcal{L}(\mathcal{H})$ be selfadjoint operators with $F\overline{\mathcal{R}(G)} \subseteq \overline{\mathcal{R}(G)}$ and let $A \in \mathcal{L}(\mathcal{H}_G)$ and $B \in \mathcal{L}(\mathcal{K}_G)$ be the selfadjoint operators constructed above. Then for all real λ, $\kappa^{\pm}(\lambda I - A) = \kappa^{\pm}(\lambda I - B)$ and*

$$\kappa^{\pm}(\lambda G - F) = \kappa^{\pm}(\lambda I - A) + \kappa^{\mp}(P_{\ker G}F| \ker G).$$

Proof. We only prove the equality for κ^-. In view of (5.1) we can assume without loss of generality that $\ker G = 0$, that is, $\mathcal{R}(G)$ is dense in $\mathcal{H}$.

Let $\mathcal{L}$ be a finite dimensional subspace of $\mathcal{H}$ such that

$$((\lambda G - F)x, x) < 0, \quad x \in \mathcal{L} \setminus \{0\}.$$

By the approximation procedure used in the proof of the Pontryagin Lemma in [4], we can assume that $\mathcal{L} \subseteq \mathcal{R}(G)$. It follows that for $x \in \mathcal{L} \setminus \{0\}$,

$$((\lambda I - B)x, x)_{\mathcal{K}_G} = (G(\lambda I - B)x, x) = ((\lambda G - F)x, x) < 0$$

and hence

$$\kappa^-(\lambda G - F) \leq \kappa^-(\lambda I - B).$$

The converse inequality can be obtained by tracing the argument in the opposite direction, hence we have

$$\kappa^-(\lambda G - F) = \kappa^-(\lambda I - B) = \kappa^-(\lambda I - A).$$

The last equality holds since A and B are unitarily equivalent. ∎

THEOREM 5.2 *Let $\mathcal{H}$ be Hilbert space and let $F, G \in \mathcal{L}(\mathcal{H})$ be selfadjoint operators with $F\overline{\mathcal{R}(G)} \subseteq \mathcal{R}(G)$. Then, for all real λ the four one-sided limits $\kappa^{\pm}((\lambda \pm G - F))$ exist and the following inequalities hold:*

$$\kappa^{\pm}(\lambda G - F) \leq \min\{\kappa^{\pm}(\lambda - 0)g - F), \kappa^{\pm}((\lambda + 0)G - F)\}. \tag{5.2}$$

Proof. Let A be the selfadjoint operator in the Kreĭn space $\mathcal{K}$ defined by the diagram in Figure 1. By Lemma 5.1 we have to prove that the one-sided limits $\kappa^{\pm}(\lambda I - A)$ always exist, which actually follows from Theorem 3.4. The inequality (5.2) follows from (3.9). ∎

Note that the construction of $\mathcal{H}_G$ does not require any completion and to compute $\kappa^{\pm}(\lambda I - A)$ it is sufficient to use the values of A on the dense set $\mathcal{R}(|G|^{1/2})$. Also to compute $\kappa^{\pm}(\lambda I - B)$ it is sufficient to know the values of B on the dense set $\mathcal{R}(G)$, but the space $\mathcal{K}_G$ is obtained via a completion procedure.

5.2 Connections with the Results in [20]. In [20] P. Lancaster, A. Shkalikov, and Q. Ye consider the selfadjoint pencil $\lambda F - G$ where F and G are bounded selfadjoint operators on a Hilbert space $\mathcal{H}$ such that:

I_b. G is boundedly invertible and $\kappa^{-}(G) < \infty$.

II_b. F is compact.

III_b. For all nonzero $x \in \ker F$, $(Gx, x) > 0$.

These asssumptions imply that for each real λ, the operator $\lambda F - G$ has only a finite number of positive eigenvalues. Denote this number by $\pi(\lambda)$, in our notation we have $\pi(\lambda) = \kappa^{+}(\lambda F - G)$. In the following we denote by $\mathcal{L}_\mu$ the root subspace of the operator $\lambda F - G$ corresponding to μ. To put this into the framework of operator theory on Kreĭn spaces we consider the polar decomposition of G

$$G = J|G| = |G|^{1/2} J |G|^{1/2}, \tag{5.3}$$

and the Kreĭn space $\mathcal{K} = \mathcal{H}$ endowed with the indefinite inner product defined by the symmetry J. Then, by I_b and II_b, $A = J|G|^{-1/2} F |G|^{-1/2}$ is a compact selfadjoint operator in $\mathcal{K}$, and for any real λ we have

$$\kappa^{\pm}(\lambda F - G) = \begin{cases} \kappa^{\mp}(\frac{1}{\lambda} I - A), & \lambda > 0, \\ \kappa^{\mp}(I), & \lambda = 0, \\ \kappa^{\pm}(\frac{1}{\lambda} I - A), & \lambda < 0, \end{cases} \tag{5.4}$$

where the signatures on the righthand side are calculated with respect to the indefinite inner product of $\mathcal{K}$.

Proposition 6 in [20] states that if $\mu > 0$ is an eigenvalue of the pencil $\lambda F - G$ (and hence isolated and of finite algebraic multiplicity) the following inequality holds

$$\pi(\mu - 0) - \pi(\mu + 0) \leq \dim \mathcal{L}_\mu^{-}, \tag{5.5}$$

where $\mathcal{L}_\mu^{-}$ is a maximal G-negative subspace of $\mathcal{L}_\mu$ and that equality holds if and only if $\mathcal{L}_\mu$ is positive definite. In our transcription with $\lambda = \mu^{-1}$, the inequality (5.5) reads as

$$\kappa^{-}((\lambda + 0)I - A) - \kappa^{-}((\lambda - 0)I - A) \leq \kappa^{-}(\mathcal{S}_\lambda), \tag{5.6}$$

where $\mathcal{S}_\lambda$ is the root subspace of A at λ. This inequality immediately follows from Corollary 3.3. Indeed, this corollary implies that

$$\kappa^-((\lambda+0)I-A)-\kappa^-((\lambda-0)I-A)=\kappa^-(\mathcal{S}_\lambda)-\kappa^+(\mathcal{S}_\lambda).$$

Hence from the point of view of spectral analysis the inequality (5.5) is obvious and also the second part of the statement follows immediately. In [20] the inequality is obtained by applying perturbation theory, in particular the Rellich–Kato Theorem (see [14]).

To relate our results to one of the main theorems in [20] we recall that, by the definition, the selfadjoint pencil $L(\lambda) = \lambda F - G$ is *uniformly definitizable*, if there exists a real polynomial p such that for some $\delta > 0$

$$(Gp(G^{-1}F)x,x)\geq \delta(x,x), \quad x\in\mathcal{H},$$

and it is called *strongly definitizable*

$$(Gp(G^{-1}F)x,x)>0, \quad x\in\mathcal{H}\setminus\{0\}.$$

Theorem 1 in [20] includes the statement that if the hypotheses I_b, II_b, and III_b are valid, then the strongly definitizability of $L(\lambda)$ is equivalent to the existence of real numbers $\alpha_0 < \alpha_1 < \ldots < \alpha_{2q}$, not eigenvalues of $L(\lambda)$, such that

$$\pi(\alpha_0)=\sum_{j=1}^{q}\left(\pi(\alpha_{2j-1})-\pi(\alpha_{2j})\right). \tag{5.7}$$

In the following we prove a result that implies this equivalence. By definition, a bounded selfadjoint operator A in the Kreĭn space $\mathcal{K}$ is *uniformly definitizable (strictly definitizable)* if there exists a real polynomial p such that $p(A)$ is a positive and boundedly invertible (injective) operator. This definition plays a role in the perturbation theory of definitizable operators (see [13], where the notion of stronlgy stable is used). Clearly, if A is uniformly definitizable then it is strongly definitizable and if it is strongly definitizable then it is definitizable, in particular, the spectral function $E(t)$ of A exists. Recall that if A is definitizable then the real spectrum of A decomposes as the disjoint union

$$\sigma(A)\cap\mathbf{R}=\sigma_+(A)\cup\sigma_-(A)\cup c(A),$$

where $\sigma_\pm(A)$ is the set of all points of positive/negative type in the spectrum of A (see [22], p. 36). It is easy to see that a bounded definitizable operator A is uniformly definitizable if and only if A satisfies the following properties:

(a) $\sigma(A)\subset\mathbf{R}$,

(b) $c(A)=\emptyset$,

(c) the distance between $\sigma_+(A)$ and $\sigma_-(A)$ is positive.

Also, the definitizable operator A is strongly definitizable if and only if it satisfies the following properties:

(a)' $\sigma(A)\subset\mathbf{R}$,

(b)' $c(A) \cap \sigma_p(A) = \emptyset$.

Using these characterizations it follows easily that if $\kappa^-(\mathcal{K}) < \infty$ and A is compact then A is strongly definitizable if and only if A is uniformly definitizable (see also [18] for a different approach).

PROPOSITION 5.3 *Let A be a compact selfadjoint operator in a Pontryagin space $\mathcal{K}$ with $\kappa^-(\mathcal{K}) < \infty$. Then:*

(i) For real numbers $\beta_{2j} < \beta_{2j-1} < \ldots < \beta_1 < 0 < \alpha_1 < \ldots < \alpha_{2i-1} < \alpha_{2i}$ and $\gamma \in (\beta_1, \alpha_1) \setminus \{0\}$ in the resolvent set of A, the following inequality holds

$$\sum_{k=1}^{j} \left(\kappa^+(\beta_{2k}I - A) - \kappa^+(\beta_{2k-1}I - A)\right) + \sum_{k=1}^{i} \left(\kappa^-(\alpha_{2k}I - A) - \kappa^-(\alpha_{2k-1}I - A)\right) \leq \kappa^-(\gamma I - A). \tag{5.8}$$

(ii) For real numbers $\beta'_{2j'} < \beta'_{2j'-1} < \ldots < \beta'_1 < 0 < \alpha'_1 < \ldots < \alpha'_{2i'-1} < \alpha'_{2i'}$ and $\gamma' \in (\beta'_1, \alpha'_1) \setminus \{0\}$ in the resolvent set of A, the following inequality holds

$$\sum_{k=1}^{j'} \left(\kappa^+(\beta'_{2k-1}I - A) - \kappa^+(\beta'_{2k}I - A)\right) + \sum_{k=1}^{i'} \left(\kappa^-(\alpha'_{2k-1}I - A) - \kappa^-(\alpha'_{2k}I - A)\right) \leq \kappa^+(\gamma' I - A). \tag{5.9}$$

(iii) A is uniformly definitizable (equivalently, strongly definitizable) if and only if for some real numbers $\{\alpha_k\}_{k=1}^{2i}$, $\{\beta_k\}_{k=1}^{2j}$, and γ ordered as in (i), equality holds in (5.8), or, equivalently, for some real numbers $\{\alpha'_k\}_{k=1}^{2i'}$, $\{\beta'_k\}_{k=1}^{2j'}$, and γ' ordered as in (ii), equality holds in (5.9).

By Corollary 3.10, from hypotheses I_b and II_b, $\kappa^+(\beta I - A) < \infty$ for any $\beta < 0$ and, similarly, $\kappa^-(\alpha I - A) < \infty$ for any $\alpha > 0$, hence the numbers on the lefthand sides of the inequalities (5.8) and (5.9) are finite. The same corollary shows that the conditions imposed on A and $\mathcal{K}$ are the natural ones in which these inequalities make sense.

Proof of Proposition 5.3. (i) Since the α_k's, the β_l's and γ are in the resolvent set of A, we use Corollary 3.3 to calculate the signatures in (5.8) and get

$$\begin{aligned}
\kappa^-(\gamma I - A) - \sum_{k=1}^{j} \left(\kappa^+(\beta_{2k}I - A) - \kappa^+(\beta_{2k-1}I - A)\right) \\
+ \sum_{k=1}^{i} \left(\kappa^-(\alpha_{2k}I - A) - \kappa^-(\beta_{2k-1}I - A)\right) = \kappa^+(E(\gamma, \infty)) + \kappa^-(E(-\infty, \gamma)) \\
+ \sum_{\Im\lambda > 0} \operatorname{rank} E(\lambda; A) - \sum_{k=1}^{j} \left(\kappa^-(E(\beta_{2k}, \beta_{2k-1})) - \kappa^+(E(\beta_{2k}, \beta_{2k-1}))\right) \\
- \sum_{k=1}^{i} \left(\kappa^+(E(\alpha_{2k-1}, \alpha_{2k})) - \kappa^-(E(\alpha_{2k-1}, \alpha_{2k}))\right)
\end{aligned} \tag{5.10}$$

$$= \kappa^-(E((-\infty,\gamma)\setminus \cup_{k=1}^{j}(\beta_{2k},\beta_{2k-1}))) + \kappa^+(E((\gamma,\infty)\setminus \cup_{k=1}^{i}(\alpha_{2k-1},\alpha_{2k})))$$
$$+ \sum_{\Im\lambda>0} \operatorname{rank} E(\lambda;A) + \sum_{k=1}^{j} \kappa^+ E(\beta_{2k},\beta_{2k-1})) + \sum_{k=1}^{i} \kappa^-(E(\alpha_{2k-1},\alpha_{2k})) \geq 0.$$

(ii) The proof of (ii) is similar as the proof of (i).

(iii) Assume that equality holds in (5.8). Then equality holds in (5.10) which implies that $\sigma(A) \subset \mathbf{R}$ and that the spectral function $E(t)$ is definite on each interval of the partition determined by the numbers α_k, β_l, and γ. If $\gamma > 0$ then there exists a neighbourhood V of 0 such that the points in $V \cap \sigma(A)$ are of positive type, in particular the kernel of A is positive definite. If $\gamma < 0$ then there exists a neighbourhood V of 0 such that the points in $V \cap \sigma(A)$ are of negative type, in which case $\mathcal{K}$ is finite dimensional and the kernel of A is negative definite. In either case it follows that A is uniformly definitizable.

We claim that if equality holds in (5.8) then equality holds in (5.9). Indeed, if $\gamma > 0$ then equality holds in (5.9) with $i' = i+1$, $j' = j$, $\alpha'_1 = \gamma$, $\alpha'_k = \alpha_{k-1}$ for $2 \leq k \leq 2i'+1$, $\gamma' = \beta'_1$, $\beta'_k = \beta_{k+1}$ for $1 \leq k \leq 2j'-1$, and $\alpha'_{2i'}$ and $\beta'_{2j'}$ chosen such that the interval $(\beta'_{2j'}, \alpha'_{2i'})$ containes $\sigma(A)$ and all other α'_k's and all other β'_k's; see Figure 2, where the $\pm$ sign above an interval Δ means that the spectral subspace $E(\Delta)\mathcal{K}$ is nonnegative/nonpositive, respectively.

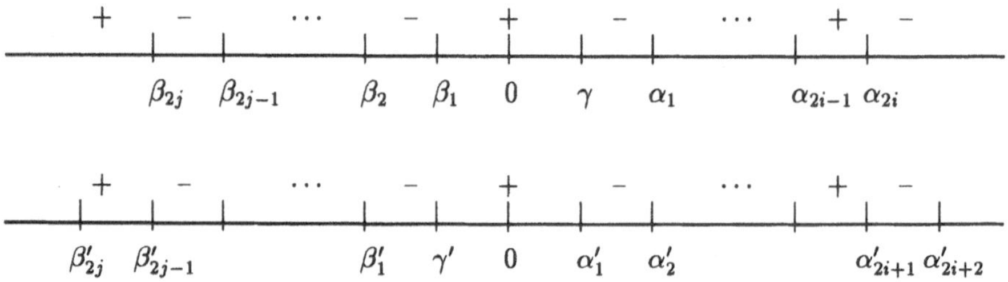

Figure 2.

If $\gamma < 0$ then equality holds in (5.9) with $i' = i$, $j' = j+1$, $\alpha'_k = \alpha_{k+1}$ for $1 \leq k \leq 2i'-1$, $\gamma' = \alpha'_1$, $\beta'_1 = \gamma$, $\beta'_k = \beta_{k-1}$ for $1 \leq k \leq 2j'+1$, and $\alpha'_{2i'}$ and $\beta'_{2j'}$ are again so large that the interval $(\beta'_{2j'}, \alpha'_{2i'})$ contains $\sigma(A)$ and all other α'_k's and all other β'_k's. (see Figure 3). Thus our claim is proved.

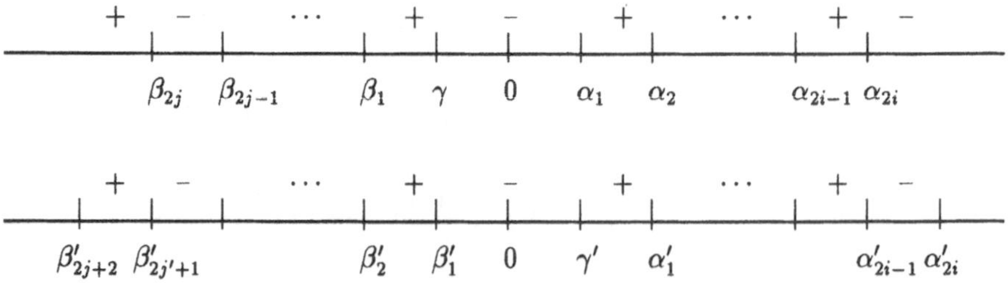

Figure 3.

If equality holds in (5.9) then a similar argument shows that one can choose real numbers $\{\alpha_k\}_{k=1}^{2i}$, $\{\beta_k\}_{k=1}^{2j}$, and γ ordered as in (i), such that equality holds in (5.8).

Conversely, if A is uniformly definitizable then there exists a partition of $\mathbf{R}$ with points in $\rho(A)$ such that $E(t)$ is definite on each interval of the partition. These points can be numbered α_k, β_l, and γ such that equality holds in (5.8). ∎

Using the decomposition (5.3), the formulae (5.4), and Proposition 5.3 we obtain the characterization of the strong definitizability of the pencil $L(\lambda)$ in [20], Theorem 1. Finally we mention that the assumption III_b is not needed for this characterization, as its validity is actually absorbed in the equality (5.7).

References

[1] T. Ando: *Linear Operators in Kreĭn Spaces*, Lecture Notes, Hokkaido University 1979.

[2] N. I. Achieser und I. M. Glasmann: *Theorie der linearen Operatoren im Hilbert-Raum*, 8., erweiterte Auflage, Akademie-Verlag, Berlin 1981.

[3] T. Ya. Azizov and I. S. Iokhvidov: *Foundations of the Theory of Linear Operators in Spaces with Indefinite Metric* [Russian], Nauka, Moscow 1986 (English Translation: *Linear Operators in Spaces with an Indefinite Metric*, John Wiley, New York 1989).

[4] J. Bognár: *Indefinite Inner Product Spaces*, Springer Verlag, Berlin-Heidelberg-New York 1974.

[5] P. A. Binding and K. Seddighi: On Root Vectors of Self-Adjoint Pencils, *J. Functional Analysis*, **70**(1987), 117-125.

[6] T. Constantinescu, A. Gheondea: The Negative Signature of Some Hermitian Matrices, *Linear Alg. Appl.*, **178**(1993), 17-42.

[7] T. Costantinescu, A. Gheondea: Elementary Rotations of Linear Operators in Kreĭn Spaces, *J. Operator Theory*, (to appear)

[8] A. Dijksma, H. Langer, and H. S. V. de Snoo: Unitary Colligations in Kreĭn Spaces and their Role in the Extension Theory of Isometric and Symmetric Linear Relations in Hilbert Spaces, in *Functional Analysis II, Proceedings Dubrovnik 1985*, Lecture Notes in Mathematics, Vol. **1242**, Springer Verlag, Berlin-Heidelberg-New York 1987, pp.1-42.

[9] A. Gheondea: Quasi-Contractions on Kreĭn Spaces, in *Operator Theory: Addvances and Applications*, Vol. **61**, Birkhäuser Verlag, Basel - Boston - Berlin 1993, pp. 123-148.

[10] I. Gohberg, P. Lancaster, and L. Rodman: *Matrices in Indefinite Scalar Products*, Birkhäuser Verlag, Basel-Boston-Stuttgart 1983.

[11] I. S. Iokhvidov, M. G. Kreĭn, and H. Langer: *Introduction to the Spectral Theory of Operators in Spaces with an Indefinite Metric*, Akademie-Verlag, Berlin 1983.

[12] P. Jonas: On the Functional Calculus and the Spectral Function for Definitizable Operators in Kreĭn space, *Beiträge zur Analysis*, **16**(1981), 121-135.

[13] P. Jonas and H. Langer: Compact Perturbations of Definitizable Operators, *J. Operator Theory*, **2**(1979), 63-77.

[14] T. Kato: *Perturbation Theory of Linear Operators*, Springer Verlag, Berlin-Heidelberg-New York 1966.

[15] M. G. Kreĭn; H. Langer: On the Spectral Function of a Selfadjoint Operator in a Space with Indefinite Metric [Russian], *Dokl. Akad. Nauk SSSR*, **152**(1963), 39-42.

[16] M. G. Kreĭn and Yu. L. Shmulyan: Plus-operators in Spaces with Indefinite Metric [Russian], *Mat. Issled*, **1**(1966),131–161.

[17] M. G. Kreĭn and Yu. L. Shmulyan: J-Polar Representations of Plus-Operators [Russian], *Mat. Issled.* **1**(1966), 172-210.

[18] P. Lancaster, A. S. Markus, and V. I. Matsaev: Definitizable Operators and Quasihyperbolic Operator Polynomials, preprint 1993.

[19] P. Lancaster and Q. Ye: Definitizable Hermitian Matrix Pencils, *Aequationes Mathematicae,* (to appear).

[20] P. Lancaster, A. Shkalikov, and Q. Ye: Strongly Definitizable Linear Pencils in Hilbert Space, *Integral Equations Operator Theory,* **17**(1993), 338-360.

[21] H. Langer: *Spektraltheorie linearer Operatoren in J-Räumen und einige Anwendungen auf den Schar* $L(\lambda) = \lambda^2 + \lambda B + C$, Habilitationsschrift, Dresden 1965.

[22] H. Langer: Spectral Functions of Definitizable Operators in Kreĭn Spaces, in *Lecture Notes in Mathematics*, Vol. **948**, Springer Verlag, Berlin–Heidelberg–New York 1983.

Aad Dijksma

University of Groningen
Department of Mathematics
P. O. Box 800
9700 AV Groningen
The Netherlands

e-mail: `a.dijksma@math.rug.nl`

Aurelian Gheondea

Institutul de Matematică al
Academiei Române,
C.P. 1–764,
70700 Bucureşti,
România

e-mail: *gheondea@imar.ro*

AMS Subject Classification: 47A56, 47B50.

Operator Theory:
Advances and Applications, Vol. 80

ON THE SPECTRAL THEORY OF AN ELLIPTIC BOUNDARY VALUE PROBLEM INVOLVING AN INDEFINITE WEIGHT

M. Faierman

We are concerned here with the spectral theory pertaining to an elliptic boundary value problem involving an indefinite weight function, or equivalently, the spectral theory for a pencil of the form $A - \lambda T$ acting in a Hilbert space $L_2(\Omega)$, where $\Omega \subset \mathbb{R}^n$ is a bounded region and $n \geq 2$. Here A is a non–selfadjoint operator and T is a multiplication operator in $L_2(\Omega)$ induced by a real–valued weight function which assumes both positive and negative values. Results are given concerning the completeness of the principal vectors of the pencil in certain function spaces as well as concerning the angular and asymptotic distribution of the eigenvalues. Furthermore, a new result is also derived pertaining to the asymptotic distribution of the eigenvalues.

1. INTRODUCTION

Although there is a relatively large literature devoted to the spectral theory for linear elliptic boundary value problems involving an indefinite weight function, most of the work to date has been concerned with either selfadjoint problems or non–selfadjoint problems arising from perturbations of selfadjoint ones. We refer to [6], [8–11], [13], [15,16], [22,23], and [26–28] for further information. With this in mind, the author [17–21] has recently initiated an investigation into the spectral theory of quite general non–selfadjoint problems, and accordingly it is the object of this paper to collect the known results for the case of second order elliptic operators and then derive some new results concerning the eigenvalue asymptotics for such operators.

We shall be concerned here with the eigenvalue problem

$$Lu = \lambda\omega(x)u \text{ in } \Omega, \tag{1.1}$$

$$Bu = 0 \text{ on } \Gamma, \tag{1.2}$$

where L is a linear elliptic operator of the second order defined in a bounded region $\Omega \subset \mathbb{R}^n$, $n \geq 2$, with boundary Γ, B is a linear differential operator of the first order defined on Γ, and ω is a real–valued function in $L_\infty(\Omega)$ which assumes both positive and negative values. Our assumptions concerning the problem (1.1–2) will be made precise in the sequel. Observe that if A denotes the operator in $L_2(\Omega)$ induced by L and the boundary condition (1.2) and T denotes the operator of multiplication in $L_2(\Omega)$ induced by ω, then the eigenvalue problem (1.1–2) can be formulated from a purely operator theoretic point of view, namely as the spectral problem for the pencil $S(\lambda) = A - \lambda T$, $\lambda \in \mathbb{C}$. By mean of certain a priori estimates for solutions of elliptic equations as well as under certain assumptions concerning the problem (1.1–2) (e.g., we require that the resolvent set of $S(\lambda)$ is not empty), we have been able to derive some important information concerning the completeness of the principal vectors of $S(\lambda)$ in certain function spaces and the angular and asymptotic distribution of the eigenvalues. Furthermore, by means of these a priori estimates, we will establish in this paper a new result pertaining to the asymptotic distribution of the eigenvalues.

In §2 of this paper we introduce some of our basic assumptions and present some known results concerning the problem (1.1–2) which we require in the sequel. In §3 we introduce the last of our basic assumptions, state our main results (see *Theorems 3.2–3*), and then, in order to give some insight into how these results were arrived at, we state a new result (see *Theorem 3.4*) and devote the remainder of the paper to its proof. In §4 we introduce a certain compact operator $\mathcal{K}$ whose eigenvalues are precisely those of the problem (1.1–2) and study some of its basic properties. The results of §4 are then used in §5 to prove *Theorem 3.4*.

2. PRELIMINARIES

In this section we are going to introduce our basic assumptions concerning the problem (1.1–2) and present some known results which we require in the sequel. To begin with, we let $x = (x_1, \ldots, x_n) = (x', x_n)$ denote a generic point in $\mathbb{R}^n$ and use the notation $D_j = \partial/\partial x_j$, $D = (D_1, \ldots, D_n)$, $D^\alpha = D_1^{\alpha_1} \cdots D_n^{\alpha_n}$, where $\alpha = (\alpha_1, \ldots, \alpha_n)$ is a multi-index whose length $\sum_{j=1}^n \alpha_j$ is denoted by $|\alpha|$. We henceforth suppose that $1 < p < \infty$ and for $0 \leq s < \infty$ and G an open set in $\mathbb{R}^n$ or $\mathbb{R}^{n-1}$ we let $W_p^s(G)$ denote the usual Sobolev–Slobodeckii space of order s related to $L_p(G)$ and denote by $\| \ \|_{s,G,p}$ the norm in this space (see [25, p.17] for details). Also for $s < 0$, we let $\| \ \|_{s,\mathbb{R}^{n-1},p}$ denote the norm in the Besov space $B_{p,p}^s(\mathbb{R}^{n-1})$ (see [34, p.169]). Turning to the problem (1.1–2), we suppose from now on that

ASSUMPTION 2.1. (1) Ω is of class $C^{3,1}$; (2) $L(x, D) = \sum_{|\alpha| \leq 2} a_\alpha D^\alpha$ is uniformly strongly elliptic in Ω with a_α real–valued if $|\alpha| = 2$ and complex–valued otherwise and such that $a_\alpha \in C^{|\alpha|,1}(\overline{\Omega})$ for $|\alpha| \geq 1$, $a_\alpha \in L_\infty(\Omega)$ otherwise, where ¯

denotes closure; (3) $B(x,D) = \sum_{|\alpha|\leq 1} b_\alpha D^\alpha$, with b_α real-valued if $|\alpha| = 1$ and complex-valued otherwise, while $b_\alpha \in C^{|\alpha|+1,1}(\Gamma)$ for $|\alpha| \geq 0$; (4) Γ is non-characteristic to B at each of its points.

REMARK 2.1. Note from Assumption 2.1 that $L_0(x, i\xi) \geq c|\xi|^2$ for $x \in \overline{\Omega}$ and $\xi \in \mathbb{R}^n$, where $L_0(x,D)$ denotes the principal part of $L(x,D)$ and c denotes a positive constant. Note also that by employing a known extension procedure, we may suppose from now on that b_α is defined in all of $\overline{\Omega}$ with $b_\alpha \in C^{|\alpha|+1,1}(\overline{\Omega})$.

It is clear that apart from certain smoothness conditions, Assumption 2.1 ensures that the boundary value problem:

$$Lu = f \text{ in } \Omega \tag{2.1}$$

together with (1.2), is a regular elliptic problem in the sense of [2], [30]. Note that if L^* denotes the formal adjoint of L and C denotes a boundary operator adjoint to B with respect to the problem (2.1), (1.2) (see [30], [33]), then the formal adjoint problem of (2.1), (1.2)

$$\begin{aligned} L^*u &= f \text{ in } \Omega, \\ Cu &= 0 \text{ on } \Gamma, \end{aligned} \tag{2.2}$$

is also a regular elliptic problem (see [17, §2]). Now putting $p' = p/(p-1)$ let us introduce in $L_p(\Omega)$ (resp. $L_{p'}(\Omega)$) the operator A_p (resp. $A'_{p'}$) with domain $D(A_p)$ (resp. $D(A'_{p'})$) as follows: we let $D(A_p)$ (resp. $D(A'_{p'})$) denote the closure in $W_p^2(\Omega)$ (resp. $W_{p'}^2(\Omega)$) of the class of functions in $C^2(\overline{\Omega})$ satisfying the boundary condition (1.2) (resp. (2.2)) and put $A_p u = Lu$ for $u \in D(A_p)$ (resp. $A'_{p'}u = L^*u$ for $u \in D(A'_{p'})$). Then we know from [2] that

THEOREM 2.1. *If $u \in D(A_p)$, then $\|u\|_{2,\Omega,p} \leq c\left[\|A_p u\|_{0,\Omega,p} + \|u\|_{0,\Omega,p}\right]$, where the constant c does not depend upon u.*

It follows from the theorem that A_p is semi-Fredholm and dim ker $A_p < \infty$. Analogous results also hold for $A'_{p'}$. Furthermore, we know from [20] that

THEOREM 2.2. *If A_p^* denotes the Banach space adjoint of A_p, then $A_p^* = A'_{p'}$.*

In light of *Theorems 2.1–2*, we conclude from [2] that

THEOREM 2.3. *A_p has non-empty resolvent set, compact resolvent, and hence a discrete spectrum. Moreover, the eigenvalues and principal vectors of A_p are the same for all p.*

Analogous results also hold for $A_p^\star$. Note that as a consequence of *Theorems 2.1–2* we now know that A_p and $A_p^\star$ are Fredholm operators, while it also follows from *Theorem 2.1*, [2], [29, Theorem 5.26, p.238], and [1, Theorem 6.2, p.144] that index $A_p = -$ index $A_p^\star = 0$. For later use let us note the following result from [2]: if $u \in D(A_p)$ and $A_p u \in L_q(\Omega)$, where $1 < q < \infty$, then $u \in D(A_q)$. An analogous result likewise holds for $A_p^\star$.

Turning next to our assumptions concerning $\omega(x)$, let

$$\Omega^+ = \{x \in \Omega | \omega(x) > 0\},\ \Omega^- = \{x \in \Omega | \omega(x) < 0\},\ \Omega^0 = \{x \in \Omega | \omega(x) = 0\}.$$

ASSUMPTION 2.2. In the sequel we suppose that: (1) $|\Omega^\pm| > 0$ and $|\Omega^0| \geq 0$, where $|\ |$ denotes n-dimensional Lebesgue measure; (2) $|\Omega^\pm \backslash \operatorname{int} \Omega^\pm| = 0$, where int = interior; (3) int Ω^+ (resp. int Ω^-) is the union of a finite number of non-empty disjoint regions, say $\{\Omega_r^+\}$ (resp. $\{\Omega_r^-\}$) in each of which $\omega(x)$ is continuous and such that for at least one r, Ω_r^+, (resp. Ω_r^-) contains a closed ball in which $\omega(x)$ is Lipschitz continuous; (4) each component Γ_{rj}^+ (resp. Γ_{rj}^-) of $\partial\Omega_r^+$ (resp. $\partial\Omega_r^-$), where ∂ = boundary, is either a component of Γ or is contained in Ω and is either a component of $\partial\Omega_s^+$ (resp. $\partial\Omega_s^-$) for some $s \neq r$, or a component of $\partial\Omega_s^-$ (resp. $\partial\Omega_s^+$) for some s, or a component of $\partial\Omega_0$ if $|\Omega^0| > 0$, where $\Omega_0 = \Omega \backslash \overline{\Omega}_1$ and $\Omega_1 =$ int $\Omega^+ \cup$ int Ω^-; (5) for each component $\Gamma_{rj}^\pm$ of $\partial\Omega_r^\pm$ either (i) $\Gamma_{rj}^\pm$ is of class $C^{1,1}$ and there is a neighbourhood of $\Gamma_{rj}^\pm$ such that in the intersection of this neighbourhood with $\Omega_r^\pm$, $\omega(x)$ is uniformly continuous and $|\omega(x)|$ has a positive infimum or (ii) $\Gamma_{rj}^\pm$ is of class $C^{2,1}$ and there is a neighbourhood of $\Gamma_{rj}^\pm$ such that in the intersection of this neighbourhood with $\Omega_r^\pm$, $\omega(x) = \omega_{rj}^\pm(x)\left(d_{rj}^\pm(x)\right)^{\gamma_{rj}^\pm}$, where $\omega_{rj}^\pm(x)$ is uniformly continuous and $|\omega_{rj}^\pm(x)|$ has a positive infimum, $d_{rj}^\pm(x) = \operatorname{dist}\{x, \Gamma_{rj}^\pm\}$, and $\gamma_{rj}^\pm \geq 2$; (6) $\omega(x)$ has been modified on a set of measure zero, if necessary, so that if $|\Omega^0| > 0$, then $\omega(x) = 1$ for $x \in \Gamma_{rj}^\pm$ if $\Gamma_{rj}^\pm \subset \Omega \backslash \Omega_0$ and $\omega(x) = 0$ for $x \in \Omega_0$, while if $|\Omega^0| = 0$, then $\omega(x) = 1$ for $x \in \Gamma_{rj}^\pm$ if $\Gamma_{rj}^\pm \subset \Omega$.

It is an immediate consequence of Assumption 2.2 that if $|\Omega^0| > 0$, then Ω_0 is the union of a finite number of non-empty disjoint regions, say $\{\Omega_{0,j}\}$, and each component of $\partial\Omega_{0j}$ is either a component of Γ or is contained in Ω and is either a component of $\partial\Omega_s^+$ or of $\partial\Omega_s^-$ for some s. Also fixing our attention upon condition (5) of Assumption 2.2, we henceforth define $\gamma_{rj}^\pm = 0$ if alternative (i) is valid.

ASSUMPTION 2.3. It will be supposed from now on that: (1) if Γ^{+}_{rj} (resp. Γ^{-}_{rj}) coincides with a $\Gamma^{\pm}_{sk}$, then $\gamma^{+}_{rj} = \gamma^{\pm}_{sk}$ (resp. $\gamma^{-}_{rj} = \gamma^{\pm}_{sk}$); (2) if $|\Omega^0| > 0$ and $\Gamma^{\pm}_{rj}$ coincides with a component of $\partial\Omega_0$, then $\Gamma^{\pm}_{rj}$ is of class $C^{3,1}$; (3) if $|\Omega^0| > 0$, then the boundary value problem: $Lu = 0$ in Ω_0, $Bu = 0$ (resp. $u = 0$) on each component of $\partial\Omega_0$ which is contained in Γ (resp. in Ω), admits, in $W^2_2(\Omega_0)$, only the trivial solution.

In the sequel we will require some further terminology. Accordingly, with this in mind, we now introduce the following

DEFINITION 2.1. Let X be a complex Banach space and S a linear operator in X. Then the set $\{\lambda \in \mathbb{C} \,|\, \lambda \neq 0, (I - \lambda S)^{-1}$ exists and is in $\mathcal{L}(X)\}$ is denoted by $\rho_m(S)$. For $\lambda \in \rho_m(S)$ we let $S_\lambda = S(I - \lambda S)^{-1}$ and call S_λ the modified resolvent of S. A complex number $\lambda \neq 0$ is called a characteristic value of S if λ^{-1} is an eigenvalue of S, and if λ is a characteristic value of S, then we denote by $\mathcal{G}_\lambda(S, X)$ the principal subspace of S corresponding to the eigenvalue λ^{-1}. Lastly, the ray arg $\lambda = 0$ in the complex λ-plane is said to be a ray of minimal growth of S_λ if for all λ on the ray, with $|\lambda|$ sufficiently large, we have $\lambda \in \rho_m(S)$ and $\|S_\lambda\|_{\mathcal{L}(X)} \leq c|\lambda|^{-1}$, where c denotes a positive constant.

Finally for later use we introduce the space $\mathcal{H}_T$ in the following way. Recalling the definition of T given in §1, we observe that when $L_2(\Omega)$, considered only as a vector space, is equipped with the inner product $(\ ,\ )_T = (T.,.)$, where $(.,.)$ denotes the usual inner product in $L_2(\Omega)$, then it becomes an indefinite inner product space [12, p.4]; and in the sequel we shall denote this latter space by $\mathcal{H}_T$. Recall also from [12] that if $\mathcal{M}, \mathcal{N}$ are any two subspaces of $L_2(\Omega)$, then $\mathcal{M}, \mathcal{N}$ are said to form a dual pair of subspaces of $\mathcal{H}_T$ if for each $u \neq 0$ in $\mathcal{M}$ there is a $v \in \mathcal{N}$ such that $(u, v)_T \neq 0$ and for each $v \neq 0$ in $\mathcal{N}$ there is a u in $\mathcal{M}$ such that $(u, v)_T \neq 0$.

3. MAIN RESULTS

In this section we are going to present our main results (see *Theorems 3.2–4* below). However, we must firstly introduce one further assumption. Accordingly, recalling from §1 the definitions of the operators A, T, and $S(\lambda)$ and noting that A is just the operator A_2 introduced in §2, we have already stated that the eigenvalue problem (1.1–2) is to be interpreted as the spectral problem for the pencil $S(\lambda)$. Observe that for each λ, $S(\lambda)$ is a closed operator in $L_2(\Omega)$ with domain $D(A)$. Now employing the usual terminology associated with the pencil $S(\lambda)$ (see [31, pp.56–57 and 102]), suppose that $\mu \in \mathbb{C}$ is an eigenvalue of $S(\lambda)$. Then we let $N_\mu = \ker S(\mu)$ and let M_μ denote the subspace of $L_2(\Omega)$ spanned by the eigenvectors of $S(\mu)$ together with their associated vectors. We henceforth call dim M_μ the algebraic multiplicity of μ and refer to a vector $u \neq 0$ in M_μ as a principal vector for the eigenvalue μ of $S(\lambda)$.

We have seen in §2 that A (as well as $A^* = A_2^*$) is a Fredholm operator with index zero, and hence it follows from *Theorem 2.1*, [29, *Theorem 5.26*, p.238], and *Rellich's theorem* [3, p.30] that $S(\lambda)$ is a Fredholm operator with index zero for every $\lambda \in \mathbf{C}$. Thus we conclude from [29, *Theorem 5.31*, p.241] that for $\lambda \in \mathbf{C}$, nul $S(\lambda) =$ def $S(\lambda) =$ constant, with the possible exception of certain isolated points. For our purposes we require that this constant be zero, and this leads us to introduce

ASSUMPTION 3.1. We suppose from now on that $\rho(S) \neq \emptyset$, where $\rho(S)$ denotes the resolvent set of $S(\lambda)$.

Observing that $0 \in \sigma(S) =$ spectrum of $S(\lambda)$ if and only if $0 \in \sigma(A)$, and if $0 \in \sigma(A)$, then 0 is an eigenvalue of $S(\lambda)$ and $N_0 = \ker A$, we might mention at this point that in [19] it was shown that (see the last paragraph of §2 for terminology)

THEOREM 3.1. *In order that $\rho(S) \neq \emptyset$, it is necessary and sufficient that either $0 \in \rho(A)$ or $0 \in \sigma(A)$ and either: (i) N_0 and $N_0^* = \ker A^*$ form a dual pair of subspaces of $\mathcal{H}_T$ or (ii) N_0 and N_0^* do not form a dual pair in $\mathcal{H}_T$, but* dim $M_0 < \infty$.

<u>Terminology.</u> In the sequel when we speak of the spectrum, eigenvalues, principal vectors, etc. of the problem (1.1–2), this will always be meant with respect of the pencil $S(\lambda)$.

We are now in a position to present the main results concerning the spectral theory for the problem (1.1–2). Indeed, from [19] we have

THEOREM 3.2. *The spectrum of the problem (1.1–2) consists solely of eigenvalues of finite algebraic multiplicity which form a denumerably infinite subset of* $\mathbf{C}$ *having no finite points of accumulation. Moreover, for any ϵ satisfying $0 < \epsilon < \pi/2$, there are infinitely many eigenvalues in each of the sectors $|\arg \lambda| < \epsilon$ and $|\arg \lambda - \pi| < \epsilon$, while there are at most a finite number of eigenvalues in each of the sectors $\epsilon \leq \arg \lambda \leq \pi - \epsilon$ and $-\pi + \epsilon \leq \arg \lambda \leq -\epsilon$. Finally, the principal vectors of the problem (1.1–2) are complete in each of the function spaces $L_2(\Omega^+ \cup \Omega^-)$ and $L_2(\Omega^+ \cup \Omega^-;\ |\omega(x)|dx)$.*

Of course, when we speak of the completeness of the principal vectors of (1.1–2) in the spaces just cited, we mean that the restrictions of the principal vectors to the set $\Omega^+ \cup \Omega^-$ are complete. Note from [19] that for the case where $|\Omega^0| > 0$ and $\Gamma_{rj}^{\pm}$ is a component of both $\partial\Omega_r^{\pm}$ and $\partial\Omega_0$ (see Assumption 2.2), it was assumed in the proof of Theorem 3.2 that certain smoothness conditions were satisfied by $\Gamma_{rj}^{\pm}$ and by the a_α, $|\alpha| = 2$, in a neighbourhood of $\Gamma_{rj}^{\pm}$. However, we have shown in [21] that such conditions are actually not necessary.

Let $\{\lambda_j\}_1^\infty$ denote the eigenvalues of the problem (1.1–2) counted according to algebraic multiplicity, and for $\lambda \geq 0$ let $N_+(\lambda)$ denote the number of eigenvalues λ_j for which $0 \leq Re\,\lambda_j \leq \lambda$, while for $\lambda > 0$ let $N_-(\lambda)$ denote the number of eigenvalues λ_j for which $-\lambda \leq Re\,\lambda_j < 0$ and put $N_-(0) = 0$. Also for $\lambda \geq 0$ let $N(\lambda) = N_+(\lambda) + N_-(\lambda)$, so that $N(\lambda)$ is the number of eigenvalues λ_j for which $|Re\,\lambda_j| \leq \lambda$ and let $N^{\#}(\lambda)$ denote the number of λ_j for which $|\lambda_j| \leq \lambda$. Then putting $\omega^+(x) = \max\{\omega(x), 0\}$, $\omega^-(x) = \max\{-\omega(x), 0\}$, and $\tau(x) = \left|\{\xi \in \mathbb{R}^n | 0 < L_0(x, i\xi) < 1\}\right|$, where $|\ |$ denotes n-dimensional Lebesgue measure, we know from [20] that

THEOREM 3.3. *Suppose that $|\Omega^0| = 0$ and that in condition (5) of Assumption 2.2 only alternative (i) can occur for each of the $\Gamma_{rj}^{\pm}$. Then it is the case that as $\lambda \to \infty$, $N_\pm(\lambda) = \kappa_\pm \lambda^{n/2} + o(\lambda^{n/2})$, $N(\lambda) = \kappa\lambda^{n/2} + o(\lambda^{n/2})$, $N^{\#}(\lambda) = \kappa\lambda^{n/2} + o(\lambda^{n/2})$, where $\kappa_\pm = (2\pi)^{-n}\int_\Omega \left(\omega^\pm(x)\right)^{n/2}\tau(x)dx$ and $\kappa = \kappa_+ + \kappa_- = (2\pi)^{-n}\int_\Omega |\omega(x)|^{n/2}\tau(x)dx$.*

In order to give some insight into how *Theorem 3.2–3* have been arrived at, we shall in this paper prove the following *new* result.

THEOREM 3.4. *Suppose that $|\Omega^0| > 0$ and that in condition (5) of Assumption 2.2 only alternative (i) can occur for each of the $\Gamma_{rj}^{\pm}$. Then the conclusions of Theorem 3.3 remain valid.*

Finally, in light of *Theorem 3.2*, it is clear that in proving *Theorem 3.4* there is no loss of generality in supposing that $0 \in \rho(S)$, and hence that $0 \in \rho(A)$, since this situation can always be achieved, if necessary, by means of a shift in the spectral parameter λ.

ASSUMPTION 3.2. It will be supposed that the hypotheses of Theorem 3.4 are valid and that $0 \in \rho(A)$.

4. SOME TECHNICAL RESULTS

In this section we are going to introduce the compact operator K_p acting in a certain Banach space and investigate the growth of the modified resolvent of K_p along certain rays emanating from the origin in $\mathbf{C}$. The results so obtained will be used in §5 to prove *Theorem 3.4*. Accordingly, let T_p denote the operator of multiplication induced in

$L_p(\Omega)$ by ω (so that T_2 is precisely the operator T introduced in §1) and let $K_p = A_p^{-1}T_p$, so that K_p is a compact operator in $L_p(\Omega)$. Next let $\Omega^\dagger = \Omega\backslash\overline{\Omega}_0$ (see Assumption 2.2 for terminology), $\mathcal{H}_p = L_p(\Omega^\dagger)$, and let us introduce the extension operator $\mathcal{E}_p\colon \mathcal{H}_p \to L_p(\Omega)$ by putting $(\mathcal{E}_p f)(x) = f(x)$ if $x \in \Omega^\dagger$, $(\mathcal{E}_p f)(x) = 0$ if $x \in \Omega\backslash\Omega^\dagger$ for $f \in \mathcal{H}_p$. Let us also introduce the restriction operator $\mathcal{R}_p\colon L_p(\Omega) \to \mathcal{H}_p$ by putting $\mathcal{R}_p f = f|\Omega^\dagger$ for $f \in L_p(\Omega)$. Then in $\mathcal{H}_p$ we introduce the compact operator $\mathcal{K}_p = \mathcal{R}_p K_p \mathcal{E}_p$ and write $\mathcal{K}_{p,\lambda}$ for $(\mathcal{K}_p)_\lambda$ if $\lambda \in \rho_m(\mathcal{K}_p)$ (see Definition 2.1). If we henceforth agree to put $\mathcal{H} = \mathcal{H}_2$, $\mathcal{K} = \mathcal{K}_2$, and $\mathcal{K}_\lambda = \mathcal{K}_{2,\lambda}$ if $\lambda \in \rho_m(\mathcal{K})$, then we know from [19] that

THEOREM 4.1. *λ is an eigenvalue of the problem (1.1–2) if and only if λ is a characteristic value of $\mathcal{K}$. Moreover, if λ is an eigenvalue of (1.1–2), then $\mathcal{R}_2$ maps M_λ onto $\mathcal{G}_\lambda(\mathcal{K},\mathcal{H})$ injectively.*

Fundamental to our work is the following

THEOREM 4.2. *If $\theta \in \mathbb{R}$ and $\theta \neq k\pi$ for $k \in \mathbb{Z}$, then the ray* $\arg\lambda = \theta$ *is a ray of minimal growth of $K_{p,\lambda}$.*

In order to prove the theorem we require the following definitions. Accordingly, let $x^0 \in \Gamma_{rj}^{\pm}$. Then by hypothesis there is an open set $U \subset R^n$ and a real valued function ϕ of $n-1$ variables such that the following conditions hold: (1) there is a Cartesian coordinate system $(y_1,\dots,y_n) = (y',y_n)$ in $\mathbb{R}^n$ about x^0, where the y_n–axis is directed along the inward normal to $\Gamma_{rj}^{\pm}$ at x^0 (i.e., pointing into $\Omega_r^{\pm}$) and the y_1–$,\dots,y_{n-1}$–axes lie in the tanget plane to $\Gamma_{rj}^{\pm}$ at x^0 such that $U = \{(y',y_n)|y' \in U',\ |y_n - \phi(y')| < \rho_1\}$, where U' is the open ball $|y'| < \rho_0$ and ρ_0, ρ_1 are positive constants, (2) $\phi \in C^{1,1}(\overline{U'})$ (see [1, pp.9,10] for notation), and (3) $U \cap \Omega_r^{\pm} = \{(y',y_n) \in U|y_n > \phi(y')\}$, $U \cap \Gamma_{rj}^{\pm} = \{(y',y_n) \in U|y_n = \phi(y')\}$, and $U \cap (\mathbb{R}^n\backslash\overline{\Omega_r^{\pm}}) = \{(y',y_n) \in U|y_n < \phi(y')\}$. We call U a neighbourhood and (y',y_n) a system of coordinates connected with the point x^0. Moreover, if we let $V = \{\eta|\eta = (\eta_1,\dots,\eta_n) = (\eta',\eta_n) \in \mathbb{R}^n,\ |\eta'| < \rho_0,\ |\eta_n| < \rho_1\}$, then U can be mapped onto V by means of the mapping $\eta_j = y_j$ for $j = 1,\dots,(n-1)$, $\eta_n = y_n - \phi(y')$, and we refer to (η',η_n) as local coordinates of $\Gamma_{rj}^{\pm}$ at the point x^0. Note that if we let e_r denote the unit vector in $\mathbb{R}^n$ parallel to and pointing in the direction of the positive y_r–axis, then in terms of the local coordinates at x^0 we have $D_j = \sum_{r=1}^n e_{rj}(\mathcal{D}_r - (\mathcal{D}_r\psi)\mathcal{D}_n)$, where $\mathcal{D}_r = \partial/\partial\eta_r$, $\psi(\eta) = \psi(\eta') = \phi(y')$, and e_{rj} is the j–th component of e_r with respect to the standard basis of $\mathbb{R}^n$. Hence if we pass to local coordinates at x^0 and restrict ourselves to the set $\{\eta \in V|\eta_n > 0\}$, then in this set (1.1) becomes

$$\mathcal{L}(\eta,\mathcal{D})v - \lambda m(\eta)v = 0, \tag{4.1}$$

where $\mathcal{D} = (\mathcal{D}_1, \ldots, \mathcal{D}_n)$, $\mathcal{L}(\eta, \mathcal{D}) = \sum_{|\alpha|\le 2} a'_\alpha(\eta)\mathcal{D}^\alpha$, $m(\eta) = \omega(x(\eta))$, and $v(\eta) = u(x(\eta))$, while if $\Gamma^\pm_{rj}$ is also a component of Γ, then, still restricting ourselves to the set V, (1.2) becomes

$$\mathcal{B}(\eta, \mathcal{D})v = 0 \text{ on } \eta_n = 0, \tag{4.2}$$

where $\mathcal{B}(\eta, \mathcal{D})$ denotes $B(x, D)$ in terms of these local coordinates and which can be expressed in the form $\mathcal{B}(\eta, \mathcal{D}) = \sum_{|\alpha|\le 1} b'_\alpha(\eta)\mathcal{D}^\alpha$.

Next for θ as in *Theorem 4.2* let $\Xi(\theta)$ denote the ray in the complex plane emanating from the origin and making an angle θ with the positive real axis, and for $\lambda \in \Xi(\theta)$ and $s \ge 0$ put $|||u|||_{s,G,p} = \|u\|_{s,G,p} + |q|^s\|u\|_{0,G,p}$ for every open set $G \subset \mathbb{R}^n$ and vector $u \in W^s_p(G)$, where $q = \lambda^{1/2}$ with arg $q = \theta/2$ if $\lambda \ne 0$. Suppose, furthermore, that the component $\Gamma^\pm_{rj}$ of $\partial\Omega^\pm_r$ is also a component of Γ, that $x^0 \in \Gamma^\pm_{rj}$, and that U is a neighbourhood connected with the point x^0. Then for $u \in W^2_p(\Omega)$, with $\mathit{supp}\, u \subset U$, where supp = support, we put $|||Bu|||'_p = \|\mathcal{B}v\|_{1/p',\mathbb{R}^{n-1},p} + |q|^{1/p'}\|\mathcal{B}v\|_{0,\mathbb{R}^{n-1},p}$, where all terms are defined above (see (4.2)) and $\mathcal{B}v$ is to be interpreted in the sense of trace on the hyperplane $\eta_n = 0$. Lastly, for $a \ge 0$ let us also put $\Xi(\theta, a) = \{\lambda \in \Xi(\theta) | |\lambda| \ge a\}$.

Let us also note that definitions and results similar to those above hold for the case where x^0 is a point of Γ which does not belong to any of the $\Gamma^\pm_{rj}$ (this situation could arise if a component of $\partial\Omega_0$ is also a component of Γ). In this case, with U denoting a neighbourhood connected with the point x^0, we put $\|Bu\|' = \|\mathcal{B}v\|_{1/p',\mathbb{R}^{n-1},p}$ for $u \in W^2_p(\Omega)$ with $\mathit{supp}\, u \subset U$.

Proof of Theorem 4.2. If $p = 2$, then the theorem has been proved in [19], and hence let us firstly prove the theorem for the case $p > 2$. To begin with we are going to establish certain a priori estimates which we require for the proof. To this end let us observe from the proof of *Lemma 4.1* of [20] that for each point $x^0 \in \Omega^\pm_r$ there exists a neighbourhood $X \subset\subset \Omega^\pm_r$ of this point and positive numbers $c^\dagger_0$, c, such that for $\lambda \in \Xi(\theta, c_1)$, $|||u|||_{2,\Omega^\dagger,p} \le c^\dagger_0\|(L - \lambda\omega)u\|_{0,\Omega^\dagger,p}$ for every $u \in W^2_p(\Omega)$ with $\mathit{supp}\, u \subset X$. Hence it follows by interpolation [34, *Theorem 4.3.1/2*, p.317, *Theorem 1.3.3*, p.25] that for such u we also have

$$|||u|||_{2-1/p',\Omega^\dagger,p} \le c_0|\lambda|^{-1/2p'}\|(L - \lambda\omega)u\|_{0,\Omega^\dagger,p} \quad \text{for } \lambda \in \Xi(\theta, c_1), \tag{4.3}$$

where the constant c_0 does not depend upon X, λ, nor u.

Fixing our attention next upon a $\Gamma^\pm_{rj}$, let us suppose firstly that $\Gamma^\pm_{rj}$ is a component of Γ. Then it follows from [20] and interpolation that for each point $x^0 \in \Gamma^\pm_{rj}$ there exists a neighbourhood X of this point, with $X \cap \Omega \subset \Omega^{\dagger}$, and positive numbers c_0, c_1 such that for $\lambda \in \Xi(\theta, c_1)$,

$$|||u|||_{2-1/p',\Omega^{\dagger},p} \leq c_0|\lambda|^{-1/2p'}\Big[\|(L-\lambda\omega)u\|_{0,\Omega^{\dagger},p} + |||Bu|||'_p\Big]$$

for every $u \in W_p^2(\Omega)$ with $supp\, u \subset X$. On the other hand, if $\Gamma_{rj}^{\pm} \subset \Omega$ and $\Gamma_{rj}^{\pm}$ coincides with a $\Gamma_{sk}^{\pm}$, then for this case it follows from [20] and interpolation that for each point $x^0 \in \Gamma_{rj}^{\pm}$ there exists a neighbourhood $X \subset\subset \Omega^{\dagger}$ of this point and positive numbers c_0, c_1 such that for $\lambda \in \Xi(\theta, c_1)$, (4.3) is valid for every $u \in W_p^2(\Omega)$ with $supp\, u \subset X$. Lastly, turning to the case where $\Gamma_{rj}^{\pm} \subset \Omega$ and $\Gamma_{rj}^{\pm}$ coincides with a component of $\partial\Omega_0$, let us recall from above that if, for $x^0 \in \Gamma_{rj}^{\pm}$ and $u \in W_p^2(\Omega)$, we pass to local coordinates at x^0, then the expression Lu goes over into $\mathcal{L}v$ (see (4.1)). Thus for this case it follows from [21] that for each point $x^0 \in \Gamma_{rj}^{\pm}$ there exists a neighbourhood $X \subset\subset \Omega$, with $X \backslash \Gamma_{rj}^{\pm} \subset \Omega^{+} \cup \Omega_0$, and positive numbers $c_0^{\dagger}$, c_1 such that for $\lambda \in \Xi(\theta, c_1)$,

$$|||u|||_{2,\Omega^{\dagger},p} + \|u\|_{2,\Omega_0,p} + |\lambda|^{1/2p'}\|u\|_{2-1/p',\Omega_0,p} \leq c_0^{\dagger}\Big[\|(L-\lambda\omega)u\|_{0,\Omega^{\dagger},p} + \|Lu\|_{0,\Omega_0,p} + |\lambda|^{1/2p'}\Big(\int_{-\infty}^{0} \|\mathcal{L}v\|^p_{-1/p',\mathbb{R}^{n-1},p}\, d\eta_n\Big)^{1/p}\Big]$$

for every $u \in W_p^2(\Omega)$ with $supp\, u \subset X$. Hence it follows from [34, Theorem 2.3.2, p.172] and interpolation that for such u we also have

$$|||u|||_{2-1/p',\Omega^{\dagger},p} + \|u\|_{2-1/p',\Omega_0,p} \leq c_0\Big[|\lambda|^{-1/2p'}\|(L-\lambda\omega)u\|_{0,\Omega^{\dagger},p} + \|Lu\|_{0,\Omega_0,p}\Big]$$

for $\lambda \in \Xi(\theta, c_1)$, where the constant c_0 does not depend upon X, λ, nor u.

Lastly, it follows immediately from [5] and the Poincaré inequality that: (i) for each point $x^0 \in \Omega_0$ there is a neighbourhood $X \subset\subset \Omega_0$ of this point and a positive number c_0 such that $\|u\|_{2,\Omega_0,p} \leq c_0\|Lu\|_{0,\Omega_0,p}$ for every $u \in W_p^2(\Omega)$ with $supp\, u \subset X$, (ii) if Γ_0 is a component of $\partial\Omega_0$ which is also a component of Γ, then for each point $x^0 \in \Gamma_0$ there exists a neighbourhood X of this point, with $X \cap \Omega \subset \Omega_0$, and a positive number c_0 such that $\|u\|_{2,\Omega_0,p} \leq c_0[\|Lu\|_{0,\Omega_0,p} + \|Bu\|']$ for every $u \in W_p^2(\Omega)$ with $supp\, u \subset X$.

Let $\mathcal{V} = \Big\{u | u \in D(A_p),\ \|\omega^{-1}Lu\|_{0,\Omega^{\dagger},p} < \infty,\ \|Lu\|_{0,\Omega_0,p} = 0\Big\}$. Then by considering a suitable covering of $\overline{\Omega}$ by means of a finite number of the open sets X described above and a partition of unity subordinate to this open covering, it follows from the above results and a standard argument (see the proof of *Theorem 4.1* of [20]) that there exist positive constants k_0, k_1 such that if $\lambda \in \Xi(\theta, k_1)$, then

$$|||u|||_{2-1/p',\Omega^{\dagger},p} + \|u\|_{2-1/p',\Omega_0,p} \leq k_0\Big[|\lambda|^{-1/2p'}\|T_p\mathcal{E}_pf\|_{0,\Omega^{\dagger},p} + \|u\|_{0,\Omega_0,p}\Big] \tag{4.4}$$

for every pair $u \in \mathcal{V}$, $f \in \mathcal{H}_p$ for which $(A_p - \lambda T_p)u = T_p \mathcal{E}_p f$.

Fixing our attention upon (4.4), let us now show that there exists the constant $k_1^\dagger \geq k_1$ such that if $\lambda \in \Xi(\theta, k_1^\dagger)$, then

$$|||u|||_{2-1/p',\Omega^\dagger,p} \leq 2k_0|\lambda|^{-1/2p'} \|T_p\mathcal{E}_pf\|_{0,\Omega^\dagger,p} \tag{4.5}$$

for every pair u, f as defined above. Indeed if this is not the case, then it follows from (4.4) that there exists a $u \in W_p^{2-1/p'}(\Omega)$ and sequences $\{u_i\}_1^\infty$ in $\mathcal{V}$ and $\{\lambda_i\}_1^\infty$ in $\Xi(\theta)$, where $\|u\|_{0,\Omega,p} = \|u_i\|_{0,\Omega,p} = 1$ and $|\lambda_i| \to \infty$ as $i \to \infty$, such that for each i, $u_i \to u$ weakly in $W_p^{2-1/p'}(\Omega)$ and $u_i \to u$ strongly in $W_p^1(\Omega)$ as $i \to \infty$, while $\|u\|_{0,\Omega^\dagger,p} = 0$. But then it follows from arguments similar to those used in [30, pp.118–119] that $\int_{\Omega_0} u\overline{L^* v}dx = 0$ for every $v \in C^2(\overline{\Omega}_0)$ satisfying the boundary conditions: $Cv = 0$ (resp. $v = 0$) on each component of $\partial\Omega_0$ which is contained in Γ (resp. Ω). On the other hand it is clear from [20, §2] that results analogous to those given in §2 for the boundary value problem (2.1), (1.2) also hold for the boundary value problem: $Lu = f$ in Ω_0, $Bu = 0$ (resp. $u = 0$) on each component of $\partial\Omega_0$ which is contained in Γ (resp. Ω), and hence, in light of Assumption 2.3, we arrive at the contradiction that $u = 0$.

Suppose next that $\lambda \in \Xi(\theta, k_1^\dagger)$, $f \in \mathcal{H}_p$, and $(I - \lambda K_p)f = 0$. Then a simple argument shows that there exists a $u \in \mathcal{V}$ such that $\mathcal{R}_p u = f$ and $(A_p - \lambda T_p)u = 0$, and so we see from (4.5) that $f = 0$. Thus we conclude that $\Xi(\theta, k_1^\dagger) \subset \rho_m(K_p)$. Moreover, if $\lambda \in \Xi(\theta, k_1^\dagger)$, $f \in \mathcal{H}_p$, and $K_{p,\lambda} f = u$, then it is not difficult to verify that there exists a $v \in \mathcal{V}$ such that $\mathcal{R}_p v = u$ and $(A_p - \lambda T_p)v = T_p\mathcal{E}_p f$, and hence it follows from (4.5) that $\|K_{p,\lambda} f\|_{0,\Omega^\dagger,p} \leq 2k_0\|w\|_{L_\infty(\Omega)}|\lambda|^{-1}\|f\|_{0,\Omega^\dagger,p}$. Since this last inequality is precisely the result we wanted, the proof of the theorem is complete for the case $p > 2$.

Finally the truth of the theorem for the case $p < 2$ is an immediate consequence of the facts that $\lambda \in \rho_m(K_p)$ if and only if $\bar{\lambda} \in \rho_m(K_{p'}^\dagger)$, where $K_{p'}^\dagger = \mathcal{R}_{p'} A_p^{*-1} T_{p'} \mathcal{E}_{p'}$, and if $K_{p,\lambda}^*$ denotes the Banach space adjoint of $K_{p,\lambda}$, then $K_{p,\lambda}^* = \mathcal{T}_{p'} K_{p',\bar{\lambda}}^\dagger \mathcal{T}_{p'}^{-1}$, where $\mathcal{T}_{p'}$ denotes the operator of multiplication induced in $\mathcal{H}_{p'}$ by ω. ∎

It is a simple matter to deduce from the results of §2 that if $1 < q < s < \infty$, then $K_q|\mathcal{H}_s = K_s$, while if $\lambda \in \rho_m(K_q)$, then $\lambda \in \rho_m(K_s)$ and $K_{q,\lambda}|\mathcal{H}_s = K_{s,\lambda}$. Furthermore, by appealing to Theorem 4.2 and the interpolation theorem [1, p.79], we can also show that

THEOREM 4.3. *Suppose that $p > 2$ and $\lambda \in \Xi(\theta, a)$, where θ satisfies the hypothesis of Theorem 4.2 and a is chosen large enough so that $\Xi(\theta, a) \subset \rho_m(K)$. Then K_λ induced a bounded linear operator from $\mathcal{H}_p$ to $W_p^2(\Omega^\dagger)$, and moreover, $\|K_\lambda u\|_{j,\Omega^\dagger,p} \leq c|\lambda|^{-1+j/2}\|u\|_{0,\Omega^\dagger,p}$, $0 \leq j \leq 2$, for every $u \in \mathcal{H}_p$ when $|\lambda|$ is sufficiently large, where c denotes a positive constant.*

Next for $x^0 \in \Omega_r^{\pm}$, let $L_0^0(D) = \sum_{|\alpha|=2} a_\alpha(x^0)D^\alpha$, $\omega_0 = \omega(x^0)$, and suppose that θ satisfies the hypothesis of *Theorem 4.2.* If $\mathcal{A}_p$ denotes the realization of $\omega_0^{-1}L_0^0(D)$ as an operator in $L_p(\mathbb{R}^n)$, where $D(\mathcal{A}_p) = W_p^2(\mathbb{R}^n)$ and $R_{p,\lambda}$ denotes the resolvent of $\mathcal{A}_p$ if $\rho(\mathcal{A}_p) \neq \emptyset$, then we have

THEOREM 4.4. *For any fixed positive number a, $\Xi(\theta, a) \subset \rho(\mathcal{A}_p)$, and for $\lambda \in \Xi(\theta, a)$ we have $R_{p,\lambda} \in \mathcal{L}(L_p(\mathbb{R}^n), W_p^2(\mathbb{R}^n))$ and $\|R_{p,\lambda}u\|_{j,\mathbb{R}^n,p} \leq c|\lambda|^{-1+j/2}\|u\|_{0,\mathbb{R}^n,p}$, $0 \leq j \leq 2$, for $u \in L_p(\mathbb{R}^n)$, where c denotes a positive constant.*

Proof. The theorem is well known when $p = 2$ (see [4]), and the proof, for arbitrary p, can be established by appealing to Michlin's multiplier theorem (see the proof of *Lemma 4.1* of [20]). ■

Finally, for $x^0 \in \Omega_r^{\pm}$, let δ satisfy $0 < \delta < dist\{x^0, \partial\Omega_r^{\pm}\}$, let ϕ, ψ be functions in $C^\infty(\mathbb{R}^n)$ whose supports are contained in the ball $|x| < 1$, and put $\phi^\delta(x) = \phi(\delta^{-1}(x - x^0))$, $\psi^\delta(x) = \psi(\delta^{-1}(x - x^0))$. Then with all terms as defined in *Theorem 4.4* and for $\lambda \in \Xi(\theta, a)$, let us introduce in $\mathcal{H}_p$ the operator $R_{p,\lambda,\delta} = \phi^\delta R_{p,\lambda}\psi^\delta$ (here we use ϕ^δ and ψ^δ to denote multiplication operators in $\mathbb{R}^n$ and $\Omega^\dagger$, respectively, and ϕ^δ (resp. ψ^δ) is to be interpreted as $\Gamma_{\Omega^\dagger} \circ \phi^\delta$ (resp. $i_{\Omega^\dagger} \circ \psi^\delta$), where $r_{\Omega^\dagger}$ denotes the natural restriction: $\mathbb{R}^n \to \Omega^\dagger$ and $i_{\Omega^\dagger}$ denotes the natural imbedding: $\Omega^\dagger \to \mathbb{R}^n$) and let $R_{\lambda,\delta} = R_{2,\lambda,\delta}$.

THEOREM 4.5. *If $p > 2$ and $\lambda \in \Xi(\theta, a)$, then $R_{\lambda,\delta}$ induces a bounded linear operator from $\mathcal{H}_p$ to $W_p^2(\Omega^\dagger)$ and $\|R_{\lambda,\delta}u\|_{j,\Omega^\dagger,p} \leq c|\lambda|^{-1+j/2}\|u\|_{0,\Omega^\dagger,p}$, $0 \leq j \leq 2$, for $u \in \mathcal{H}_p$, where c denotes a positive constant.*

Proof. The theorem is an immediate consequence of *Theorem 4.4* if we bear in mind that $R_{p,\lambda}f = R_{2,\lambda}f$ for $f \in L_p(\mathbb{R}^n) \cap L_2(\mathbb{R}^n)$. ■

5. PROOF OF THEOREM 3.4.

If $n > 4$, then let s denote the smallest integer exceeding $n/4$, let $p_1 = q_1 = 2$ and let $\{p_j\}_2^{s+1}$, $\{q_j\}_2^s$ be real numbers satisfying $p_1 < \cdots p_s < n/2 < p_{s+1}$, $p_{j+1} < np_j/(n - 2p_j)$ for $j = 1, \ldots, s, q_1 < \cdots < q_{s-1} < n/2 < q_s$, $q_{j+1} < nq_j/(n - 2q_j)$ for $j = 1, \ldots, (s-1)$. If $n \leq 4$, then let $s = 2$, put $p_1 = q_1 = 2$, and choose p_2, p_3, and q_2 in $\mathbb{R}$ so that $p_1 < p_2 < p_3$, $q_1 < q_2$. Hence if we let $m = 2s + 1$, then $2m > n$. Moreover, since it follows from [2] and [24, p.27] that $\mathcal{K}$ is a compact operator in $\mathcal{H}$ of class C_ν for any ν satisfying $\nu > n/2$, it follows from [24, p.92] that $\mathcal{K}^m$ is of trace class, and hence we conclude from *Theorem 4.1* that $\sum_{j=1}^\infty |\lambda_j^m|^{-1} < \infty$.

Suppose next that $\theta \in \mathbb{R}$, $\theta \neq k\pi$ for $k \in \mathbb{Z}$, and that $\lambda \in \Xi(\theta, 1)$. Let $\mu(j) = \lambda^{1/m}\exp\{2\pi i(j-1)/m\}$ for $j = 1, \ldots, m$, where the principal value of the root is taken. Then for $|\lambda|$ sufficiently large we have

$$\mathcal{K}_\lambda^m = \left(\prod_{j=1}^{s+1} \mathcal{K}_{\mu(j)}\right) \mathcal{T}^{-1} \left(\prod_{j=s+2}^{m} \mathcal{K}_{\bar{\mu}(j)}^{\dagger}\right)^{*} \mathcal{T},$$

where S^* denotes the adjoint of S in $\mathcal{H}$, $\mathcal{T} = \mathcal{T}_2$, $\mathcal{K}^\dagger = \mathcal{R}_2 A^{*-1} \mathcal{T}\mathcal{E}_2$, and $\bar{\mu}(j)$ denotes the complex conjugate of $\mu(j)$. If $S_\lambda^{\#} = \left(\prod_{j=1}^{s+1} \mathcal{K}_{\mu(j)}\right) \mathcal{T}^{-1}$, then we see from *Theorem 4.3* and the Sobolev imbedding theorem that $S_\lambda^{\#} \in \mathcal{L}(\mathcal{H}, L_\infty(\Omega^\dagger))$ and for $u \in \mathcal{H}$ and $x \in \Omega^\dagger$ we have

$$|(S_\lambda^{\#}u)(x)| \leq \prod_{j=1}^{s+1} \|\mathcal{K}_{\mu(s+2-j)}\|_{(\mathcal{H}_{p_j}, \mathcal{H}_{p_{j+1}})} \|\mathcal{T}^{-1}\|_{(\mathcal{H},\mathcal{H})} \|u\|_{0,\Omega^\dagger,2},$$

where $\mathcal{H}_{p_{s+2}} = L_\infty(\Omega^\dagger)$ and $\| \ \|_{(\mathcal{H}_p, \mathcal{H}_q)}$ denotes the norm in $\mathcal{L}(\mathcal{H}_p, \mathcal{H}_q)$. It now follows from *Theorem 4.3* and [32, *Corollary 2*, p.68] that for $\lambda \in \Xi(\theta, a)$, with a sufficiently large, $|(S_\lambda^{\#}u)(x)| \leq c|\lambda|^{[n-4(s+1)]/4m}\|u\|_{0,\Omega^\dagger,2}$, where the constant c does not depend upon x, λ, nor u. Thus we conclude from an argument similar to that used in the proof of *Lemma 2.1* of [4] that $S_\lambda^{\#}$ is an integral operator with kernel $G_\lambda^{\#}(x,y)$ satisfying $\left(\int_{\Omega^\dagger} |G_\lambda^{\#}(x,y)|^2 dy\right)^{1/2} \leq c|\lambda|^{[n-4(s+1)]/4m}$ for $x \in \Omega^\dagger$, and that the mapping: $\Omega^\dagger \to \mathcal{H}$ given by $x \to G_\lambda^{\#}(x,.)$ is continuous. On the other hand, if we let $(S_\lambda^{\#})^*$ denote the adjoint of $S_\lambda^{\#}$ when considered as a mapping from $\mathcal{H}$ to $L_\infty(\Omega^\dagger)$, then $(S_\lambda^{\#})^*$ induces a bounded linear transformation from $L_1(\Omega^\dagger)$ to $\mathcal{H}$, and hence we see from [14, Problem 59, p.519] that $(S_\lambda^{\#})^*|L_1(\Omega^\dagger)$ is an integral operator with a Hilbert–Schmidt kernel. It follows immediately that $S_\lambda^{\#}$ is also generated by a Hilbert–Schmidt kernel, say $H_\lambda^{\#}(x,y)$, and that for almost every $x \in \Omega^\dagger$, $H_\lambda^{\#}(x,y) = G_\lambda^{\#}(x,y)$ for almost every $y \in \Omega^\dagger$. Similarly we can show that for $\lambda \in \Xi(\theta, a)$, $S_\lambda^\dagger = \prod_{j=s+2}^{m} \mathcal{K}_{\bar{\mu}(j)}^\dagger$ is an integral operator with kernel $G_\lambda^\dagger(x,y)$ satisfying $\left(\int_{\Omega^\dagger} |G_\lambda^\dagger(x,y)|^2 dy\right)^{1/2} \leq c|\lambda|^{(n-4s)/4m}$ for $x \in \Omega^\dagger$, where the constant c does not depend upon x nor λ, that the mapping: $\Omega^\dagger \to \mathcal{H}$ given by $x \to G_\lambda^\dagger(x,.)$ is continuous, and that $S_\lambda^\dagger$ is also generated by a Hilbert–Schmidt kernel $H_\lambda^\dagger(x,y)$ such that for almost every $x \in \Omega^\dagger$, $H_\lambda^\dagger(x,y) = G_\lambda^\dagger(x,y)$ for almost every $y \in \Omega^\dagger$. It follows immediately from these results and those of [24, p.27] that for $\lambda \in \Xi(\theta, a)$, $S_\lambda^{\#}(S_\lambda^\dagger)^*$ is an integral operator of trace class with kernel $G_\lambda(x,y)$ which is continuous in $\Omega^\dagger \times \Omega^\dagger$ and

$$(5.1) \qquad |G_\lambda(x,y)| \leq c|\lambda|^{-1+n/2m} \text{ for } (x,y) \in \Omega^\dagger \times \Omega^\dagger,$$

where the constant c does not depend upon λ. Hence we conclude that for $\lambda \in \Xi(\theta, a)$, $\mathcal{K}_\lambda^m$ is an integral operator of trace class with kernel $G_\lambda(x, y)\omega(y)$.

Let $x^0 \in \Omega_r^\pm$. Then recalling the definitions of $\omega_0^{-1}L_0^0(D)$ and $R_{p,\lambda}$ given in §4 and writing R_λ for $R_{2,\lambda}$, it is clear that for $\lambda \in \Xi(\theta, a)$

$$F_\lambda = \prod_{j=1}^{m} R_{\mu(j)} = \left(\prod_{j=1}^{s+1} R_{\mu(j)}\right)\left(\prod_{j=s+2}^{m} R_{\bar{\mu}(j)}^\dagger\right)^*$$

is an integral operator in $L_2(\mathbb{R}^n)$ with kernel

$$Q_\lambda(x, y) = (2\pi)^{-n} \int_{\mathbb{R}^n} \exp\{i(x-y)\cdot\xi\}\left[(\omega_0^{-1}L_0^0(i\xi))^m - \lambda\right]^{-1} d\xi,$$

where $\cdot$ denotes inner product, S^* denotes the adjoint of S in $L_2(\mathbb{R}^n)$, and $R_\mu^\dagger$ denotes the resolvent of the operator induced in $L_2(\mathbb{R}^n)$ by the formal adjoint of $\omega_0^{-1}L_0^0(D)$. Let $\phi(x)$, $\psi(x)$, $\chi(x)$, and $\{\phi_j(x)\}_1^{s+1}$ be functions in $C^\infty(\mathbb{R}^n)$ such that for $0 \le k \le s+3$, $0 \le \phi_k(x) \le 1$, $\phi_k(x) = 1$ in a neighbourhood of $x = 0$, and supp ϕ_k is contained in the ball $|x| < 1$, and $\phi_k(x)\phi_{k+1}(x) = \phi_k(x)$ for $0 \le k \le s+2$, where we have written $\phi_0(x)$ for $\phi(x)$, $\phi_{s+2}(x)$ for $\psi(x)$, and $\phi_{s+3}(x)$ for $\chi(x)$. For $0 < \delta < dist\ \{x^0, \partial\Omega_r^\pm\}$, let $\phi^\delta(x) = \phi(\delta^{-1}(x-x^0))$, $\psi^\delta(x) = \psi(\delta^{-1}(x-x^0))$, $\chi^\delta(x) = \chi(\delta^{-1}(x-x^0))$, $\phi_j^\delta(x) = \phi_j(\delta^{-1}(x-x^0))$ for $j = 1, \ldots, (s+1)$, and put:

$$\mathcal{K}_{\mu(j),\delta} = \phi_{j-1}^\delta \mathcal{K}_{\mu(j)}\phi_j^\delta,\ R_{\mu(j),\delta} = \phi_{j-1}^\delta R_{\mu(j)}\phi_j^\delta \ \text{ for } j = 1, \ldots, (s+1),$$

$$\mathcal{K}_{\bar{\mu}(j),\delta}^\dagger = \phi_{j-s-2}^\delta \mathcal{K}_{\bar{\mu}(j)}^\dagger \phi_{j-s-1}^\delta,\ R_{\bar{\mu}(j),\delta}^\dagger = \phi_{j-s-2}^\delta R_{\bar{\mu}(j)}^\dagger \phi_{j-s-1}^\delta \ \text{ for } j = (s+2), \ldots, m,$$

where $\phi_0^\delta = \phi^\delta$ and where we recall from §4 that ϕ_j^δ is used in the sense of a multiplication operator. Then it is not difficult to verify that

$$P_\lambda^\delta = \phi^\delta \mathcal{K}_\lambda^m \phi^\delta - \left(\prod_{j=1}^{s+1} \mathcal{K}_{\mu(j),\delta}\right)\mathcal{T}^{-1}\left(\prod_{j=s+2}^{m} \mathcal{K}_{\bar{\mu}(j),\delta}^\dagger\right)^*\mathcal{T}$$

is a finite sum of operators of the form

$$-\left(\prod_{j=1}^{r} \mathcal{K}_{\mu(j),\delta}\right)\left(\phi_r^\delta[\mathcal{K}_{\mu(r+1)}, \phi_{r+1}^\delta]\eta_{r+1}\right)\left(\prod_{j=r+1}^{s+1} \mathcal{K}_{\mu(j)}\right)\mathcal{T}^{-1}\left(\prod_{j=s+2}^{m} \mathcal{K}_{\bar{\mu}(j)}^\dagger\right)^*\phi^\delta\mathcal{T}$$

(5.2) and

$$-\left(\prod_{j=1}^{s+1} \mathcal{K}_{\mu(j),\delta}\right)\mathcal{T}^{-1}\left(\prod_{j=s+2}^{s+r-1} \mathcal{K}_{\bar{\mu}(j),\delta}^\dagger\right)\left(\phi_{r-2}^\delta[\mathcal{K}_{\bar{\mu}(s+r)}^\dagger, \phi_{r-1}^\delta]\eta_{r-1}\right)\left(\prod_{j=s+r+1}^{m} \mathcal{K}_{\bar{\mu}(j)}^\dagger\right)^*\mathcal{T},$$

where $[S, \phi_j^\delta]$ denotes the commutator $S\phi_j^\delta - \phi_j^\delta S$ and either $\eta_j = I$ or $\eta_j = \phi_j^\delta$. Observing that if we let $A^\dagger$ denote the restriction of the differential operator A to $\Omega^\dagger$, then $(\mathcal{T}^{-1}A^\dagger - \mu(j)I)\mathcal{K}_{\mu(j)} = \mathcal{K}_{\mu(j)}(\mathcal{T}^{-1}A^\dagger - \mu(j)I) = I$, it follows that $[\mathcal{K}_{\mu(j)}, \phi_j^\delta] = \mathcal{K}_{\mu(j)}\mathcal{T}^{-1}(\phi_j^\delta A^\dagger - A^\dagger\phi_j^\delta)\mathcal{K}_{\mu(j)}$. Hence we can appeal to *Theorem 4.3* and to the arguments used in the proof of *Lemma 6* of [7], and then argue as we did with $\mathcal{K}_\lambda^m$ above to deduce that the first expression in (5.2) is an integral operator such that in $\Omega^\dagger \times \Omega^\dagger$ the modulus of its kernel is bounded by

$$c|\lambda|^{-1+(2n-1)/4m} \quad \text{for} \quad |\lambda| \geq \max\{a, \delta^{-4m}\}, \tag{5.3}$$

where the constant c does not depend upon λ or δ. Since a similar result holds for the second expression in (5.2), we conclude that P_λ^δ is an integral operator such that in $\Omega^\dagger \times \Omega^\dagger$ the modulus of its kernel satisfies a bound of the form (5.3). Moreover, by appealing to *Theorem 4.5* we can also show that an analogous result holds for the operator

$$\phi^\delta F_\lambda \phi^\delta - \left(\prod_{j=1}^{s+1} R_{\mu(j),\delta}\right)\mathcal{T}^{-1}\left(\prod_{j=s+2}^{m} R_{\bar{\mu}(j),\delta}^\dagger\right)^* \mathcal{T}.$$

On the other hand since

$$\begin{aligned}\mathcal{K}_{\mu(j),\delta} - R_{\mu(j),\delta} &= \phi_{j-1}^\delta \mathcal{K}_{\mu(j)}\psi^\delta(\mathcal{A}_2 - \mathcal{T}^{-1}A^\dagger)\chi^\delta R_{\mu(j)}\phi_j^\delta \\ &\quad + \phi_{j-1}^\delta \mathcal{K}_{\mu(j)}\mathcal{T}^{-1}(\psi^\delta A^\dagger - A^\dagger\psi^\delta)\mathcal{K}_{\mu(j)}(\mathcal{T}^{-1}A^\dagger - \mu(j)I)\chi^\delta R_{\mu(j)}\phi_j^\delta,\end{aligned}$$

with a similar result holding for $\mathcal{K}_{\bar{\mu}(j),\delta}^\dagger - R_{\bar{\mu}(j),\delta}^\dagger$, it follows immediately from Theorems 4.3–5 and the arguments used above for dealing with $\mathcal{K}_\lambda^m$ that $\phi^\delta \mathcal{K}_\lambda^m \phi^\delta - \phi^\delta F_\lambda \phi^\delta$ is an integral operator with kernel $H_\lambda^\delta(x,y) = \phi^\delta(x)(G_\lambda(x,y)\omega(y) - Q_\lambda(x,y))\phi^\delta(y)$ satisfying

$$|H_\lambda^\delta(x,y)| \leq c|\lambda|^{-1+n/2m}[\Phi(\delta) + |\lambda|^{-1/4m}] \quad \text{for} \quad (x,y) \in \Omega^\dagger \times \Omega^\dagger \quad \text{and} \quad |\lambda| \geq \max\{a, \delta^{-4m}\},$$

where the constant c does not depend upon λ nor δ and $\Phi(\delta) \to 0$ as $\delta \to 0$. Since δ is arbitrary, we conclude that

$$G_\lambda(x^0, x^0)\omega(x^0) = |\lambda|^{-1+n/2m}[\rho_\theta(x^0) + o(1)] \quad \text{as} \quad |\lambda| \to \infty,\ \lambda \in \Xi(\theta), \tag{5.4}$$

where

$$\rho_\theta(x^0) = (2\pi)^{-n} \int_{\mathbb{R}^n} \left[\left(\omega(x^0)^{-1}L_0(x^0, i\xi)\right)^m - e^{i\theta}\right]^{-1} d\xi.$$

Recalling from above that K_λ^m is of trace class, it follows from *Theorem 4.1* that $tr\, K_\lambda^m = \sum_{j=1}^\infty (\lambda_j^m - \lambda)^{-1}$ for $\lambda \in \Xi(\theta, a)$, where tr denotes trace. On the other hand, since we know from above and [24, p.27] that K_λ^m is the product of two Hilbert–Schmidt operators, we conclude from [3, *Theorem 12.21*, p.205] that $\int_{\Omega^\dagger} G_\lambda(x,x)\omega(x)dx = \sum_{j=1}^\infty (\lambda_j^m - \lambda)^{-1}$ for $\lambda \in \Xi(\theta, a)$. Hence it follows from (5.1) and (5.4) that

$$(5.5) \qquad \sum_{j=1}^\infty (\lambda_j^m - \lambda)^{-1} = |\lambda|^{-1+n/2m} \left[\int_{\Omega^\dagger} \rho_\theta(x)dx + o(1) \right] \quad \text{as } |\lambda| \to \infty,\ \lambda \in \Xi(\theta).$$

Taking $\theta = \pi/2$ and putting $\lambda = it$, we see from (5.5) that

$$(5.6) \qquad \sum_{j=1}^\infty (\lambda_j^m - it)^{-1} = \left(\int_{\Omega^\dagger} \rho_{\pi/2}(x)dx \right) t^{-1+n/2m} + o(t^{-1+n/2m}) \quad \text{as } t \to \infty.$$

Hence if we let $\lambda_j = \mu_j + i\nu_j$ for $j \geq 1$, then it is not difficult to deduce from (5.6) that

$$(5.7) \qquad \sum_{j=1}^\infty (\mu_j^m - it)^{-1} = \left(\int_{\Omega^\dagger} \rho_{\pi/2}(x)dx \right) t^{-1+n/2m} + o(t^{-1+n/2m}) \quad \text{as } t \to \infty.$$

We can now appeal to a Tauberian theorm of Hardy and Littlewood and argue with (5.7) precisely as in the proof of *Theorem 14.6* of [3, p.250] to establish the assertions of the theorem for $N(\lambda)$ and the $N_\pm(\lambda)$. Moreover, since it follows from (5.6) that

$$(5.8) \qquad \sum_{j=1}^\infty \left(|\lambda_j^m| - it\right)^{-1} = \left(\int_{\Omega^\dagger} \rho_{\pi/2}(x)dx \right) t^{-1+n/2m} + o(t^{-1+n/2m}) \quad \text{as } t \to \infty,$$

we can argue with (5.8) as we argued with (5.7) to establish the assertion of the theorem concerning $N^\#(\lambda)$.

REFERENCES

1. R.A. Adams, *Sobolev spaces*, Academic, New York, 1975.
2. S. Agmon, *On the eigenfunctions and on the eigenvalues of general elliptic boundary value problems*, Comm. Pure Appl. Math. **15** (1962), 119–147.
3. S. Agmon, *Lectures on elliptic boundary value problems*, Van Nostrand, Princeton, N.J., 1965.
4. S. Agmon, *On kernels, eigenvalues, and eigenfunctions of operators related to elliptic problems*, Comm. Pure Appl. Math. **18** (1965), 627–663.
5. S. Agmon, A. Douglis, and L. Nirenberg, *Estimates near the boundary for solutions of elliptic partial differential equations satisfying general boundary conditions. I*, Comm. Pure Appl. Math. **12** (1959), 623–727.

6. W. Allegretto and A.B. Mingarelli, *Boundary problems of the second order with an indefinite weight-function*, J. Reine Angew. Math. **398** (1989), 1–24.
7. R. Beals, *Asymptotic behaviour of the Green's function and spectral function of an elliptic operator*, J. Funct. Anal. **5** (1970), 484–503.
8. R. Beals, *Indefinite Sturm-Liouville problems and half-range completeness*, J. Differential Equations **56** (1985), 391–407.
9. P. Binding and B. Najman, *A variational principle in Krein spaces* (preprint).
10. M.S. Birman and M.Z. Solomjak, *Asymptotic behaviour of the spectrum of differential equations*, J. Soviet Math. **12** (1974), 247–282.
11. M.S. Birman and M.Z. Solomjak, *Asymptotics of the spectrum of variational problems on solutions of elliptic equations*, Siberian Math. J. **20** (1979), 1–15.
12. J. Bognár, *Indefinite inner product spaces*, Springer, Berlin, 1974.
13. B. Ćurgus and B. Najman, *A Krein space approach to elliptic eigenvalue problems with indefinite weights* (preprint).
14. N. Dunford and J.T. Schwartz, *Linear operators*, part I, Wiley, New York, 1988.
15. M. Faierman, *On the eigenvalues of nonselfadjoint problems involving indefinite weights*, Math. Ann. **282** (1988), 369–377.
16. M. Faierman, *Elliptic problems involving an indefinite weight*, Trans. Amer. Math. Soc. **320** (1990), 253–279.
17. M. Faierman, *Non-selfadjoint elliptic problems involving an indefinite weight*, Comm. Partial Differential Equations **15** (1990), 939–982.
18. M. Faierman, *Eigenvalue asymptotics for a non-selfadjoint elliptic problem involving an indefinite weight*, Rocky Mountain J. Math. (to appear).
19. M. Faierman, *On an oblique derivative problem involving an indefinite weight*, Arch. Math. (Brno) (to appear).
20. M. Faierman, *On the eigenvalue asymptotics for a non-selfadjoint elliptic problem involving an indefinite weight* (submitted).
21. M. Faierman, *On an a priori estimate for solutions of an elliptic equation* (submitted).
22. J. Fleckinger and M.L. Lapidus, *Eigenvalues of elliptic boundary value problems with an indefinite weight function*, Trans. Amer. Math. Soc. **295** (1986), 305–324.
23. J. Fleckinger and M.L. Lapidus, *Remainder estimates for the asymptotics of elliptic eigenvalue problems with indefinite weights*, Arch. Rational Mech. Anal. **98** (1987), 329–356.
24. I.C. Gohberg and M.G. Krein, *Introduction to the theory of linear nonselfadjoint operators*, Amer. Math. Soc., Providence, R.I., 1969.
25. P. Grisvard, *Elliptic problems in nonsmooth domains*, Pitman, London, 1985.
26. P. Hess, *On the relative completeness of the generalized eigenvectors of elliptic eigenvalue problems with indefinite weight functions*, Math. Ann. **270** (1985), 467–475.
27. P. Hess, *On the asymptotic distribution of eigenvalues of some non-selfadjoint problems*, Bull. London Math. Soc. **18** (1986), 181–184.
28. P. Hess, *On the spectrum of elliptic operators with respect to indefinite weights*, Linear Algebra Appl. **84** (1986), 99–109.

29. T. Kato, *Perturbation theory for linear operators*, 2nd edn., Springer, Berlin, 1976.
30. J.L. Lions and E. Magenes, *Non-homogeneous boundary value problems and applications*, Vol. I, Springer, Berlin, 1972.
31. A.S. Markus, *Introduction to the spectral theory of polynomial operator pencils*, Amer. Math. Soc., Providence, R.I., 1988.
32. V.G. Maz'ja, *Sobolev spaces*, Springer, Berlin, 1985.
33. M. Schechter, *General boundary value problems for elliptic partial differential equations*, Comm. Pure Appl. Math. **12** (1959), 457–482.
34. H. Triebel, *Interpolation theory, function spaces, differential operators*, North-Holland, Amsterdam, 1978.

Department of Mathematics
University of the Witwatersrand
Johannesburg, WITS 2050
South Africa

MSC: Primary 35P10, 35P20; Secondary 47F05

Operator Theory:
Advances and Applications, Vol. 80

NONLINEARITY IN H^∞-CONTROL THEORY, CAUSALITY IN THE COMMUTANT LIFTING THEOREM, AND EXTENSION OF INTERTWINING OPERATORS

Ciprian Foias, Caixing Gu and Allen Tannenbaum

The problems studied in this note have been motivated by our work in generalizing linear H^∞ control theory to nonlinear systems. These ideas have led to a design procedure applicable to analytic nonlinear plants. Our technique is a generalization of the linear H^∞ theory. In contrast to previous work on this topic ([9], [10]), we now are able to explicitly incorporate a causality constraint into the theory. In fact, we show that it is possible to reduce a causal optimal design problem (for nonlinear systems) to a classical interpolation problem solvable by the commutant lifting theorem [8]. Here we present the complete operator theoretical background of our research together with a short control theoretical motivation.

INTRODUCTION

In this paper, we present the operator theoretical background of all our work on an implementable nonlinear extension of the powerful linear H^∞ design methodology [8]. In what follows, we will just consider discrete–time systems, even though the techniques explained below carry over to the continuous–time setting as well.

* This work was supported in part by grants from the Research Fund of Indiana University, the National Science Foundation DMS-8811084 and ECS-9122106, by the Air Force Office of Scientific Research F49620-94-1-0098DEF, and by the Army Research Office DAAL03-91-G-0019 and DAAH04-93-G-0332

Our approach is based on previous work ([9], [10]) in which we considered systems described by analytic input/output operators. A key idea here involved the expression of each n–linear term of a suitable Taylor expansion of the given operator as an equivalent linear operator acting on a certain associated tensor space which allowed us to iteratively apply the classical commutant lifting theorem in designing a compensator.

More precisely, in such an approach one is reduced to applying the classical (linear) commutant lifting theorem to an H^2–space defined on some D^n (where D denotes the unit disc). Now when one applies the classical result to D^n ($n \geq 2$), even though time–invariance is preserved (that is, commutation with the appropriate shift), causality may be lost. Indeed, for systems described by analytic functions on the disc D (these correspond to stable, discrete–time, 1–D systems), time–invariance (that is, commutation with the basic unilateral shifts) implies causality. For analytic functions on the polydisc ($n > 1$), this is not necessarily the case. For dynamical system control design and for any physical application, this is of course a major drawback for such an approach. (The compensators we previously obtained were "weakly causal" and causal approximations were discussed [9], [10].)

Hence for a dilation result in $H^2(D^n)$ we need to include the causality constraint explicitly in the set–up of the dilation problem. It is precisely this problem which motivated the mathematical operator–theoretic work of [11] and [7] which incorporated Arveson theory [1] into the dilation, commutant lifting framework.

While, the general method explicated in this paper is based on a causal extension of the commutant lifting theorem, for the purposes of the operators and spaces which appear in control, we give a direct simple method for finding the optimal causal compensators. In fact, we show that the computation of an optimal causal nonlinear compensator can be reduced to a known interpolation problem.

1. SYSTEMS

We recall that a discrete–time system can be viewed as a procedure transferring an input sequence $a = (a_0, a_1, \ldots)$ into an output sequence $b = (b_0, b_1, \ldots)$:

$$\xrightarrow{a=(a_0,a_1,a_2,\ldots)} \square \xrightarrow{b=(b_0,b_1,b_2,\ldots)} \tag{1.0}$$

We recall the following basic definitions:

(1.1) Time invariance: if $(a'_0, a'_1, a'_2, \ldots) = (0, a_0, a_1, \ldots)$ then $(b'_0, b'_1, b'_2, \ldots) = (0, b_0, b_1, \ldots)$.

(1.2) Causality: b_k = independent of $a_{k+1}, a_{k+2}, \ldots$ $(\forall k)$.

(1.3) Linearity: b_k depends linearly on the a_j's $(\forall k)$.

Clearly (1.1) and (1.3) imply

(1.4) $\exists (f_0, f_1, \ldots)$ such that $b_k = f_0 a_k + f_1 a_{k-1} + \cdots + f_k a_0 \quad \forall a \equiv (a_0, a_1, \ldots)$,

so

(1.5) Time invariance and linearity $\Longrightarrow$ causality.

(1.6) "Energy" of $(a_0, a_1, \ldots) := \sum_{n=0}^{\infty} |a_n|^2 = \|\hat{a}\|_{H^2}^2$ where

$$\hat{a} := \sum_{n=0}^{\infty} a_n z^n \,, \; z \in \mathbf{D} = \{|z| < 1\}$$

and H^2 is the usual Hardy space, viewed as formed either by analytic functions in $\mathbf{D}$ or by their boundary values on $\partial\mathbf{D}$. The relation (1.4) becomes

$$\hat{b} = f\hat{a} \text{ where } f(z) = \sum_{n=0}^{\infty} f_n z^n \,. \tag{1.7}$$

Recall also the property:

(1.8) Stability (linear case): the mapping $\hat{a} \mapsto \hat{b} = f\hat{a}$ defines a bounded operator on H^2, i.e. $f \in H^\infty :=$ space of all bounded analytic functions on $\mathbf{D}$.

In general, for stability, first we require

$$\hat{a} \mapsto \hat{b} \text{ continuous on } H^2 \,. \tag{1.9a}$$

If $\hat{b}$ = homogeneous polynomial in $\hat{a}$ of degree k, say

$$\hat{b} = F(\hat{a}, \hat{a}, \ldots, \hat{a}) \tag{1.9b}$$

where F is linear in each argument, then

(1.9c) $$\hat{b} = F_{\text{lin}}(\hat{a} \otimes \hat{a} \otimes \ldots \otimes \hat{a}) ,$$

where F_{lin} is the linear map associated to F on $(H^2)^{\otimes k} := H^2 \otimes \ldots \otimes H^2(k\text{–times})$. In this case, we require that F_{lin} be continuous with respect to the Hilbert space norm on $(H^2)^{\otimes k}$; the corresponding Hilbert space can be identified with the usual Hardy space $H^2(\mathbf{D}^k)$ on the polydisc $\mathbf{D}^k$. Thus F_{lin} is linear and continuous on that space:

(1.9d) $$F_{\text{lin}} : H^2(\mathbf{D}^k) \Longrightarrow H^2(= H^2(\mathbf{D})) \text{ continuous} .$$

Note

(1.9e) $$(\hat{a} \otimes \ldots \otimes \hat{a})(z_1, z_2, \ldots, z_k) = \hat{a}(z_1)\hat{a}(z_2)\ldots\hat{a}(z_k) , \ z_1, \ldots, z_k \in \mathbf{D} .$$

(1.10) In this representation of (1.0), time invariance means $UF_{\text{lin}} = F_{\text{lin}} S^{\otimes k}$ where $S^{\otimes k}$ = multiplication by $z_1 z_2 \ldots z_k$ in $H^2(\mathbf{D}^k), U$ = multiplication by z in H^2.

(1.11) Example. (Time–invariant stable noncausal system)

$$b_n = a_n^2 + 2a_n a_{n+1} \ (n = 0, 1, \ldots)$$

Here, degree $k = 2$, and

$$F_{\text{lin}} \left(\sum \alpha_{n_1 n_2} z_1^{n_1} z_2^{n_2} \right) = \sum (\alpha_{nn} + \alpha_{n,n+1} + \alpha_{n+1,n}) z^n .$$

2. TRACKING

One of the simplest and important problems in H^∞–Control Theory, is that of tracking. Namely in the diagram

(2.1)

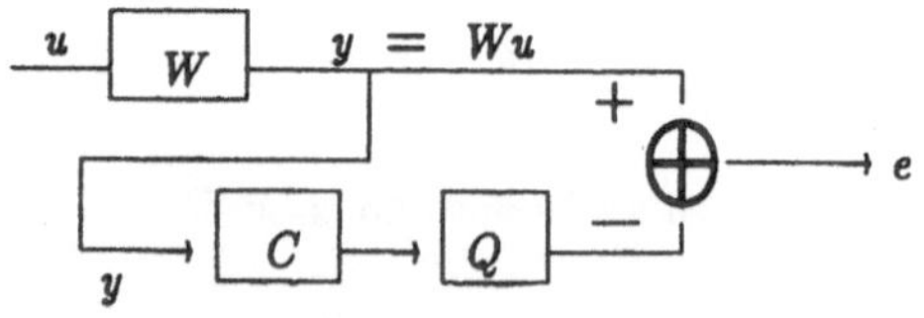

we are interested in minimizing the energy of the error e

(2.2) $$e = Wu - QCy = Wu - QCWu$$

by a suitable design of the controller C. Mathematically, we are required to find μ_δ, where

(2.3) $$\mu_\delta := \inf_C \sup_{\|u\|\le\delta} \|e\| = \inf_C \sup_{\|u\|\le\delta} \|Wu - QCWu\| ,$$

as well as a minimizing C. Note that if $W =$ invertible, then

(2.4) $$\mu_\delta = \inf_C \sup_{\|u\|\le\delta} \|Wu - QCu\| .$$

Here W, Q and C are stable, time invariant, causal maps; if W, P are linear, then C is linear too, and

(2.5) $$\mu_\delta = \delta \cdot \inf_C \|W - QC\|$$

where the $\|\cdot\|$ denotes the operator norm. If

$$W = W_1 + W_2 + \cdots ,$$

$$Q = Q_1 + Q_2 + \cdots ,$$

$$C = C_1 + C_2 + \cdots ,$$

with W_k, P_k and C_k homogeneous polynomials of degree k, then

(2.6) $$\mu_\delta = \delta \cdot \inf_{C_1} \|W_1 - Q_1C_1\| + 0(\delta^2) .$$

So even in this case, the optimization up to order 2 in δ, is identical to that in the linear case (see (2.5)).

3. THE LINEAR OPTIMIZATION PROBLEM AND THE COMMUTANT LIFTING THEOREM

We consider the linear bounded operators $W : \mathcal{G} \mapsto \mathcal{K}$, $Q : \mathcal{L} \mapsto \mathcal{K}$, where $\mathcal{G}, \mathcal{K}, \mathcal{L}$ are Hilbert spaces and isometries S on $\mathcal{G}, U$ on $\mathcal{K}, V$ on $\mathcal{L}$, which satisfy

(3.1) $$WS = UW , \ QV = UQ .$$

In all applications, it is generic that Q is bounded from below, i.e. for some $q > 0$:

(3.2) $$\|Q\ell\| \ge q\|\ell\| , \quad \forall \ell \in \mathcal{L} .$$

The previous discussion motivates the following (see [12]):

The Linear Optimization Problem (LOP). Find

$$C_0 : \mathcal{G} \mapsto \mathcal{L}\,,\ C_0 S = V C_0 \tag{3.3a}$$

such that

$$\|W - QC_0\| = \mu := \inf\{\|W - QC\| : C : \mathcal{G} \mapsto \mathcal{L}\,,\ CS = VC\}\,. \tag{3.3b}$$

One way to give a useful answer to this problem is to refer to the following theorem [14], [15], [6]:

THE COMMUTANT LIFTING THEOREM (CLT). *Let $\mathcal{H} \subset \mathcal{K}$ be a closed subspace invariant to U^*. Set $T = (U^*|\mathcal{H})^*$, and let $A : \mathcal{G} \mapsto \mathcal{H}$ be a linear bounded operator such that*

$$AS = TA\,. \tag{3.4a}$$

Then there exists a linear operator $B : \mathcal{G} \mapsto \mathcal{K}$ such that

$$BS = UB\,, \tag{3.4b}$$

$$A = PB\,, \tag{3.4c}$$

where $P = P_{\mathcal{H}}^{\mathcal{K}}$ is the orthogonal projection of $\mathcal{K}$ onto $\mathcal{H}$, and

$$\|B\| = \|A\|\,. \tag{3.4d}$$

(Any B satisfying (3.4b,c) is called an intertwining dilation of A.)

The connection (between LOP and CLT).
Let $\mathcal{H} = \mathcal{K} \ominus Q\mathcal{L}\,,\ A = PW$. Then

$$\|A\| \le \mu\,,\ TA = PUW = PWS = AS\,.$$

Hence there exists B as in (3.4c,d). Thus

$$P(W - B) = 0 ,$$

and so for $g \in \mathcal{G}$, $Wg - Bg = Q\ell(g)$ with a unique $\ell(g) \in \mathcal{L}$. Then

$$C_0 : g \mapsto \ell(g) \quad (\forall g \in \mathcal{G}) \text{ is the optimal operator .}$$

Note that in the linear case in §2, $\mathcal{G} = \mathcal{L} = \mathcal{K} = H^2$, $S = U = V =$ usual multiplication by z in H^2.

4. THE NONLINEAR OPTIMIZATION PROBLEM AND THE ITERATIVE COMMUTANT LIFTING PROCEDURE

We return to the general case in §2 as treated in [10], [8]. Let C_{10} be the optimal operator for the norm of the linear part in (2.6), i.e.

$$\mu_\delta = \delta \| W_1 - Q_1 C_{10} \| + 0(\delta^2) .$$

Then

$$W(u) - Q(C(u)) = (W_1 - Q_1 C_{10})(u) + (W_2(u) - Q_2(C_{10}u) - Q_1 C_2(u)) + \cdots$$

so

$$\begin{aligned} &W(u) - Q(C(u)) - (W_1 - Q_1 C_{10}(u)) = \\ &W_2(u) - Q_2(C_{10}(u)) - Q_1 C_2(u) + \text{higher order terms;} \end{aligned}$$

therefore

$$\begin{aligned} &\sup_{\|u\| \le \delta} \|(W - QC)(u) - (W_1 - Q_1 C_{10})u\| = \\ &\| \underbrace{W_{2\text{lin}} - Q_{2\text{lin}} \cdot C_{10} \otimes C_{10}}_{W_{2,0} = \text{given}} - Q_1 C_{2\text{lin}} \| \delta^2 + \sigma(\delta^3) . \end{aligned}$$

We obtain again the LOP, but for $\mathcal{G} = H^2(\mathrm{D}^2)$, $S = S^{\otimes 2}$, while $\mathcal{L} = \mathcal{K}(= H^2)$, $V = U(=$ multiplication by $z)$ are as in the linear case.

Applying the CLT gives $C_{2,\text{lin},0}$, and so on, we obtain $C_{3,\text{lin},0,\ldots}$. However this iterative procedure faces the following:

Serious Problem: $C_{k,\text{lin},0}$ may not be causal.

5. THE CAUSAL LOP AND THE CAUSAL CLT

In order to treat the problem of causality, we must first find the characterization of causal stable homogeneous polynomial (of degree k) operators from $H^2(= H^2(\mathbb{D}))$ into H^2 in terms of their associated bounded linear operators from $H^2(\mathbb{D}^k) \cong (H^2)^{\otimes k}$ into H^2.

For this purpose, recall that if

$$\Phi_{\text{lin}}\left(\sum a_m z^{(m)}\right) = \sum b_n z^n \left(z^{(m)} = z^{m_1} \ \dots \ z_k^{m_k}\right),$$

then Φ_{lin} is causal iff

(5.1) $$b_n = \text{ depends only of } a_m \text{ with } \max_j m_j \leq n\,.$$

Therefore if

(5.2) $$\mathcal{G}_n = \text{ closed linear span of } z^{(m)} \text{ with } \max_j m_j \geq n \quad (n = 0, 1, \dots)\,,$$

then (5.1) is equivalent to

(5.3) $$\Phi_{\text{lin}}\mathcal{G}_n \subset U^n H^2 \quad (n = 0, 1, \dots)\,,$$

where U is the multiplication by z on H^2.

Properties of $(\mathcal{G}_n)_{n=0}^{\infty}$

(5.4a) $$\mathcal{G}_0 \supset \mathcal{G}_1 \supset \dots (\mathcal{G}_0 \text{ denotes } H^2(\mathbb{D}^k))\,,$$

(5.4b) $$\mathcal{G}_n \supset S^n\mathcal{G}_0 (S \text{ denotes } S^{\otimes k}) \ (n = 0, 1, \dots)\,,$$

(5.4c) $$S\mathcal{G}_n \subset \mathcal{G}_{n+1} \ (n = 0, 1, \dots)\,.$$

So in the framework of §3, the operators W, C and C_0 are causal with respect to a system of subspaces satisfying (5.4a-c). With this set–up, in the sequel $(\mathcal{G}_n)_{n=0}^{\infty}$ will be a fixed sequence of subspaces of $\mathcal{G}$ satisfying (5.4a-c), and $\mathcal{G}_0 = \mathcal{G}$. Operators $B : \mathcal{G} \mapsto \mathcal{K}$ and $D : \mathcal{G} \mapsto \mathcal{L}$ will be called causal (w.r.t. $(\mathcal{G}_n)_{n=0}^{\infty}$) if

(5.5) $$B\mathcal{G}_n \subset U^n\mathcal{K}\,, \text{ resp. } D\mathcal{G}_n \subset V^n\mathcal{L} \quad (\forall n \geq 0)\,.$$

We can now state:

The Causal Linear Optimization Problem (CLOP). Let W, Q be as in LOP. Moreover, let W be causal. Find

$$C_0 : \mathcal{G} \mapsto \mathcal{L}\ ,\ C_0 S = V C_0\ ,\ C_0 \text{ causal} \tag{5.5a}$$

such that

$$\|W - QC_0\| = \mu_c := \inf\{\|W - QC\| : C : \mathcal{G} \mapsto \mathcal{L}\ ,\ CS = VC\ ,\ C \text{ causal}\}\ . \tag{5.5b}$$

Causal Commutant Lifting Problem (CCLP). Hypotheses and notation as in the CLT (§3). When does there exist a causal intertwining dilation B of A? If such a B exists, find

$$\|A\|_c := \inf\{\|B\| : B \text{ causal intertwining dilation of } A\}\ . \tag{5.6a}$$

To give an answer to this question, we consider the minimal unitary extension $\hat{S}$ of S on $\hat{\mathcal{G}} \supset \mathcal{G}$; define

$$\mathcal{G}_c = \left(\bigcup_{n=0}^{\infty} \hat{S}^{*n} \mathcal{G}_n\right)^{-} \subset \hat{\mathcal{G}}\ ,\ S_c = \hat{S}|\mathcal{G}_c\ . \tag{5.7a}$$

If T is invertible, we define

$$A_c \hat{g} = T^{-n} A g_n \tag{5.7b}$$

for $\hat{g} = \hat{S}^{*n} g_n$ $(g_n \in \mathcal{G}_n\ ,\ n = 0, 1, \ldots)$.

An answer to CCLP is given by the following theorem [11], [7]:

CAUSAL COMMUTANT LIFTING THEOREM (CCLT).

(i) *If the condition*

$$T \textit{ is invertible} \tag{5.8}$$

holds, then CLOP is solvable iff $\|A_c\| < \infty$. *In this case* $A_c S_c = T A_c$ *and* $\|A\|_c = \|A_c\|$.

(ii) *If condition* (5.8) *is dropped, then CLOP is solvable iff there exists a bounded linear operator* $A' : \mathcal{G}_c \mapsto \mathcal{H}$ *such that*

$$A' S_c = T A' \textit{ and } A'|\mathcal{G} = A\ . \tag{5.9a}$$

If such an A' *exists, then*

$$\|A\|_c = \min\{\|A'\| : A' \text{ as in } (5.9a)\}\ . \tag{5.9b}$$

COROLLARY. *If* (5.8) *holds, then in CLOP we have*

(5.10) $$\mu_c = \|PW\|_c = \|(PW)_c\| \, .$$

6. CLOP AND CLT

Let W, Q be as in the CLOP (§5). Then

(6.1) $$W|\mathcal{G}_n = U^n W_n \, , \; W_n : \mathcal{G}_n \mapsto \mathcal{K} \, ,$$

(6.2a) $$W_n S|\mathcal{G}_n = UW_n \, ,$$

(6.2b) $$W_n = W_{n+1} S|\mathcal{G}_n \, , \text{ for } n = 0, 1, 2, \dots \, .$$

Define

(6.3) $$W_c \hat{g} = W_n g_n \text{ for } \hat{g} = \hat{S}^{*n} g_n \, , \; g_n \in \mathcal{G}_n \, , \; n = 0, 1, \dots \, .$$

Then

(6.4a) $$\|W_c\| = \|W\| \, , \; W_c|\mathcal{G} = W \, ,$$

(6.4b) $$W_c S_c = UW_c \, .$$

If $C : \mathcal{G} \mapsto \mathcal{L}$, $CS = VS$, $C =$ causal, then

(6.5) $$\|W_c - QC_c\| = \|(W - QC)_c\| = \|W - QC\| \, .$$

Thus we can easily conclude the following:

THEOREM. *CLOP for W and LOP for W_c are equivalent; namely*

(i) $\mu_c(W) = \mu(W_c)$;

(ii) C_{0c} *is a causal optimal operator for CLOP iff* $(C_{0c})_c$ *is a optimal operator for LOP;*

(iii) *if C_0 is a optimal operator for LOP, then $C_{0c} = C_0|\mathcal{G}$ is a causal optimal operator for CLOP.*

This theorem reduces the CLOP directly to the CLT.

COROLLARY. $\mu_c(W) = \|(PW)_c\|$.

Note that unlike (5.10), the above equality holds even if (5.8) is not true.

Example.

$$W(a_0, a_1, \ldots, a_n, \ldots) = \left(\frac{a_0 - a_1}{2}, \frac{a_1 - a_2}{2}, \ldots\right) \tag{6.6a}$$

$$Q(a_0, a_1, \ldots, a_n, \ldots) = (a_0^2, a_1^2, a_0 + a_2^2, a_1 + a_3^2, \ldots) \tag{6.6b}$$

In the function representation

$$W = \text{ multiplication by } (1-z)/2\,,\ Q_1 = \text{ multiplication by } z^2\,, \tag{6.6c}$$

$$\|W_1 - Q_1 C_{10}\| = (\sqrt{5}+1)/2 \tag{6.6c}$$

$$C_{10} = \text{ multiplication by } \frac{\beta}{2(1-\beta z)}\,, \text{ where } \beta = \frac{\sqrt{5}-1}{2}\,.$$

Recall that for the 2nd order optimization, we solve CLOP for $W_{2,0} = -Q_{2\text{lin}} \circ C_{10} \otimes C_{10}$. Easy, but long, computations then lead to the following:

Fact:

$$\|W_{2,0}\| \approx 1.902 \tag{6.8a}$$

$$\mu(W_{2,0}) \approx 1.431 \tag{6.8b}$$

and

$$\mu_c(W_{2,0})(= \mu((W_{2,0})_c) \approx 1.807. \tag{6.8c}$$

This shows that the causal optimal bound can be worse than the noncausal optimal bound (compare (6.8b) with (6.8c)), but that the causal optimal bound leads to an improvement in the design of the controller (compare (6.8a) with (6.8c)).

7. CONCLUSION

From the applied operator theoretic point of view, the results in §6 give a complete solution to the causal linear optimal problem by reducing it to the commutant lifting theorem. In this way, the corresponding iterative causal commutant lifting procedure (see §4) solves also the nonlinear analytic control problem. However from the purely operator theoretic point of view, the causal commutant lifting theorem (general case (ii)) in §5, is not yet a definitive result. Indeed a workable characterization of the existence of the intertwining extensions A' (see (5.9a)) is still not available. The finding of such a characterization is an interesting open problem in Operator Theory.

REFERENCES

[1] W. Arveson, "Interpolation problems in nest algebras," *J. Funct. Anal.*, **20**(1975), 208–233.

[2] J. Ball and J.W. Helton, "Sensitivity bandwidth optimization for nonlinear feedback systems," Technical Report, Department of Mathematics, University of California at San Diego, 1988.

[3] J. Ball and J.W. Helton, "H^∞ control for nonlinear plants: connections with differential games," *IEEE Conference on Decision and Control* Tampa, Florida, 1989, 956-962.

[4] J. Doyle, B. Francis and A. Tannenbaum, *Feedback Control Theory*, MacMillan, New York, 1991.

[5] C. Foias, "Contractive intertwining dilations and waves in layered media", *Proc. International Congress Math.*, Helsinki **2**(1978), 605-613.

[6] C. Foias and A. Frazho, *The Commutant Lifting Approach to Interpolation Problems*, Birkhauser–Verlag, Boston, 1990.

[7] C. Foias, C. Gu and A. Tannenbaum, "Intertwining dilations, intertwining extensions and causality," *Acta Sci. Math. (Szeged)*, **57**(1993), 101-123.

[8] C. Foias, C. Gu and A. Tannenbaum, "On a causal linear optimization theorem," to appear in *J. of Math. Anal. and Appl.*.

[9] C. Foias and A. Tannenbaum, "Iterated commutant lifting for systems with rational symbol," *Oper. Theory: Adv. and Appl.*, **41**(1989), 255-277.

[10] C. Foias and A. Tannenbaum, "Weighted optimization theory for nonlinear systems," *SIAM J. on Control and Optimiz.*, **27**(1989), 843-860.

[11] C. Foias and A. Tannenbaum, "On a causal commutant lifting theorem," *J. of Funct. Anal.*, **118**(1993), 407-441.

[12] B. Francis, *A Course in H^∞ Control Theory*, McGraw–Hill, New York, 1981.

[13] B. Francis and A. Tannenbaum, "Generalized interpolation theory in control," *Mathematical Intelligencer*, **10**(1988), 48-43.

[14] D. Sarason, "Generalized interpolation in H^∞," *Trans. AMS*, **127**(1967), 179-203.

[15] B. Sz.-Nagy and C. Foias, *Harmonic Analysis of Operators on Hilbert Space*, North–Holland Publishing Company, Amsterdam, 1970.

[16] G. Zames, "Feedback and optimal sensitivity: model reference transformations, multiplicative semi-norms, and approximate inverses," *IEEE Trans. Auto. Control*, **AC-26**(1981), 301-320.

Ciprian Foias and Caixing Gu
Department of Mathematics
Indiana University
Bloomington, IN 47405

Allen Tannenbaum
Department of Electrical Engineering
University of Minnesota
Minneapolis, MN 55455

1980 *Mathematics Subject Classifications* (1985 Revision). Primary 47A20, Secondary 47A99, 93B35, 93C05.

Operator Theory:
Advances and Applications, Vol. 80

Analysis of the radiation loss: asymptotics beyond all orders

Jishan Hu and Wing-Cheong Cheng

1. INTRODUCTION

Kath and Kriegsmann recently studied a model in bent fibre-optic tunnelling (see [6]). An interesting singular perturbation problem on the half axis:

$$\epsilon^2 y'' + Q(x;\lambda)y = 0, \qquad x \in (0,+\infty), \tag{1.1}$$

arises. Here $0 < \epsilon \ll 1$ is a parameter and λ is an eigenvalue. One boundary condition associated with equation (1.1) is

$$\epsilon y'(0) + hy(0) = 0, \tag{1.2}$$

for some $h > 0$. The other boundary condition is imposed at $x = +\infty$. Several authors have computed a desired quantity $\mathrm{Im}Q(0;\lambda)$, which is called the radiation loss, for several special cases. The radiation loss problem is an nonlinear eigenvalue problem. In this paper, we try to have a general discussion for a variety of functions Q.

Let us assume

$$Q(x;\lambda) \sim Q_+(x) > 0, \qquad x \to +\infty. \tag{1.3}$$

One boundary condition used at $x = +\infty$ is that $y(x)$ has controlling behavior $e^{ip(x)/\epsilon}$ for some $p(x) > 0$. In particular, if Q goes to constant $Q_+(+\infty) > 0$ fast enough as $x \to +\infty$, the boundary condition at $x = +\infty$ is that y has controlling behavior $e^{iQ_+(+\infty)x/\epsilon}$. Since the equation under consideration is linear, after rescaling the solution of equation (1.1), we can easily see that this only condition determines one solution, except a constant multiplier. Because of the homogeneity of condition (1.2), this multiplier cannot be determined at $x = 0$. Therefore, we have an over-determined boundary value problem if $h > 0$ is given. This generates a relation between $Q(0;\lambda)$ and h, thereafter, the radiation loss $\mathrm{Im}Q(0;\lambda)$, in terms of h. Primary results indicates that for many functions of Q, for instance, $Q(x;\lambda) = \lambda + x^n$, the radiation loss is a transcendentally small quantity of ϵ if $h > 0$ (see [7], [10], [12]).

Comparing the above problem with adiabatic invariance problem, or equivalently, reflection coefficient problem (see [2], [4], [5], [8], [9], [11]), we demonstrate that these problems share not only similar equations, but also a very similar method to compute the radiation loss as well as the reflection coefficient.

In section 2, we discuss the asymptotic solutions of equation (1.1) by using the work by Gingold (see [3]). His results allow us to write an expression of the general solution of equation (1.1). It is "invariant" in the sense that it is valid even as x approaches turning points. In section 3, we formulate radiation loss problem for general functions Q and show how to compute it for functions Q being analytic and having critical points on the nearest critical line.

2. ASYMPTOTIC FORMULAS

It is well known that WKB approximation gives valid asymptotic solutions of equation (1.1) for problems without turning points. For problems with turning points, WKB approximation becomes invalid near those points. In order to obtain global representations of solutions, various corrections have been developed to overcome the difficulty. A typical treatment is to establish additional so-called connection formulas. Recent work by Gingold made a significant progress toward turning point problems. His representations of solutions are valid all the way up to the turning points and WKB approximation can be extracted from them. Gingold's formulas are particularly useful for problems with turning points at infinity. In this section, we use his formulas to express solutions of equation (1.1).

Consider a second ordinary differential equation of the form (1.1):

$$\epsilon^2 y'' + Q(x)y = 0, \qquad x \in (0, +\infty). \tag{2.1}$$

In this section, we assume that $Q \in C^\infty([a,b])$, with $0 \le a < b \le +\infty$. On (a,b), we assume

$$Q(x) \neq 0, \tag{2.2}$$

i.e., equation (2.1) has no turning point on (a,b). For fixed and sufficient small $\epsilon > 0$, denote

$$\varphi(x) := -\,Q(x)/\epsilon^2, \tag{2.3}$$

$$\lambda(x) := \sqrt{\varphi(x) + \left(\frac{1}{4}\frac{\varphi'(x)}{\varphi(x)}\right)^2} = \sqrt{-\frac{Q(x)}{\epsilon^2} + \left(\frac{1}{4}\frac{Q'(x)}{Q(x)}\right)^2}, \tag{2.4}$$

$$\ell(x) := \frac{1}{4}\frac{\varphi'(x)}{[\varphi(x)]^{3/2}} = -\frac{\epsilon}{4}\frac{Q'(x)}{(-Q(x))^{3/2}}, \tag{2.5}$$

$$\theta(x) := \ln\left[\frac{1 - i\ell(x)}{1 + i\ell(x)}\right]^{1/4} = \ln\left[\frac{1 + \dfrac{i\epsilon}{4}\dfrac{Q'(x)}{(-Q(x))^{3/2}}}{1 - \dfrac{i\epsilon}{4}\dfrac{Q'(x)}{(-Q(x))^{3/2}}}\right]^{1/4}, \tag{2.6}$$

$$\begin{aligned} r(x) &:= -\frac{i}{2}\frac{\ell'(x)}{1 + \ell^2(x)} \\ &= \frac{i\epsilon}{16} \cdot \frac{3(Q'(x))^2 - 2Q(x)Q''(x)}{(-Q(x))^{5/2}} \cdot \frac{1}{1 - \dfrac{\epsilon^2}{16}\dfrac{Q'^2(x)}{Q^3(x)}}, \end{aligned} \tag{2.7}$$

and

$$e(\tau_1,\tau_2) := \exp\left\{2\int_{\tau_2}^{\tau_1}\lambda(s)ds\right\}$$
$$= \exp\left\{2\int_{\tau_2}^{\tau_1}\sqrt{-\frac{Q(x)}{\epsilon^2}+\left(\frac{1}{4}\frac{Q'(x)}{Q(x)}\right)^2}\,ds\right\}. \tag{2.8}$$

A point $x \in [a,b]$ is called an induced turning point if $\ell^2(x) = -1$. If we assume that

$$\left|Q'^2(x)/Q^3(x)\right| \leq M, \qquad x \in [a,b], \tag{2.9}$$

then, by (2.2), equation (2.1) has no induced turning point on (a,b). Furthermore, we assume that

$$\int_a^b \left|\left(\frac{Q'(x)}{Q^{3/2}(x)}\right)'\right| dx < +\infty. \tag{2.10}$$

Under assumptions (2.2), (2.9) and (2.10), two linearly independent solutions and there derivatives can be expressed as follows:

$$y_1 = [Q(x)]^{-1/4}\Big\{[\cosh\theta(x)+i\sinh\theta(x)]\,(1+p_{11}(x,\alpha_{11},\alpha_{21}))$$
$$-i\,[\cosh\theta(x)-i\sinh\theta(x)]\,p_{21}(x,\alpha_{11},\alpha_{21})\Big\}\cdot\exp\left\{+\int_{x_0}^x\lambda(s)ds\right\}, \tag{2.11}$$

$$y_2 = [Q(x)]^{-1/4}\Big\{[\cosh\theta(x)+i\sinh\theta(x)]\,p_{12}(x,\alpha_{12},\alpha_{22})$$
$$-i\,[\cosh\theta(x)-i\sinh\theta(x)]\,(1+p_{22}(x,\alpha_{12},\alpha_{22}))\Big\}\cdot\exp\left\{-\int_{x_0}^x\lambda(s)ds\right\}, \tag{2.12}$$

$$y_1' = i\epsilon^{-1}\,[Q(x)]^{+1/4}\Big\{[\cosh\theta(x)-i\sinh\theta(x)]\,(1+p_{11}(x,\alpha_{11},\alpha_{21}))$$
$$+i\,[\cosh\theta(x)+i\sinh\theta(x)]\,p_{21}(x,\alpha_{11},\alpha_{21})\Big\}\cdot\exp\left\{+\int_{x_0}^x\lambda(s)ds\right\}, \tag{2.13}$$

and

$$y_2' = -i\epsilon^{-1}\,[Q(x)]^{+1/4}\Big\{[\cosh\theta(x)-i\sinh\theta(x)]\,p_{12}(x,\alpha_{12},\alpha_{22})$$
$$+i\,[\cosh\theta(x)+i\sinh\theta(x)]\,(1+p_{22}(x,\alpha_{12},\alpha_{22}))\Big\}\cdot\exp\left\{-\int_{x_0}^x\lambda(s)ds\right\}, \tag{2.14}$$

where $x_0 \in (a,b)$ is fixed and $p_{jk}(x)$ are convergent series for any $x \in [a,b]$:

$$p_{11}(x,\alpha_{11},\alpha_{21}) = \sum_{m=0}^{+\infty}\int_{\alpha_{11}}^x r(\hat{t}_0)d\hat{t}_0\cdot\int_{\alpha_{21}}^{\hat{t}_0} r(t_0)e(t_0,\hat{t}_0)dt_0$$
$$\cdot\prod_{n=1}^m\int_{\alpha_{11}}^{t_{n-1}} r(\hat{t}_n)d\hat{t}_n\int_{\alpha_{21}}^{\hat{t}_n} r(t_n)e(t_n,\hat{t}_n)dt_n, \tag{2.15}$$

$$p_{22}(x,\alpha_{12},\alpha_{22}) = \sum_{m=0}^{+\infty} \int_{\alpha_{22}}^{x} r(\hat{t}_0)d\hat{t}_0 \cdot \int_{\alpha_{12}}^{\hat{t}_0} r(t_0)e(\hat{t}_0,t_0)dt_0 \cdot \prod_{n=1}^{m} \int_{\alpha_{22}}^{t_{n-1}} r(\hat{t}_n)d\hat{t}_n \int_{\alpha_{12}}^{\hat{t}_n} r(t_n)e(\hat{t}_n,t_n)dt_n, \tag{2.16}$$

$$p_{12}(x,\alpha_{12},\alpha_{22}) = \sum_{m=0}^{+\infty} \int_{\alpha_{12}}^{x} r(\hat{t}_0)e(x,\hat{t}_0)d\hat{t}_0 \cdot \prod_{n=1}^{m} \int_{\alpha_{22}}^{\hat{t}_{n-1}} r(\hat{t}_n)d\hat{t}_n \int_{\alpha_{12}}^{\hat{t}_n} r(t_n)e(\hat{t}_n,t_n)dt_n, \tag{2.17}$$

and

$$p_{21}(x,\alpha_{11},\alpha_{21}) = \sum_{m=0}^{+\infty} \int_{\alpha_{21}}^{x} r(\hat{t}_0)e(\hat{t}_0,x)d\hat{t}_0 \cdot \prod_{n=1}^{m} \int_{\alpha_{11}}^{\hat{t}_{n-1}} r(\hat{t}_n)d\hat{t}_n \int_{\alpha_{21}}^{\hat{t}_n} r(t_n)e(t_n,\hat{t}_n)dt_n. \tag{2.18}$$

with α_{jk}, $j,k=1,2$, being any arbitrary constants in $[a,b]$.

In particular, for equation (1.1), if condition (1.3) holds, and if the function Q_+ satisfies (2.9) and (2.10) for some finite a and $b=+\infty$, then the general solution of equation (1.1) behaves in the following form as $x\to\pm\infty$:

$$\begin{cases} y(x) \sim C_1[Q_+(x)]^{-1/4}\cdot\exp\left\{+\int^x \lambda_+(s)ds\right\} \\ \qquad +C_2[Q_+(x)]^{-1/4}\cdot\exp\left\{-\int^x \lambda_+(s)ds\right\}, \\ y'(x) \sim iC_1\epsilon^{-1}[Q_+(x)]^{-1/4}\cdot\exp\left\{+\int^x \lambda_+(s)ds\right\} \\ \qquad -iC_2\epsilon^{-1}[Q_+(x)]^{-1/4}\cdot\exp\left\{-\int^x \lambda_+(s)ds\right\}. \end{cases} \tag{2.19}$$

Here λ_+ is defined in (2.4) with $Q=Q_+$.

As indicated by Gingold, we can extend the above discussion to the complex-x plane.

3: RADIATION LOSS PROBLEMS

From the discussions in section 2, we know that under conditions:

$$\begin{cases} Q(x;\lambda) \sim Q_+(x), \qquad x\to+\infty, \\ \left|Q'^2(x;\lambda)/Q^3(x;\lambda)\right| \le M, \\ \displaystyle\int^{+\infty} \left|\left(\frac{Q'_+(x)}{Q_+^{3/2}(x)}\right)'\right| dx < +\infty, \end{cases} \tag{3.1}$$

we have a well-defined nonlinear eigenvalue problem

$$\begin{cases} \epsilon^2 y'' + Q(x;\lambda)y = 0, & x \in (0,+\infty), \\ \epsilon y'(0) + hy(0) = 0, \\ y(x) \sim [Q_+(x)]^{-1/4} \cdot \exp\left\{+\int_{\bar{x}}^{x} \lambda_+(s)ds\right\}, & x \to +\infty, \end{cases} \tag{3.2}$$

where $\bar{x}$ is a sufficiently large real number and

$$\lambda_+(x) = \sqrt{-\frac{Q_+(x)}{\epsilon^2} + \left(\frac{1}{4}\frac{Q'_+(x)}{Q_+(x)}\right)^2}. \tag{3.3}$$

The radiation loss problem is to solve the problem (3.2) in order to obtain the radiation loss $\mathrm{Im}Q(0;\lambda)$.

The method to compute the radiation loss is similar to that was used to solve the reflection coefficient problems, studied by Gingold and Hu (see [4]). The reflection coefficient problems under different conditions were well studied by many authors (see [2], [5], [8], [9], [11]). To find the radiation loss, we need to solve the problem (3.2) along the nearest critical level line from the x-complex upper plane, of

$$\mathrm{Re}\left\{\int^{x} \lambda(s)ds\right\} = \text{const}, \tag{3.4}$$

on which there exists at least one critical point of the differential equation. Here

$$\lambda(x) = \sqrt{-\frac{Q(x;\lambda)}{\epsilon^2} + \left(\frac{1}{4}\frac{Q'(x;\lambda)}{Q(x;\lambda)}\right)^2}. \tag{3.5}$$

Let us assume that the nearest level line of (3.4) is L_1. Here we presume that the value of the eigenvalue λ is known. Let us assume that $x = x_c$ is a critical point on L_1 and near $x = x_c$,

$$Q(x;\lambda) \sim b_c(x - x_c)^{2\gamma_c - 2}, \qquad x \to x_c, \tag{3.6}$$

with $\gamma_c > 0$. In general, b_c, x_c and γ_c all depend on the value of λ. On L_1, away from the critical point $x = x_c$, we assume that the leading order of y is given by

$$y \sim Q(x;\lambda)^{-1/4}\left[r_1 \exp\left\{+\int_{x_c}^{x} \lambda(s)ds\right\} + r_2 \exp\left\{-\int_{x_c}^{x} \lambda(s)ds\right\}\right]. \tag{3.7}$$

The values of r_1, r_2 can be determined by the continuation of the boundary behavior (3.2) of y near $+\infty$, which gives

$$\begin{cases} r_1 = \exp\left\{-\int_{x_c}^{\bar{x}} \lambda(s)ds\right\}, \\ r_2 = 0. \end{cases} \tag{3.8}$$

In a neighborhood of $x = x_c$, the leading term of y satisfies

$$\epsilon^2 y'' + b_c(x - x_c)^{2\gamma_c - 2} y = 0, \tag{3.9}$$

whose general solution can be expressed in terms of Hankel functions

$$y = (x - x_c)^{1/2} \left\{ T_1 H^{(1)}_{1/2\gamma_c} \left(\frac{b_c^{1/2}}{\epsilon\gamma_c} (x - x_c)^{\gamma_c} \right) + T_2 H^{(2)}_{1/2\gamma_c} \left(\frac{b_c^{1/2}}{\epsilon\gamma_c} (x - x_c)^{\gamma_c} \right) \right\}. \tag{3.10}$$

Matching (3.10) with (3.7) gives

$$\begin{cases} T_1 = r_1 \sqrt{\frac{\pi}{2\epsilon\gamma_c}} e^{+\pi i/4\gamma_c + \pi i/4}, \\ T_2 = r_2 \sqrt{\frac{\pi}{2\epsilon\gamma_c}} e^{-\pi i/4\gamma_c - \pi i/4}. \end{cases} \tag{3.11}$$

We omit the estimates of the overlapping region for the matching since it is a routine work. Similar discussions can be found in [7] or [1]. Since $x = x_c$ is a critical point of equation (3.9), near $x = x_c$, the level line of (3.4) consists of hyperbola-like curves with the angle of each leaf being π/γ_c. We extend the solution (3.10) passing through $x = x_c$ from the branch L_1 to the next branch L_2 at the same level in the clockwise direction. The passage is equivalent to a change of the argument of $(x - x_c)^{\gamma_c}$ by $-\pi$. Hence, on L_2, to the leading order, the function y has the form

$$y = (x - x_c)^{1/2} e^{\pi i/2\gamma_c} \left\{ \left[2T_1 \cos(\pi/2\gamma_c) - T_2 e^{\pi i/2\gamma_c} \right] H^{(1)}_{1/2\gamma_c} \left(\frac{b_c^{1/2}}{\epsilon\gamma_c} (x - x_c)^{\gamma_c} \right) + T_1 e^{-\pi i/2\gamma_c} H^{(2)}_{1/2\gamma_c} \left(\frac{b_c^{1/2}}{\epsilon\gamma_c} (x - x_c)^{\gamma_c} \right) \right\}. \tag{3.12}$$

By the discussions in section 2, on L_2 away from the critical point $x = x_c$, the function y has the form

$$y \sim Q(x; \lambda)^{-1/4} \left[r_1' \exp\left\{ + \int_{x_c}^{x} \lambda(s) ds \right\} + r_2' \exp\left\{ - \int_{x_c}^{x} \lambda(s) ds \right\} \right]. \tag{3.13}$$

Matching (3.13) with (3.12) near $x = x_c$ gives

$$\begin{cases} r_1' = T_1 \sqrt{\frac{2\epsilon\gamma_c}{\pi}} e^{-\pi i/4\gamma_c - \pi i/4}, \\ r_2' = e^{\pi i/2\gamma_c} \left[2T_1 \cos(\pi/2\gamma_c) - T_2 e^{\pi i/2\gamma_c} \right] \sqrt{\frac{2\epsilon\gamma_c}{\pi}} e^{+\pi i/4\gamma_c + \pi i/4}. \end{cases} \tag{3.14}$$

Now by using the boundary condition at $x = 0$ in (3.2), we have

$$iQ(0;\lambda)^{1/4}\left[r_1'e^{\tau} - r_2'e^{-\tau}\right] + hQ(0;\lambda)^{-1/4}\left[r_1'e^{\tau} + r_2'e^{-\tau}\right] \to 0,$$

as $\epsilon \to +\infty$, i.e.,

$$Q(0;\lambda) \sim -h^2\left(\frac{1+(r_1'/r_2')e^{-2\tau}}{1-(r_1'/r_2')e^{-2\tau}}\right)^2, \qquad \epsilon \to +\infty, \tag{3.15}$$

where

$$\tau = i\int_{x_c}^{0}[Q(s;\lambda)]^{1/2}ds/\epsilon. \tag{3.16}$$

Here we use the fact that

$$\int_{x_c}^{0}\lambda(s)ds - \tau \to 0, \qquad \epsilon \to 0+.$$

If we presume that $e^{-2\tau} \ll 1$, then, by combining (3.14) and (3.15), we have

$$Q(0;\lambda) \sim -h^2\left(1 - \frac{2ie^{-\pi i/\gamma_c}}{\cos(\pi/2\gamma_c)}e^{-2\tau}\right), \qquad \epsilon \to 0+. \tag{3.17}$$

Hence, we have the radiation loss

$$\mathrm{Im}Q(0;\lambda) \sim \frac{2h^2}{\cos(\pi/2\gamma_c)}\cos(\pi/\gamma_c - 2\mathrm{Im}\tau)\cdot e^{-2\mathrm{Re}\tau}, \qquad \epsilon \to 0+. \tag{3.18}$$

Example 3.1 If $Q(x;\lambda) = x^n + \lambda$, this is a linear eigenvalue problem. Liu & Wood (see [7]) showed that the radiation loss in this case is

$$\mathrm{Im}\lambda \sim \begin{cases} -\dfrac{2h^2}{e}\exp\left\{\dfrac{-4h^2}{3\epsilon}\right\}, & \text{if } n = 1,\\[2ex] -2h^2\exp\left\{-\dfrac{2h^{(n+2)/n}}{\epsilon}\cdot\dfrac{\Gamma(1+1/n)\Gamma(3/2)}{\Gamma(3/2+1/n)}\right\} & \text{if } n \geq 2. \end{cases}$$

This result can be recovered from the formula (3.17), with $\gamma_c = 3/2$ and a formula given by Liu & Wood:

$$\lambda = -h^2 - n!\epsilon^n/(2h)^n + o(\epsilon^n).$$

Example 3.2 If $Q(x;\lambda) = 1/(x^2+\lambda^2)$, the classical WKB approximation cannot be used to compute the radiation loss, since $x = +\infty$ is a turning point of the original problem. However, we can use formula (3.18) to calculate the radiation loss

$$\mathrm{Im}Q(0;\lambda) = \mathrm{Im}\lambda^{-2} \sim 2h^2\exp\left\{-\frac{\pi}{\epsilon}\right\}, \qquad \epsilon \to 0+.$$

References

1. C. M. Bender & Orszag, S. A., *Advanced mathematical methods for scientists and engineers*, McGraw-Hill Book Co., New York, 1978.

2. M. V. Berry, *Semiclassically week reflections above analytic and non-analytic potential barriers*, J. Phys. A: Math. Gen., **15**, 3693–3704, 1982.

3. H. Gingold, *An invariant asymptotic formula for solutions of second-order linear ODE's*, Asym. Anal., **1**, 317–350, 1988.

4. H. Gingold and J. Hu, *Transcendentally small reflection of waves for problems with /without turning points near infinity: a new uniform approach*, J. Math. Phys., **32**(12), 3278–3284, 1991.

5. J. Hu and M. Kruskal, *Reflection coefficient beyond all orders for singular problems(I)*, J. Math. Phys., *32*(10), 2400–2405, 1991.

6. W. L. Kath & G. A. Kriegsmann, *Optical tunnelling: radiation losses in bent fibre-optic waveguides*, IMA J. Appl. Math., **41**, 85–103, 1988.

7. J. Liu & A. D. Wood, *Matched asymptotics for a generalisation of a model equation for optical tunnelling*, Euro. J. Appl. Math., **2**, 223–231, 1991.

8. J. J. Mahony, *The reflection of short waves in a variable medium*, Quart. Appl. Math., **25**, 313–316, 1967.

9. R. E. Meyer, *Quasiclassical scattering above barriers in one dimension*, J. Math. Phys., **17**, 1039–1041, 1976.

10. R. B. Paris & A. D. Wood, *A model equation for optical tunnelling*, IMA J. Appl. Math., **43**, 273–284, 1989.

11. V. L. Pokrovskii & I. M. Khalatnikov, *On the problem of above-barrier reflection of high-energy particles*, Soviet Phys. JETP, **13**, 1207–1210, 1961.

12. A. D. Wood, *Exponential asymptotics and spectral theory for curved optical waveguides*, Asymptotics beyond all orders, Edited by H. Segur *et al*, Plenum Press, New York, 1991.

Department of Mathematics
The Hong Kong University of Science and Technology
Clear Water Bay, Kowloon, Hong Kong

AMS classification numbers: 34B05, 34E20, 78A40.

Operator Theory:
Advances and Applications, Vol. 80

SELFADJOINT EXTENSIONS OF A CLOSED LINEAR RELATION OF DEFECT ONE IN A KREIN SPACE

Peter Jonas and **Heinz Langer**

In this paper we study the selfadjoint and the nonnegative selfadjoint extensions of a nonnegative closed linear relation (c.l.r.) A_0 of defect one in a Krein space $(\mathcal{H}, [\cdot,\cdot])$. These extensions are described by their resolvents, that is, M. G. Krein's formula for the resolvents of the extensions of a symmetric densely defined operator with defect $(1,1)$ is generalized to the situation considered here. The main difficulties which arise with this generalization are the following.

It may happen that all the selfadjoint extensions of a nonnegative c.l.r. in a Krein space have empty resolvent sets. This case will be excluded from our consideration: We restrict ourselves to the case of a nonnegative c.l.r. A_0 of *regular* defect one, which means that for sufficiently many complex numbers z the ranges of $A_0 - z$ are closed and of codimension one. This notion is introduced in Section 3.

Even under this assumption there can exist at most one selfadjoint extension of A_0 which has an empty resolvent set. This situation is described in detail in Sections 4 and 6. In Section 5 a connection between the signature of the defect subspaces and the nonreal spectrum of the selfadjoint extensions of A_0 is established, which generalizes corresponding results for densely defined operators in a Pontryagin space.

The main results are the generalization of M. G. Krein's formula to a nonnegative c.l.r. A_0 of regular defect one (Theorem 7.2) and the characterization of the nonnegative selfadjoint extensions of A_0 (Theorem 8.2).

This paper can be considered as a continuation of our note [10]. In a subsequent paper we shall study the nonreal eigenvalue of the selfadjoint extensions of a nonnegative c.l.r. A_0 of regular defect one and its dependence on the real parameter in formula (7.8) below. Also, here we restrict ourselves to selfadjoint extensions of A_0 which act in the originally given space. Corresponding results for extensions with exit, that is, a description of all the generalized resolvents of A_0, will be considered in another note.

Finally we should like to mention that similar problems were considered in [5]. There for a densely defined hermitian operator A in a Krein space such that the form $[Ax, y]$ has a finite number of negative squares on $\mathcal{D}(A)$, the selfadjoint extensions $\widetilde{A}$ and generalized resolvents are described under the assumption that the form $[\tilde{A}x, y]$ also has a finite number of negative squares on $\mathcal{D}(\tilde{A})$.

The authors thank Professor A. Dijksma for carefully reading the manuscript and valuable remarks.

1. Selfadjoint closed linear relations with a finite number of negative squares

Let $(\mathcal{H}, [\cdot,\cdot])$ be a (complex) Krein space. Recall that a closed linear relation (c.l.r.) T in $\mathcal{H}$ is a closed linear subspace of $\mathcal{H}^2$; a closed linear operator T in $\mathcal{H}$ is viewed as a c.l.r. via its graph in $\mathcal{H}^2$. For the usual definitions of the linear operations with c.l.r.'s of the inverse, kernel, range, etc., we refer to [6], [8]. Sums of vectors in $\mathcal{H}^2$ and also the linear span of two linear subspaces of $\mathcal{H}^2$ will be denoted by $\dot{+}$. The *resolvent set* $\rho(T)$ of a c.l.r. T is the set of all $z \in \mathbf{C}$ such that $R(z;T) := (T-z)^{-1}$ is an everywhere defined (and hence bounded) operator, the spectrum $\sigma(T)$ is the complement of $\rho(T)$ in $\mathbf{C}$. The point spectrum $\sigma_p(T)$ of T is the set of all $z \in \mathbf{C}$ such that $\binom{f}{zf} \in T$ for some $f \neq 0$. If the c.l.r. T is not an operator, that is $T(0) \neq \{0\}$, then ∞ is also called an eigenvalue of T with the nonzero elements of $T(0)$ being corresponding eigenvectors. The set

$$\tilde{\sigma}_p(T) = \begin{cases} \sigma_p(T) & \text{if } T(0) = \{0\} \\ \sigma_p(T) \cup \{\infty\} & \text{if } T(0) \neq \{0\} \end{cases}$$

is called the *extended point spectrum* of T. The point $z \in \mathbf{C}$ is said to be of *regular type* for T if there exists a $C > 0$ such that

$$\|g - zf\| \geq C\|f\| \qquad \text{for all} \qquad \binom{f}{g} \in T;$$

the set of all points of regular type for T is denoted by $r(T)$. Here $\|\cdot\|$ denotes an arbitrary Hilbert majorant of $[\cdot,\cdot]$. It follows immediately from the definition that for every $z \in r(T)$ the range $\mathcal{R}(T-z)$ is closed and that $r(T)$ is an open set. Finally, the c.l.r. $\tilde{T}$ is called an *extension* of T if $\tilde{T} \supset T$.

With the c.l.r. T there is associated an adjoint c.l.r. T^+. In order to define it we introduce in $\mathcal{H}^2$ the inner product $[\![\cdot,\cdot]\!]$ by the relation

$$(1.1) \qquad [\![\binom{x_1}{y_1}, \binom{x_2}{y_2}]\!] := i([x_1, y_2] - [y_1, x_2]), \qquad \binom{x_1}{y_1}, \binom{x_2}{y_2} \in \mathcal{H}^2.$$

Then $T^+ = T^{[\![\perp]\!]}$, the $[\![\cdot,\cdot]\!]$-orthogonal companion of T in $\mathcal{H}^2$. In other words

$$T^+ = \{\binom{h}{k} : [g,h] = [f,k] \qquad \text{for all} \qquad \binom{f}{g} \in T\}.$$

The c.l.r. T is called *symmetric* if $T \subset T^+$ and *selfadjoint* if $T = T^+$. In terms of the inner product $[\![\cdot,\cdot]\!]$, T is symmetric if and only if it is a $[\![\cdot,\cdot]\!]$-neutral subspace of $\mathcal{H}^2$ and selfadjoint if and only if it coincides with its $[\![\cdot,\cdot]\!]$-orthogonal companion. It is easy to see that the c.l.r. T is symmetric if and only if $[f,g]$ is real for all $\binom{f}{g} \in T$; it is called *nonnegative* ($T \geq 0$), if $[f,g] \geq 0$ for all $\binom{f}{g} \in T$. More generally, we say that the symmetric c.l.r. T has κ *negative squares* (for some $\kappa \in \mathbf{N}_0$) if the hermitian inner product $< \cdot,\cdot >$ on T, defined by

$$< \binom{x_1}{y_1}, \binom{x_2}{y_2} > := [x_1, y_2], \quad \binom{x_1}{y_1}, \binom{x_2}{y_2} \in T,$$

has this property. Evidently, $\kappa = 0$ means that T is nonnegative.

Theorem 1.1. *If A is a selfadjoint c.l.r. which has κ negative squares then either $\sigma(A) = \mathbf{C}$, or $\sigma(A) \setminus \mathbf{R}$ consists of an at most finite number of pairs α_j, $\bar{\alpha}_j$, $j = 1, 2, ..., n$, which are eigenvalues of A. In the second case, the linear span of the algebraic eigenspaces corresponding to $\alpha_1, ..., \alpha_n$ chosen in $\mathbf{C}^+$ is a neutral subspace of $\mathcal{H}$ of a dimension not greater than κ.*

Remark If $\alpha, \bar{\alpha}$ is a pair of nonreal isolated eigenvalues of finite algebraic multiplicity of the selfadjoint c.l.r. A with $\sigma(A) \neq \mathbf{C}$, then the algebraic eigenspaces of A corresponding to α and $\bar{\alpha}$ are skewly linked and their Jordan structures coincide as in the case of a selfadjoint operator, see [2, Theorem VI.6.5], [15, Proposition 3.2].

Proof. Assume that $\sigma(A) \neq \mathbf{C}$ and choose $z_0 \in \mathbf{C}^+ \cap \rho(A)$. Then also the bounded selfadjoint operator $B := R(\bar{z}_0; A)(I + z_0 R(z_0; A))$ has κ negative squares. Therefore its nonreal spectrum and the corresponding algebraic eigenspaces have the properties stated in Theorem 1.1 for the c.l.r. A (see [1]). Application of a version of the spectral mapping theorem yields the claim (comp. [9, §3]).

Corollary 1.2. *If A is a nonnegative selfadjoint c.l.r. then either $\sigma(A) = \mathbf{C}$ or $\sigma(A) \subset \mathbf{R}$.*

For an example of a nonnegative selfadjoint (unbounded) operator with empty resolvent set see [14]. Here is an example of a nonnegative selfadjoint c.l.r. A with $\sigma(A) = \mathbf{C}$ in a two-dimensional Krein space (comp. [8, §3]). Take $\mathcal{H} := (\mathbf{C}^2, [\cdot,\cdot])$ where $[\binom{x_1}{y_1}, \binom{x_2}{y_2}] := x_1\bar{x}_2 - y_1\bar{y}_2$, $e := \binom{1}{1}$ and

$$A := \{\binom{\alpha e}{\beta e} : \quad \alpha, \beta \in \mathbf{C}\}. \tag{1.2}$$

Then $[f, g] = 0$ for all $\binom{f}{g} \in A$, hence A is nonnegative. As A is two-dimensional it is selfadjoint. Moreover, $\mathcal{D}(A) = \mathcal{N}(A) = A(0) = \operatorname{span}\{e\}$, and $\sigma(A) = \mathbf{C}$ since $\binom{e}{ze} \in A$ for all $z \in \mathbf{C}$.

Theorem 1.3. *If A is a nonnegative selfadjoint c.l.r. in the Krein space $(\mathcal{H}, [\cdot,\cdot])$ then for $z \in \mathbf{C} \setminus \mathbf{R}$ it holds*

$$\mathcal{R}(A - z)^{[\perp]} = \mathcal{N}(A) \cap A(0) \quad (\subset A(0) \cap A(0)^{[\perp]}). \tag{1.3}$$

In particular, the set on the left hand side is independent of z and neutral.

Proof. If $y \in \mathcal{R}(A - z)^{[\perp]}$ then $y \in \mathcal{N}(A - \bar{z})$, or $\binom{y}{\bar{z}y} \in A$. As $\bar{z}[y, y]$ is real and $z \in \mathbf{C} \setminus \mathbf{R}$ it follows that $[y, y] = 0$. For arbitrary $\binom{u}{v} \in A$ the nonnegativity of A implies

$$|[\bar{z}y, u]|^2 \leq [\bar{z}y, y][v, u] = 0,$$

hence $[y, u] = 0$ for all $\binom{u}{v} \in A$, and the selfadjointness of A yields $\binom{0}{y} \in A$. Together with $\binom{y}{\bar{z}y} \in A$ it follows that also $\binom{y}{0} \in A$, and $y \in \mathcal{N}(A) \cap A(0)$ is proved.

Conversely, if $y \in \mathcal{N}(A) \cap A(0)$, $\binom{u}{v} \in A$ and $z \in \mathbf{C}$ then

$$[v - zu, y] = [v, y] - z[u, y] = 0,$$

or $y \in \mathcal{R}(A-z)^{[\perp]}$. The Theorem 1.3 is proved.

If for the nonnegative selfadjoint c.l.r. A the set in (1.3) contains nonzero elements, the following "reduction" of A is possible. Let $\mathcal{L}_0 := \mathcal{N}(A) \cap A(0)$. Then the factor space $\mathcal{H}_1 := \mathcal{L}_0^{[\perp]}/\mathcal{L}_0$, equipped with the inner product

$$[x+\mathcal{L}_0, y+\mathcal{L}_0] := [x,y], \qquad x, y \in \mathcal{L}_0^{[\perp]},$$

is a Krein space and

$$A_1 := \{\begin{pmatrix} x+\mathcal{L}_0 \\ y+\mathcal{L}_0 \end{pmatrix} : \begin{pmatrix} x \\ y \end{pmatrix} \in A, \quad x, y \in \mathcal{L}_0^{[\perp]}\}.$$

is a nonnegative selfadjoint c.l.r. in $\mathcal{H}_1$ with the property $\mathcal{N}(A_1) \cap A_1(0) = \{0\}$. That is, $\mathcal{R}(A_1 - z)$ is dense in $\mathcal{H}_1$ if $z \neq \bar{z}$.

2. Nonnegative closed linear relations of defect one

Let A_0 be a symmetric c.l.r. in the Krein space $(\mathcal{H}, [\cdot,\cdot])$. On each component Δ of the open set $r(A_0)$ the defect number dim $\mathcal{R}(A_0 - z)^{[\perp]}$, $z \in \Delta$, is constant (see [8, Section 2]). However, the upper and lower half planes $\mathbf{C}^+$ and $\mathbf{C}^-$, respectively, do in general not belong to $r(A_0)$ or can contain more than one component of $r(A_0)$. For this reason, in [18, §4] Yu. L. Shmul'yan has introduced the defect numbers of a symmetric operator A_0 in the Krein space $\mathcal{H}$ as the defect numbers of the neutral subspace A_0 (more precisely: the graph of A_0) in $(\mathcal{H}^2, [\![\cdot,\cdot]\!])$ (see (1.1)). For an arbitrary fundamental symmetry J of $\mathcal{H}$, these numbers coincide with the usual defect numbers of the symmetric operator $H_0 := JA_0$ in the Hilbert space $(\mathcal{H}, (\cdot,\cdot)_J)$ where $(x,y)_J := [Jx, y]$, x, $y \in \mathcal{H}$. In this paper we consider only a nonnegative c.l.r. of defect one:

Definition 2.1. The nonnegative c.l.r. A_0 in the Krein space $(\mathcal{H}, [\cdot,\cdot])$ is said to be of *defect one* if for some fundamental symmetry J of $\mathcal{H}$ the nonnegative c.l.r. JA_0 in the Hilbert space $(\mathcal{H}, (\cdot,\cdot)_J)$ is of defect one:

$$n_\pm(JA_0) := \dim \mathcal{R}(JA_0 \mp i)^\perp = 1. \tag{2.1}$$

As the c.l.r. A is a selfadjoint extension of A_0 in $(\mathcal{H}, [\cdot,\cdot])$ if and only if JA is a selfadjoint extension of JA_0 in $(\mathcal{H}, (\cdot,\cdot)_J)$, the following lemma is an immediate consequence of a particular case of the well-known von Neumann formula for the selfadjoint extensions of a symmetric relation (see [3], [7, Corollary 6.4]).

Lemma 2.2. *Let A_0 be a nonnegative c.l.r. in the Krein space $\mathcal{H}$. A_0 is of defect one if and only if there exists a selfadjoint extension A of A_0 such that*

$$\dim A/A_0 = 1.$$

In this case, if J is any fundamental symmetry and we choose

$$e \in \mathcal{R}(JA_0 + i)^\perp, \qquad f \in \mathcal{R}(JA_0 - i)^\perp, \qquad \|e\|_J = \|f\|_J = 1,$$

the selfadjoint extensions of A_0 *can be parametrized as*

$$(2.2)\qquad A_s := A_0 \dotplus \operatorname{span}\{\boldsymbol{e} + s\boldsymbol{f}\}, \qquad s \in \mathbf{T},$$

where $\boldsymbol{e} := \binom{e}{iJe}$, $\boldsymbol{f} := \binom{f}{-iJf}$.

By Lemma 2.2 the numbers $n_\pm(JA_0)$ in Definition 2.1 do not depend on the choice of the fundamental symmetry J and for two different selfadjoint extensions A, A' of A_0 it holds $A \cap A' = A_0$. The description (2.2) of the selfadjoint extensions of A_0 implies that an extension of A_0 is either nonnegative or has one negative square.

Here is a simple example which shows that for a nonnegative c.l.r. A_0 of defect one in a Krein space the set $r(A_0)$ of points of regular type can be empty. Take a Hilbert space $(\mathcal{H}_0, (\cdot,\cdot))$ and a closed nonnegative hermitian operator C_0 in $\mathcal{H}_0$ of defect one. In the Krein space $\mathcal{H} = \mathcal{H}_0 \oplus \mathcal{H}_0$ with the inner product generated by the Gram operator $G = \begin{pmatrix} 0 & I \\ I & 0 \end{pmatrix}$ we consider the nonnegative operator $A_0 = \begin{pmatrix} 0 & 0 \\ C_0 & 0 \end{pmatrix}$. It is easy to check that for $z \neq 0$ the range $\mathcal{R}(A - z)$ is the dense subset $\mathcal{D}(C_0) \oplus \mathcal{H}$. In the next section we introduce the notion of a nonnegative c.l.r. of regular defect one which excludes this possibility.

If the c.l.r. A_0 is nonnegative in $(\mathcal{H}, [\cdot,\cdot])$ and, hence, $H_0 := JA_0$ is nonnegative in the Hilbert space $(\mathcal{H}, (\cdot,\cdot)_J)$, there exists at least one nonnegative selfadjoint extension H of H_0. There are two extremal nonnegative selfadjoint extensions of H_0, the Friedrichs extension H_F and the von Neumann-Krein extension H_K. These extremal extensions coincide if and only if H_0 has only one nonnegative selfadjoint extension ([4]). Recall that

$$H_F := \{\tbinom{f}{g} \in H_0^* : f \in \mathcal{D}[H_0]\},$$

where $\mathcal{D}[H_0]$ is the completion of $\mathcal{D}(H_0)$ with respect to the norm $(\|x\|_J^2 + (y,x)_J)^{\frac{1}{2}}$ $(\binom{x}{y} \in H_0)$. The extension H_F is also characterized by the property $\mathcal{D}(H_F) \subset \mathcal{D}[H_0]$. Then $A_F := JH_F$ is called the *Friedrichs extension* of A_0 (in the Krein space $(\mathcal{H}, [\cdot,\cdot])$). It can be characterized by the relation

$$(2.3)\qquad A_F := \{\tbinom{f}{g} \in A_0^+ : \quad f \in \mathcal{D}[A_0]\},$$

where $\mathcal{D}[A_0]$ is the completion of $\mathcal{D}(A_0)$ with respect to the norm $(\|x\|_J^2 + [y,x])^{\frac{1}{2}}$ $(\binom{x}{y} \in A_0)$. Therefore, A_F is independent of the choice of the fundamental symmetry J. The *von Neumann-Krein extension* A_K of A_0 is the inverse of the Friedrichs extension of the nonnegative symmetric c.l.r. A_0^{-1}, or, what amounts to the same:

$$(2.4)\qquad A_K = \{\tbinom{f}{g} \in A_0^+ : \quad g \in \mathcal{D}[A_0^{-1}]\}.$$

The nonnegative c.l.r. A_0 has a unique nonnegative selfadjoint relation extension if and only if A_F and A_K coincide.

Next we recall M. G. Krein's formula describing the resolvents of all selfadjoint extensions of a symmetric c.l.r. H_0 in a Hilbert space $(\mathcal{H}, (\cdot,\cdot))$, see [16]. To this end we fix a selfadjoint extension H of H_0 in $(\mathcal{H}, (\cdot.\cdot))$ and a defect vector $e \in \mathcal{R}(H_0 + i)^\perp$. With the resolvent $R(z) := (H - z)^{-1}$ $(z \in \mathbf{C}^+ \cup \mathbf{C}^-)$ we define the vector function g:

$$g(z) := e + (z - i)R(z)e$$

and the scalar function

$$Q(z) := i(g(i), g(i)) + (z - i)(g(z), g(-i)) = z(e, e) + (z^2 + 1)(R(z)e, e).$$

Then the relation

$$R(z; H_{(\gamma)}) = R(z) - (\gamma + Q(z))^{-1}(\cdot, g(\bar{z}))\, g(z), \qquad z \neq \bar{z}, \tag{2.5}$$

defines a bijection between the set of all $\gamma \in \bar{\mathbf{R}} := \mathbf{R} \cup \{\infty\}$ and all resolvents $R(z; H_{(\gamma)}) := (H_{(\gamma)} - z)^{-1}$ of the selfadjoint extensions of H_0 in $\mathcal{H}$. Here, of course, if $\gamma = \infty$ the second term on the right hand side of (2.5) is supposed to be zero, that is, $H_{(\infty)} = H$.

If A_0 and hence also H_0 are nonnegative, we describe the set of all parameters γ in (2.5) for which the corresponding selfadjoint c.l.r. $H_{(\gamma)}$ is also nonnegative. To this end we choose the fixed extension H in (2.5) nonnegative, denote by P_0 and P_∞ the orthogonal projections onto $\overline{\mathcal{D}(H)}$ and $\overline{\mathcal{R}(H)}$, respectively, and by H_{op} and $(H^{-1})_{op}$ the operator parts of H and H^{-1}:

$$H_{op} := H \cap (P_0\mathcal{H})^2, \qquad (H^{-1})_{op} := H^{-1} \cap (P_\infty\mathcal{H})^2.$$

If $x \in \mathcal{H}$ we define

$$\begin{aligned} |x|_+ &:= \begin{cases} \|H_{op}^{\frac{1}{2}} x\| & \text{if} \quad x \in \mathcal{D}(H_{op}^{\frac{1}{2}}) \\ \infty & \text{otherwise} \end{cases} \\ |x|_- &:= \begin{cases} \|(H^{-1})_{op}^{\frac{1}{2}} x\| & \text{if} \quad x \in \mathcal{D}((H^{-1})_{op}^{\frac{1}{2}}) \\ \infty & \text{otherwise} \end{cases} \end{aligned} \tag{2.6}$$

and write, if necessary, more fully $|x|_\pm^{(H)}$ instead of $|x|_\pm$. Put

$$\gamma_+ := |g(i)|_+^2, \qquad \gamma_- := -|g(i)|_-^2. \tag{2.7}$$

Observe that $\mathcal{D}(H_{op}^{\frac{1}{2}}) = \mathcal{D}[H]$ and $\mathcal{D}((H^{-1})_{op}^{\frac{1}{2}}) = \mathcal{D}[H^{-1}]$. It is easy to see that the unitary operator $I - 2iR(-i)$ in $\mathcal{H}$ maps $\mathcal{D}(H_{op}^{\frac{1}{2}})$ $(\mathcal{D}((H^{-1})_{op}^{\frac{1}{2}}))$ isometrically with respect to $\|H_{op}^{\frac{1}{2}} \cdot \|$ (resp. $\|(H^{-1})_{op}^{\frac{1}{2}} \cdot \|$) onto itself. Since $g(-i) = (I - 2iR(-i))g(i)$ it follows that

$$\gamma_+ = |g(-i)|_+^2, \qquad \gamma_- = -|g(-i)|_-^2. \tag{2.8}$$

Lemma 2.3. *Assume that H is nonnegative in $(\mathcal{H}, (\cdot,\cdot))$. Then $H_{(\gamma)}$ is nonnegative if and only if $\gamma \notin (\gamma_-, \gamma_+)$; $H_{(\gamma_-)}$ is the von Neumann-Krein extension and $H_{(\gamma_+)}$ is the Friedrichs extension of H_0.*

Proof. One easily verifies that Q is a Nevanlinna function (that is, Q is holomorphic in $\mathbf{C}^+ \cup \mathbf{C}^-$, $\operatorname{Im} z \operatorname{Im} Q(z) \geq 0$ and $Q(\bar{z}) = \overline{Q(z)}$ if $z \in \mathbf{C}^+ \cup \mathbf{C}^-$) which is holomorphic on $(-\infty, 0)$. Moreover, Q is not constant on $\mathbf{C}^+$ and $\mathbf{C}^-$. Indeed, if Q would be constant on $\mathbf{C}^+$ and on $\mathbf{C}^-$, we would have

$$\operatorname{Im} Q'(i) = \operatorname{Im}\{(e, e) + 2i(R(i)e, e)\} = ((R(i) + R(-i))e, e) = 0$$

and, as $R(i)+R(-i)$ is a nonnegative operator,

$$0=(R(i)+R(-i))e=(R(i)+R(-i))P_0e.$$

This implies $e=e_0+e_\infty$ with some $e_0\in\mathcal{N}(H)$, $e_\infty\in H(0)$ and

$$Q(z)=z(e_0,e_0)+z(e_\infty,e_\infty)+(z^2+1)(R(z)e_0,e_0)=z(e_\infty,e_\infty)-z^{-1}(e_0,e_0).$$

Hence $e=0$, a contradiction.

It follows that Q restricted to $(-\infty,0)$ is a strictly increasing function. Set

$$M:=\sup\{Q(t):\ t\in(-\infty,0)\},\qquad m:=\inf\{Q(t):\ t\in(-\infty,0)\}.$$

By (2.5), $H_{(\gamma)}$ has a negative eigenvalue if and only if $\gamma\in(-M,-m)$.

We show that $M=|e|_-^2$ and $m=-|e|_+^2$. If $E(\cdot)$ is the spectral function of H_{op} in $P_0\mathcal{H}$ and $e':=P_0e$, then

$$M=\lim_{n\to\infty}Q(-n^{-1})=\lim_{n\to\infty}(R(-n^{-1})e',e')=\lim_{n\to\infty}\int_{-0}^{\infty}(s+n^{-1})^{-1}d(E(s)e',e').$$

Hence M is finite if and only if $e'\in\mathcal{D}(H_{op}^{-\frac{1}{2}})$ or, equivalently, $e\in\mathcal{D}((H^{-1})_{op}^{\frac{1}{2}})$. In this case $M=|e|_-^2$. Similarly,

$$\begin{aligned} m &= \lim_{n\to\infty}Q(-n)= \\ &= \lim_{n\to\infty}\{-n\|e-e'\|^2-n(e',e')+(1+n^2)\int_{-0}^{\infty}(s+n)^{-1}d(E(s)e',e')\}= \\ &= \lim_{n\to\infty}\{-n\|e-e'\|^2-\int_{-0}^{\infty}ns(s+n)^{-1}d(E(s)e',e')\}. \end{aligned}$$

It is easy to see that m is finite if and only if $e=e'$ and $e'\in\mathcal{D}(H_{op}^{\frac{1}{2}})$. If this holds, then $m=-|e|_+^2$.

The last assertion of the lemma is a consequence of [4, Theorem 5].

3. Nonnegative closed linear relations of regular defect one

For a nonnegative c.l.r. A_0 of defect one in a Krein space the ranges $\mathcal{R}(A_0-z)$, $z\neq\bar z$, need not be closed (see the example following Lemma 2.2). In order to exclude this possibility we suppose that the ranges $\mathcal{R}(A_0-z_0)$ and $\mathcal{R}(A_0-\bar z_0)$ are closed for at least one point $z_0\in\mathbf{C}^+\cup\mathbf{C}^-$.

Definition 3.1. The nonnegative c.l.r. A_0 in the Krein space $(\mathcal{H},[\cdot,\cdot])$ is said to be of *regular defect one*, if there exist points $z_0,\bar z_0\in r(A_0)$, $z_0\neq\bar z_0$, such that the subspaces $\mathcal{R}(A_0-z_0)^{[\perp]}$ and $\mathcal{R}(A_0-\bar z_0)^{[\perp]}$ are both of dimension one.

If A_0 is of regular defect one we have

(3.1) $$\mathcal{N}(A_0)\cap A_0(0)=\{0\}.$$

Indeed, if $\binom{x}{0}\in A_0$ and $\binom{0}{x}\in A_0$ it follows that $\binom{x}{zx}\in A_0$ for all $z\in\mathbf{C}$, which is impossible as $r(A_0)\neq\emptyset$.

The following theorem implies that a nonnegative c.l.r. of regular defect one is of defect one.

Theorem 3.2. *The nonnegative c.l.r. A_0 in the Krein space $(\mathcal{H}, [\cdot,\cdot])$ is of regular defect one if and only if it is of defect one and there exists a selfadjoint extension A of A_0 in $\mathcal{H}$ with $\rho(A) \neq \emptyset$.*

Proof. 1. Suppose that A_0 is of defect one and that there exists a selfadjoint relation extension A of A_0 with $\rho(A) \neq \emptyset$. According to Theorem 1.1, the nonreal spectrum of A is empty or consists of one pair $\lambda_0,\bar{\lambda}_0$. The relation $\|y - zx\| \geq C\|x\|$ for all $\binom{x}{y} \in A$ ($\|\cdot\|$ being an arbitrary Hilbert majorant of $[\cdot,\cdot]$) implies the same for all $\binom{x}{y} \in A_0$, hence $\rho(A) \subset r(A_0)$. For every $z \in \rho(A)$ the linear mapping $\binom{x}{y} \longmapsto y - zx$ is an isomorphism of A onto $\mathcal{H}$. Hence

$$\operatorname{def} \mathcal{R}(A_0 - z) = \operatorname{def} \mathcal{R}(A_0 - \bar{z}) = \operatorname{def}_A A_0 = 1$$

and A_0 is of regular defect one.

2. Suppose now that A_0 has regular defect one and let z_0, $\bar{z}_0$ be points of regular type of A_0 such that

$$\operatorname{def} \mathcal{R}(A_0 - z_0) = \operatorname{def} \mathcal{R}(A_0 - \bar{z}_0) = 1.$$

If for a selfadjoint extension A of A_0 we have z_0, $\bar{z}_0 \in \sigma(A)$, then

$$\mathcal{R}(A - z) = \mathcal{R}(A_0 - z) \qquad \text{for} \quad z = z_0, \bar{z}_0. \tag{3.2}$$

Indeed, assume that $\mathcal{R}(A - z_0) = \mathcal{H}$. Then $A - \bar{z}_0$ is injective and, hence, $\mathcal{R}(A - \bar{z}_0) = \mathcal{H}$. This implies $z_0,\bar{z}_0 \in \rho(A)$, a contradiction.

The theorem will be proved if we show that (3.2) cannot hold for all selfadjoint extensions A of A_0. Assume that (3.2) holds for all selfadjoint extensions A of A_0. Let A' be a nonnegative selfadjoint extension of A_0. Then, by Theorem 1.3,

$$\mathcal{R}(A_0 - z)^{[\perp]} = \mathcal{R}(A' - z)^{[\perp]} = \mathcal{N}(A') \cap A'(0), \qquad z = z_0, \bar{z}_0.$$

Using again (3.2) we deduce

$$\mathcal{R}(A - z_0)^{[\perp]} = \mathcal{R}(A - \bar{z}_0)^{[\perp]} = \operatorname{span}\{g\}$$

with some $g \in \mathcal{H}$, $g \neq 0$, $[g,g] = 0$, hence $\binom{g}{z_0 g}$, $\binom{g}{\bar{z}_0 g} \in A$ for all selfadjoint extensions A of A_0. On the other hand, if $\binom{x}{y} \in A$ for all selfadjoint extensions A of A_0, then $\binom{x}{y} \in A_0$, which follows easily from the corresponding fact in a Hilbert space if A_0 and A are replaced by JA_0 and JA. We conclude that $\binom{g}{z_0 g}$, $\binom{g}{\bar{z}_0 g} \in A_0$ which is impossible because of z_0, $\bar{z}_0 \in r(A_0)$. Hence there exists a selfadjoint extension A of A_0 with z_0, $\bar{z}_0 \in \rho(A)$.

Then, in view of $\operatorname{def} \mathcal{R}(A_0 - z_0) = 1$, it follows that $\dim A/A_0 = 1$ and, by Lemma 2.2, that A_0 is of defect one.

Remark 1. If A_0 is a nonnegative c.l.r. of regular defect one and A is a selfadjoint extension of A_0 in $\mathcal{H}$, then $\rho(A) \subset r(A_0)$. In particular, all nonreal z, with the possible exception of one pair λ_0, $\bar{\lambda}_0 \in \sigma(A)$, belong to $r(A_0)$ and

$$\dim \mathcal{R}(A_0 - z)^{[\perp]} = 1 \quad \text{for all} \quad z \in r(A_0).$$

Remark 2. If A_0, A are as in Remark 1 and $z,\bar{z} \in \rho(A), z \neq \bar{z}$, then the Cayley transform $U = I + (z - \bar{z})(A - z)^{-1}$ maps the defect subspace $\mathcal{R}(A_0 - z)^{[\perp]}$ onto $\mathcal{R}(A_0 - \bar{z})^{[\perp]}$. As it is an isometry in the Krein space $(\mathcal{H}, [\cdot,\cdot])$, the two defect subspaces are of the same signature (positive, negative or neutral). For the connections between the existence of selfadjoint extensions with nonempty resolvent set and the signatures of the defect spaces in more general situations see [1, §V.2].

Let again A_0 be a nonnegative c.l.r. of regular defect one and let A be a selfadjoint extension of A_0 with $\rho(A) \neq \emptyset$ as in Theorem 3.2. Then by Theorem 1.1 the nonreal spectrum of A is either empty or it consists of a pair λ_0, $\bar{\lambda}_0$ of eigenvalues. As in formula (2.5), we can introduce defect vectors of A_0, which we denote again by $g(z)$:

$$(3.3) \qquad g(z) \in \mathcal{R}(A_0 - \bar{z})^{[\perp]}, \qquad z \in \rho(A),$$

which depend holomorphically on z. To this end we fix some $z_0 \in \mathbf{C}^+ \cap \rho(A)$ and $g(z_0) \in \mathcal{R}(A - \bar{z}_0)^{[\perp]}$, $g(z_0) \neq 0$ and define with $R(z) := (A - z)^{-1}$

$$(3.4) \qquad g(z) := g(z_0) + (z - z_0)R(z)g(z_0), \qquad z \in \rho(A).$$

Then for $\binom{x}{y} \in A_0$ we have

$$[y - \bar{z}x, g(z)] = [y - \bar{z}x, g(z_0) + (z - z_0)R(z)g(z_0)] =$$

$$= [(I + (\bar{z} - \bar{z}_0)R(\bar{z}))(y - \bar{z}x), g(z_0)] = [y - \bar{z}x + (\bar{z} - \bar{z}_0)x, g(z_0)] = 0.$$

Moreover, $g(z) \neq 0$ as otherwise we would have

$$g(z_0) = -(z - z_0)R(z)g(z_0),$$

that is $\binom{g(z_0)}{-(z-z_0)g(z_0)} \in A - z$ or $\binom{g(z_0)}{z_0 g(z_0)} \in A$ which is impossible as $z_0 \in \rho(A)$. It is easy to check that for z, $\zeta \in \rho(A)$ it holds

$$(3.5) \qquad g(z) - g(\zeta) = (z - \zeta)R(z)g(\zeta).$$

In the following the set

$$\Delta_0 := \{z \in \mathbf{C}^+ \cup \mathbf{C}^- : \ \mathcal{R}(A_0 - z)^{[\perp]} \text{ is neutral}\}$$

plays some role. Analogously to the operator case (see [12], [14]) we have

Theorem 3.3. *If A_0 is a nonnegative c.l.r. of regular defect one, then either Δ_0 does not contain an inner point or it is all of $\mathbf{C}^+ \cup \mathbf{C}^-$ with possible exception of at most one pair of points λ_0, $\bar{\lambda}_0$.*

Proof. Let A be a selfadjoint extension of A_0 with $\rho(A) \neq \emptyset$, and assume that Δ_0 contains an open disc. In this disc we find a point $z_0 \in \rho(A)$, and define $g(z)$ according to (3.4) for all $z \in \rho(A) \cap (\mathbf{C}^+ \cup \mathbf{C}^-)$, that is for all nonreal z with the possible exception of one pair of points λ_0, $\bar{\lambda}_0$ which are eigenvalues of A. Then the function

$$f(z, \zeta) := [g(z), g(\zeta)], \qquad z, \zeta \in \rho(A),$$

has the property $f(z, z) = 0$ for all z in some neighbourhood of z_0. This implies (see [17]) that $f(z, \zeta) = 0$ for all z, ζ in this neighbourhood, and by analytic continuation we find $f(z, z) = 0$ for all $z \in \rho(A) \cap \mathbf{C}^+$ and, according to the relation

$$[g(z), g(z)] = [g(\bar{z}), g(\bar{z})], \qquad z \in \rho(A) \cap \mathbf{C}^+$$

(comp. Remark 2 following Theorem 3.2) for all $z \in \mathbf{C} \setminus \{\lambda_0, \bar{\lambda}_0\}$, and Theorem 3.3 is proved.

Here is an example of a nonnegative operator A_0 of regular defect one such that an (arbitrarily chosen) $\lambda_0 \in \mathbf{C}$ belongs to $\sigma_p(A_0)$. Consider the two-dimensional Krein space $\mathcal{K}$ spanned by a pair of neutral vectors g, g' which are skewly linked. Then the operator

$$A_0 := \operatorname{span}\{\begin{pmatrix} g \\ \lambda_0 g \end{pmatrix}\} \tag{3.6}$$

is of regular defect one, $\lambda_0 \in \sigma_p(A_0)$ and, as $[\lambda_0 g, g] = 0$, it is nonnegative. Further, $\mathcal{R}(A_0 - z)^{[\perp]} = \{(\lambda_0 - z)g\}^{[\perp]}$ is a neutral subspace for all $z \neq \lambda_0$, hence

$$\Delta_0 = (\mathbf{C}^+ \cup \mathbf{C}^-) \setminus \{\lambda_0\}.$$

It is easy to see that

$$\hat{A} := \operatorname{span}\{\begin{pmatrix} g \\ 0 \end{pmatrix}, \begin{pmatrix} 0 \\ g \end{pmatrix}\}$$

is a selfadjoint extension of A_0 with $\sigma_p(\hat{A}) = \mathbf{C}$.

In the following section we shall show, that a nonnegative c.l.r. A_0 in $\mathcal{K}$ with a nonreal eigenvalue is of a simple structure: It is the $[\cdot,\cdot]$-orthogonal sum of a nonnegative selfadjoint relation with real spectrum and a c.l.r. A_0 of the form (3.6) (with $\lambda_0 \neq \bar{\lambda}_0$) in a two-dimensional Krein space.

4. Nonnegative closed linear relations with a nonreal eigenvalue

The main result of this section is the following

Theorem 4.1. *Let A_0 be a nonnegative c.l.r. of regular defect one in the Krein space $\mathcal{H}$ which has a nonreal eigenvalue $\lambda_0 \in \sigma_p(A_0)$; denote by g_0 a corresponding eigenvector. Then the space $\mathcal{L}_0 := \mathcal{R}(A_0 - \lambda_0)^{[\perp]}$ is nondegenerate,* $\dim \mathcal{L}_0 = 2$ *and*

$$A_0 = (A_0 \cap (\mathcal{L}_0^{[\perp]})^2) \dotplus A_0 \cap \mathcal{L}_0^2, \tag{4.1}$$

where $A_0 \cap (\mathcal{L}_0^{[\perp]})^2$ is a nonnegative selfadjoint c.l.r. in the Krein space $\mathcal{L}_0^{[\perp]}$ with real spectrum and $A_0 \cap \mathcal{L}_0^2$ is a nonnegative c.l.r. of regular defect one in the two-dimensional Krein space $\mathcal{L}_0$, in fact

$$A_0 \cap \mathcal{L}_0^2 = \operatorname{span}\{\begin{pmatrix} g_0 \\ \lambda_0 g_0 \end{pmatrix}\}. \tag{4.2}$$

In particular, we have $\bar{\lambda}_0 \in r(A_0)$, $\mathcal{R}(A_0 - z)^{[\perp]} = \operatorname{span}\{g_0\}$ $(z \neq \lambda_0)$ and $\Delta_0 = (\mathbf{C}^+ \cup \mathbf{C}^-) \setminus \{\lambda_0\}$.

First we prove two lemmas; A_0 is always supposed to satisfy the assumptions of Theorem 4.1.

Lemma 4.2. *The subspace $\mathcal{R}(A_0-\lambda_0)^{[\perp]}$ is of dimension two and nondegenerate, in fact, if A is any selfadjoint extension of A_0 with $\rho(A) \neq \emptyset$, $\mathcal{R}(A_0-\lambda_0)^{[\perp]}$ is the linear span of $\mathcal{N}(A-\lambda_0)$ and $\mathcal{N}(A-\bar{\lambda}_0)$.*

Proof. According to Theorem 1.1 λ_0 is the only nonreal eigenvalue of A_0 and it is of algebraic multiplicity one. Let g_0 be a corresponding eigenvector. Evidently, $[g_0, g_0] = 0$. If $\widetilde{A}$ is a nonnegative selfadjoint extension of A_0, it follows from Theorem 1.3 that $g_0 \in \mathcal{N}(\widetilde{A}-\lambda_0) = \mathcal{N}(\widetilde{A}) \cap \widetilde{A}(0)$. This implies

$$g_0 \in \mathcal{N}(\widetilde{A}-z) \quad \text{for all} \quad z \in \mathbf{C},$$

hence

$$g_0 \in \mathcal{N}(\widetilde{A}-\bar{\lambda}_0) \subset \mathcal{N}(A_0^+-\bar{\lambda}_0) = \mathcal{R}(A_0-\lambda_0)^{[\perp]}.$$

If A is any selfadjoint extension of A_0 with $\rho(A) \neq \emptyset$, then $\bar{\lambda}_0 \in \sigma_p(A)$, and a corresponding eigenvector h_0 of A, $[h_0, h_0] = 0$, can be chosen such that $[h_0, g_0] = 1$. Then

$$h_0 \in \mathcal{N}(A-\bar{\lambda}_0) \subset \mathcal{N}(A_0^+-\bar{\lambda}_0) = \mathcal{R}(A_0-\lambda_0)^{[\perp]}. \tag{4.3}$$

Thus $\dim \mathcal{R}(A_0-\lambda_0)^{[\perp]} \geq 2$. On the other hand

$$\dim \mathcal{R}(A-\lambda_0)^{[\perp]} = \dim \mathcal{N}(A-\bar{\lambda}_0) = 1$$

and $\dim \mathcal{R}(A-\lambda_0)/\mathcal{R}(A_0-\lambda_0) \leq 1$, hence

$$\dim \mathcal{R}(A_0-\lambda_0)^{[\perp]} \leq 2.$$

Lemma 4.3. *There exists a unique nonnegative selfadjoint extension $\hat{A}$ of A_0:*

$$\hat{A} = A_0 \dotplus \operatorname{span}\{\tbinom{g_0}{0}, \tbinom{0}{g_0}\}, \tag{4.4}$$

where $g_0 \in \mathcal{N}(A_0-\lambda_0)$, $g_0 \neq 0$. Evidently, $\sigma_p(\hat{A}) = \mathbf{C}$.

Proof. Let $\widetilde{A}$ be an arbitrary nonnegative selfadjoint extension of A_0. In the proof of Lemma 4.2 it was shown that

$$g_0 \in \mathcal{N}(\widetilde{A}) \cap \widetilde{A}(0). \tag{4.5}$$

Hence $g_0 \in \mathcal{N}(A_0^+) \cap A_0^+(0)$, and it follows that

$$[g_0, \mathcal{D}(A_0)] = [g_0, \mathcal{R}(A_0)] = \{0\}. \tag{4.6}$$

Therefore, if $\binom{x}{y} \in A_0$ and $\xi_1, \xi_2 \in \mathbf{C}$ we find

$$[x+\xi_1 g_0, y+\xi_2 g_0] = [x, y] \geq 0,$$

that is, the expression on the right hand side of (4.4) is a nonnegative c.l.r. which is (because of (4.5)) contained in $\widetilde{A}$. As $\mathcal{N}(A_0) \cap A_0(0) = \{0\}$ (see (3.1)), the right hand side of (4.4) is a proper extension of A_0 and hence selfadjoint. Therefore it must coincide with each nonnegative selfadjoint extension $\tilde{A}$.

Proof of Theorem 4.1. Denote by Q_0 the orthogonal projection in $\mathcal{H}$ onto $\mathcal{L}_0$ and $P_0 := I - Q_0$. According to Lemma 4.2 and with the vectors g_0, h_0 introduced in the proof of Lemma 4.2, we have

$$Q_0 = [\cdot, h_0]g_0 + [\cdot, g_0]h_0.$$

If $\binom{x}{y} \in A_0$ write

$$\binom{x}{y} = \binom{x-[x,h_0]g_0}{y-[x,h_0]\lambda_0 g_0} + \binom{[x,h_0]g_0}{[x,h_0]\lambda_0 g_0}. \tag{4.7}$$

The relations (4.6) imply $Q_0 x = [x, h_0]g_0$, $Q_0 y = [y, h_0]g_0$, hence

$$Q_0(y - [x, h_0]\lambda_0 g_0) = [y, h_0]g_0 - \lambda_0[x, h_0]g_0 = [y - \lambda_0 x, h_0]g_0 = 0$$

(observe (4.3)). Therefore the first vector on the right hand side of (4.7) belongs to $(\mathcal{L}_0^{[\perp]})^2$, the second one to $\mathcal{L}_0^2$, which proves the relations (4.1) and (4.2).

It remains to prove that $A_0 \cap (\mathcal{L}_0^{[\perp]})^2$ is a selfadjoint c.l.r. in $\mathcal{L}_0^{[\perp]}$ with real spectrum. From (4.3) it follows that the c.l.r.

$$A := A_0 \dotplus \operatorname{span}\{\binom{h_0}{\bar{\lambda}_0 h_0}\}$$

is symmetric. As $\bar{\lambda}_0 \notin \sigma_p(A_0)$, it is a proper extension of A_0 and hence selfadjoint. Consider any nonreal λ_1, $\lambda_1 \neq \lambda_0, \bar{\lambda}_0$. Then $h_0 \in \mathcal{R}(A - \lambda_1) \cap \mathcal{R}(A - \bar{\lambda}_1)$. On the other hand, on account of (4.1) and (4.2)

$$h_0 \notin \mathcal{R}(A_0 - \lambda_1) \quad \text{and} \quad h_0 \notin \mathcal{R}(A_0 - \bar{\lambda}_1). \tag{4.8}$$

Then in view of Remark 1 after Theorem 3.2 it follows that $\mathcal{R}(A - \lambda_1) = \mathcal{R}(A - \bar{\lambda}_1) = \mathcal{H}$ and, hence, $\lambda_1, \bar{\lambda}_1 \in \rho(A)$. Further,

$$A_0 \cap (P_0\mathcal{H})^2 = A \cap (P_0\mathcal{H})^2. \tag{4.9}$$

Indeed, if $\binom{x}{y} \in A \cap (P_0\mathcal{H})^2$ then $\binom{x}{y}$ is of the form $\binom{x_0}{y_0} \dotplus c\binom{h_0}{\bar{\lambda}_0 h_0}$ with $\binom{x_0}{y_0} \in A_0$, $c \in \mathbf{C}$, and

$$Q_0(x_0 + ch_0) = [x_0, h_0]g_0 + [x_0, g_0]h_0 + ch_0 = [x_0, h_0]g_0 + ch_0 = 0.$$

It follows that $c = 0$ and $\binom{x}{y} \in A_0 \cap (P_0\mathcal{H})^2$, which proves (4.9).

Since $\sigma(A) \setminus \mathbf{R} = \{\lambda_0, \bar{\lambda}_0\}$, λ_0 and $\bar{\lambda}_0$ are simple eigenvalues and g_0, h_0 are corresponding eigenvectors, the relation $A \cap (P_0\mathcal{H})^2$ is selfadjoint and its spectrum is real. The Theorem 4.1 is proved.

We mention that under the assumptions of Theorem 4.1, the holomorphic defect vectors $z \longmapsto g(z)$ of A_0 satisfying (3.5) are of the form

$$g(z) = (\lambda_0 - \zeta)(\lambda_0 - z)^{-1} g_0, \tag{4.10}$$

where ζ is an arbitrary nonreal point with $\zeta \neq \lambda_0$.

The decomposition (4.1) of the nonnegative c.l.r. A_0 of regular defect one in Theorem 4.1 reduces the problem of describing all selfadjoint extensions of A_0 to a description of all selfadjoint extensions of the nonnegative c.l.r. $A^0 := A_0 \cap \mathcal{L}_0^2$ in the two-dimensional Krein space $\mathcal{L}_0$. To this end we introduce the elements $u_s := h_0 + g_0 + s(h_0 - g_0)$, $s \in \mathbf{T}$. Then the set of all selfadjoint extensions of A^0 is the set of all

$$A_s^0 := \operatorname{span}\{\binom{g_0}{\lambda_0 g_0}, \binom{u_s}{\bar{\lambda}_0 u_s}\}, \qquad s \in \mathbf{T}.$$

The proof of this purely algebraic statement is left to the reader. We only mention that

$$\sigma(A_s^0) = \begin{cases} \{\lambda_0, \bar{\lambda}_0\} & \text{if } s \neq -1, \\ \mathbf{C} & \text{if } s = -1. \end{cases}$$

If $s \neq -1$, the eigenvector corresponding to λ_0 ($\bar{\lambda}_0$) is g_0 (u_s, respectively); if $s = -1$, g_0 is eigenvector for each $z \in \mathbf{C}$. The extension A_{-1}^0 is nonnegative and A_s^0, $s \neq -1$, is not nonnegative. Evidently, the nonnegative selfadjoint extension $\hat{A}$ of A_0 in Lemma 4.3 is given by $\hat{A} = A_0 \dotplus A_{-1}^0$.

A converse statement to the fact that for a c.l.r. A_0 as in Theorem 4.1 every selfadjoint extension has the eigenvalues λ_0, $\bar{\lambda}_0$ is contained in the

Theorem 4.4. *If A_0 is a nonnegative c.l.r. of regular defect one in the Krein space $\mathcal{H}$ and there exist two different selfadjoint extensions A, A' of A_0 such that for some nonreal λ_0 the points λ_0 and $\bar{\lambda}_0$ belong to $\sigma(A) \cap \sigma(A')$, then either λ_0 or $\bar{\lambda}_0$ belong to $\sigma_p(A_0)$.*

Proof. The points λ_0 and $\bar{\lambda}_0$ belong to the spectra of all selfadjoint extensions of A_0. Otherwise we would have $\lambda_0, \bar{\lambda}_0 \in r(A_0)$ and

$$\mathcal{R}(A_0 - \lambda_0) = \mathcal{R}(A - \lambda_0) = \mathcal{R}(A' - \lambda_0)$$

(cf. the reasoning following (3.2)). Hence $0 \neq g \in \mathcal{R}(A_0 - \lambda_0)^{[\perp]}$ implies $\binom{g}{\bar{\lambda}_0 g} \in A \cap A' \subset A_0$ and $\bar{\lambda}_0 \in \sigma_p(A_0)$. Analogously $\lambda_0 \in \sigma_p(A_0)$, a contradiction (see Theorem 4.1).

Let $\hat{A}$ be a nonnegative selfadjoint extension of A_0. Then, according to Theorem 1.1, $\rho(\hat{A}) = \emptyset$. If $z_0, \bar{z}_0 \in r(A_0)$, $z_0 \neq \bar{z}_0$, and $\operatorname{def} \mathcal{R}(A_0 - z_0) = 1$, then $\mathcal{R}(\hat{A} - z_0) = \mathcal{R}(A_0 - z_0)$ and, by Theorem 1.3,

$$\{0\} \neq \mathcal{N}(\hat{A}) \cap \hat{A}(0) =: \operatorname{span}\{g\}.$$

Since

$$\mathcal{R}(A_0 - \lambda) \subset \mathcal{R}(\hat{A} - \lambda), \qquad \lambda = \lambda_0, \bar{\lambda}_0,$$

we have

(4.11) $$g \in \mathcal{R}(A_0 - \lambda)^{[\perp]}, \qquad \lambda = \lambda_0, \bar{\lambda}_0.$$

Let A'' be a selfadjoint extension of A_0 such that $\sigma(A'') \cap (\mathbf{C} \setminus \mathbf{R}) = \{\lambda_0, \bar{\lambda}_0\}$. Then λ_0 and $\bar{\lambda}_0$ are simple eigenvalues of A'', the subspaces $\mathcal{R}(A'' - \lambda_0)$ and $\mathcal{R}(A'' - \bar{\lambda}_0)$ are closed and their defects are equal to one.

Suppose that $A_0 - \lambda_0$ and $A_0 - \bar{\lambda}_0$ are invertible. If $\mathcal{R}(A_0 - \lambda_0)$ is closed, then $\lambda_0 \in r(A_0)$ and it follows that $\operatorname{def} \mathcal{R}(A_0 - \lambda_0) = 1$, i.e.

$$\mathcal{R}(A_0 - \lambda_0) = \mathcal{R}(A'' - \lambda_0).$$

If $\mathcal{R}(A_0 - \lambda_0)$ is not closed, then the fact that

$$\mathcal{R}(A_0 - \lambda_0) + x_0 = \mathcal{R}(A'' - \lambda_0)$$

for some $x_0 \in \mathcal{H}$ implies that

$$\overline{\mathcal{R}(A_0 - \lambda_0)} = \mathcal{R}(A'' - \lambda_0).$$

In both cases we have, in view of (4.11), $\mathcal{R}(A_0 - \lambda_0)^{[\perp]} = \operatorname{span}\{g\}$. Analogously for λ_0 replaced by $\bar{\lambda}_0$, $\mathcal{R}(A_0 - \bar{\lambda}_0)^{[\perp]} = \operatorname{span}\{g\}$. This implies that the eigenvectors of A'' corresponding to the eigenvalues λ_0 and $\bar{\lambda}_0$ coincide, a contradiction. Hence either λ_0 or $\bar{\lambda}_0$ is an eigenvalue of A_0.

5. The nonreal spectrum of the selfadjoint extensions

If the nonnegative c.l.r. A_0 of regular defect one has a nonreal eigenvalue, the structure of A_0 was described completely in Theorem 4.1. Therefore in this and the next section we restrict ourselves to the case $\sigma_p(A_0) \subset \mathbf{R}$. Then for each nonreal λ we find a selfadjoint extension A such that $\lambda \in \rho(A)$ (see Theorem 4.4), hence

$$\mathbf{C}^+ \cup \mathbf{C}^- \subset r(A_0).$$

In [12] it was shown that for a densely defined symmetric operator A_0 with equal defect numbers in a Pontryagin space Π a point $z \in (\mathbf{C}^+ \cup \mathbf{C}^-) \cap r(A_0)$ belongs to the spectrum of some selfadjoint extension A in Π if and only if the defect subspace $\mathcal{R}(A_0 - z)^{[\perp]}$ contains a nonzero neutral vector. An analogous result holds in the situation considered here.

Theorem 5.1. *Let A_0 be a nonnegative c.l.r. of regular defect one with $\sigma_p(A_0) \subset \mathbf{R}$. Then $z \neq \bar{z}$ belongs to the spectrum of some selfadjoint extension A of A_0 if and only if $z \in \Delta_0$. Moreover, for each $z \in \Delta_0$ there is exactly one selfadjoint extension A of A_0 with $z \in \sigma(A)$.*

Proof. If A is a selfadjoint extension of A_0 and $z \in \sigma(A)$, $z \neq \bar{z}$, then

$$\mathcal{R}(A - z)^{[\perp]} = \mathcal{R}(A_0 - z)^{[\perp]}$$

(cf. the reasoning following (3.2)). For a nonzero element g of this subspace it follows then $g \in \mathcal{N}(A - \bar{z})$, that is $\binom{g}{\bar{z}g} \in A$. As $\bar{z}[g, g]$ must be real, g is neutral.

Conversely, consider $z \in \Delta_0$ and denote by g a nonzero defect vector: $\mathcal{R}(A_0 - z)^{[\perp]} = \operatorname{span}\{g\}$. Since $\bar{z} \notin \sigma_p(A_0)$ we have $\binom{g}{\bar{z}g} \notin A_0$ and the c.l.r. A':

$$A' := A_0 \dotplus \operatorname{span}\{\tbinom{g}{\bar{z}g}\},$$

is a proper extension of A_0. As $\binom{g}{\bar{z}g} \in A_0^+$ for arbitrary $\binom{x}{y} \in A_0$ and $\alpha \in \mathbf{C}$ we have

$$[y, g] = [x, \bar{z}g],$$

$$[x + \alpha g, y + \alpha \bar{z}g] = [x, y] + \alpha[g, y] + \bar{\alpha}[x, \bar{z}g] = [x, y] + 2\operatorname{Re}\{\alpha[g, y]\},$$

therefore A' is symmetric. It follows that A' is a selfadjoint extension of A_0 with $\bar{z} \in \sigma_p(A')$, hence $z \in \sigma(A')$ too. The fact that a selfadjoint extension A of A_0 with $z \in \sigma(A)$ coincides with A' follows as in the beginning of this proof in view of the fact that $\mathcal{R}(A_0 - z)^{[\perp]}$ is of dimension one.

6. Selfadjoint extensions with an empty resolvent set

In Section 4 it was shown that a nonnegative c.l.r. A_0 of regular defect one with a nonreal eigenvalue λ_0 has a selfadjoint extension $\hat{A}$ with $\rho(\hat{A}) = \emptyset$ (see (4.4) in Lemma 4.3). On the other hand, the example (3.6) shows that a selfadjoint extension $\hat{A}$ with an empty resolvent set can also exist in the case when A_0 does not have a nonreal eigenvalue. In the following theorem those nonnegative c.l.r.'s A_0 of regular defect one with $\sigma_p(A_0) \subset \mathbf{R}$ are characterized which have a selfadjoint extension with empty resolvent set.

Theorem 6.1. *Let A_0 be a nonnegative c.l.r. of regular defect one with $\sigma_p(A_0) \subset \mathbf{R}$. Then:*

(1) There exists a selfadjoint extension $\hat{A}$ of A_0 with $\rho(\hat{A}) = \emptyset$ if and only if $\Delta_0 = \mathbf{C} \setminus \mathbf{R}$. In this case $\hat{A}$ is uniquely determined.

(2) If the extension $\hat{A}$ in (1) is nonnegative then A_0 has an eigenvalue $\lambda \in \bar{\mathbf{R}}$ with a neutral eigenvector g_λ and

$$\hat{A} = A_0 \dotplus \operatorname{span}\{\binom{g_\lambda}{0}, \binom{0}{g_\lambda}\};$$

if $\hat{A}$ is not nonnegative then there exist nonzero neutral elements $g_0 \in \mathcal{N}(A_0)$, $g_\infty \in A_0(0)$ with

$$\hat{A} = A_0 \dotplus \operatorname{span}\{\binom{g_\infty}{g_0}\}.$$

Proof. 1. Let $\hat{A}$ be a selfadjoint extension of A_0 with $\rho(\hat{A}) = \emptyset$. Then for every $z \in \mathbf{C} \setminus \mathbf{R}$ we have

$$\mathcal{R}(A_0 - z) = \mathcal{R}(\hat{A} - z) \tag{6.1}$$

and, hence,

$$\mathcal{R}(A_0 - z)^{[\perp]} = \mathcal{N}(\hat{A} - \bar{z}). \tag{6.2}$$

This implies that $\mathcal{R}(A_0 - z)^{[\perp]}$ is neutral, i.e. $\Delta_0 = \mathbf{C} \setminus \mathbf{R}$. By Theorem 5.1 $\hat{A}$ is uniquely determined.

If $\hat{A}$ is nonnegative, then by (6.1) and Theorem 1.3 the subspaces $\mathcal{R}(A_0 - z)^{[\perp]}$, $z \in \mathbf{C}\setminus\mathbf{R}$, are independent of z, say $\mathcal{R}(A_0 - z)^{[\perp]} = \operatorname{span}\{g\}$, $z \in \mathbf{C}\setminus\mathbf{R}$. Then, by (6.2), g is an eigenvector of $\hat{A}$ to every nonreal z. Then $[g, g] = 0$ and $\binom{g}{0}, \binom{0}{g} \in \hat{A}$. In view of (3.1) at least one of the vectors $\binom{g}{0}, \binom{0}{g}$ does not belong to A_0. It follows that $\hat{A} = A_0 \dotplus \operatorname{span}\{\binom{g}{0}, \binom{0}{g}\}$. Since A_0 has defect one, there exist $\alpha, \beta \in \mathbf{C}$ with $\alpha\binom{g}{0} \dotplus \beta\binom{0}{g} \in A_0$, i.e. g is an eigenelement of A_0. As A_0 has no nonreal point spectrum, the eigenvalue corresponding to g is in $\bar{\mathbf{R}}$.

2. Assume now that $\Delta_0 = \mathbf{C} \setminus \mathbf{R}$. In order to prove that there exists a selfadjoint extension $\hat{A}$ of A_0 with $\rho(\hat{A}) = \emptyset$ we may assume that there exists a nonnegative selfadjoint extension A of A_0 with $\rho(A) \neq \emptyset$. Then by Corollary 1.2, $\mathbf{C}^+ \cup \mathbf{C}^- \subset \rho(A)$. Let $R(z) :=$

$(A-z)^{-1}$ ($z \in \rho(A)$), $g \in \mathcal{R}(A_0+i)^{[\perp]}$, $g \neq 0$, and $g(z) := g + (z-i)R(z)g$. As in the proof of Theorem 3.3 it follows that

$$(6.3) \qquad [g(z), g(\zeta)] = 0, \qquad z, \zeta \in \mathbf{C}^+,$$

hence

$$[R(z)g, g] = (z-i)^{-1}[g(z)-g, g] = 0, \qquad z \in \mathbf{C}^+ \setminus \{i\}$$

and

$$[R(i)g, g] = 0,$$
$$2\mathrm{Re}\,[R(i)g, g] = [(R(i)+R(-i))g, g] = 0.$$

Since $R(i)+R(-i) = 2R(i)AR(-i)$ is a nonnegative operator in $(\mathcal{H}, [\cdot,\cdot])$ we obtain

$$(6.4) \qquad (R(i)+R(-i))g = 0.$$

Let E be the spectral function of A, $E^{(0)} := E([-1,1])$, $E^{(\infty)} := E(\bar{\mathbf{R}} \setminus [-1,1])$ and $A^{(0)} := A \cap (E^{(0)}\mathcal{H})^2$, $A^{(\infty)} := A \cap (E^{(\infty)}\mathcal{H})^2$. Then (6.4) implies

$$(R(i; A^{(0)}) + R(-i; A^{(0)}))E^{(0)}g = 0.$$

It follows that

$$g_0 := E^{(0)}g \in \mathcal{N}(A^{(0)}) = \mathcal{N}(A).$$

The relation

$$(R(i; A^{(\infty)}) + R(-i; A^{(\infty)}))E^{(\infty)}g = 0$$

implies $A^{(\infty)-1}E^{(\infty)}g = 0$, i.e.

$$g_\infty := E^{(\infty)}g \in A^{(\infty)}(0) = A(0).$$

Therefore $g = g_0 + g_\infty$, $g_0 \in \mathcal{N}(A)$, $g_\infty \in A(0)$, and

$$g(z) = (I + (z-i)R(z))(g_0+g_\infty) = g_0 + g_\infty - (z-i)z^{-1}g_0 = iz^{-1}g_0 + g_\infty.$$

By (6.3),

$$0 = [g(z), g(\zeta)] = [iz^{-1}g_0 + g_\infty, i\zeta^{-1}g_0 + g_\infty] = [g_\infty, g_\infty] + z^{-1}\zeta^{-1}[g_0, g_0]$$

for all $z, \zeta \in \mathbf{C}^+$ and, hence,

$$[g_0, g_0] = [g_\infty, g_\infty] = 0.$$

Since $g(z)$ is neutral and $\mathrm{span}\{g(z)\}^{[\perp]} = \mathcal{R}(A_0 - \bar{z})$ we have $g(z) \in \mathcal{R}(A_0 - \bar{z})$ and

$$\begin{pmatrix} R(\bar{z})g(z) \\ g(z) \end{pmatrix} = \begin{pmatrix} -i|z|^{-2}g_0 \\ iz^{-1}g_0 + g_\infty \end{pmatrix} \in A_0 - \bar{z}, \qquad z \in \mathbf{C} \setminus \mathbf{R}.$$

It follows that

$$\begin{pmatrix} -i|z|^{-2}g_0 \\ g_\infty \end{pmatrix} \in A_0, \qquad z \in \mathbf{C} \setminus \mathbf{R}.$$

Considering this relation for two different nonreal points z we derive $g_0 \in \mathcal{N}(A_0)$, $g_\infty \in A_0(0)$. Then for arbitrary $\binom{x}{y} \in A_0$ we get $[g_0, y] = [g_\infty, x] = 0$ and

$$0 = [y - \bar{z}x, iz^{-1}g_0 + g_\infty] = [y, g_\infty] + i[x, g_0].$$

This implies that the c.l.r.

$$\hat{A} := A_0 \dot{+} \operatorname{span}\{\binom{g_\infty}{ig_0}\}$$

is symmetric. Indeed, if $\binom{x}{y} \in A_0$ and $\alpha \in \mathbf{C}$ the expression

$$(6.5)\qquad \begin{aligned}[x+\alpha g_\infty, y+i\alpha g_0] &= [x,y]+\alpha[g_\infty,y]-i\bar{\alpha}[x,g_0] = \\ &= [x,y]+2\mathrm{Re}\{\alpha[g_\infty,y]\}\end{aligned}$$

is real. Moreover, if $z \in \mathbf{C}\setminus\{0\}$ it holds

$$\binom{iz^{-1}g_0+g_\infty}{zg_\infty+ig_0} \in A_0 \dot{+} \operatorname{span}\{\binom{g_\infty}{ig_0}\}$$

hence

$$\mathbf{C}\setminus\{0\} \subset \sigma_p(\hat{A}).$$

It follows that $\hat{A}$ is a proper extension of A_0 and, therefore, selfadjoint. The resolvent set of this selfadjoint extension of A_0 is empty, and the first assertion of Theorem 6.1 is proved.

3. Assume that there exists a selfadjoint extension $\hat{A}$ of A_0 with $\rho(\hat{A}) = \emptyset$ which is not nonnegative. Then there exists a nonnegative selfadjoint extension A of A_0 with $\rho(A) \neq \emptyset$ and, as in part 2 of this proof we find $g_0 \in \mathcal{N}(A_0)$, $g_\infty \in A_0(0)$, $[g_0,g_0] = [g_\infty,g_\infty] = 0$ such that

$$\hat{A} = A_0 \dot{+} \operatorname{span}\{\binom{g_\infty}{ig_0}\}.$$

The selfadjointness of $\hat{A}$ implies

$$(6.6)\qquad [y,g_\infty] = -i[x,g_0] \qquad \text{for all} \qquad \binom{x}{y} \in A_0.$$

If one of the elements g_0 or g_∞ is zero, then by (6.5) and (6.6) $\hat{A}$ is nonnegative. Therefore, $g_0 \neq 0$ and $g_\infty \neq 0$ and Theorem 6.1 is proved.

Corollary 6.2. *Let A_0 be a nonnegative c.l.r. of regular defect one with $\sigma_p(A_0) \subset \mathbf{R}$, and assume that there exists a selfadjoint extension $\hat{A}$ of A_0 with $\rho(\hat{A}) = \emptyset$. Then $\hat{A}$ is nonnegative if and only if there exists a neutral eigenvector $g_\lambda \neq 0$ of A_0 corresponding to some eigenvalue $\lambda \in \bar{\mathbf{R}}$ such that*

$$(6.7)\qquad \mathcal{R}(A_0-z)^{[\perp]} = \operatorname{span}\{g_\lambda\}, \qquad z \in \mathbf{C}\setminus\mathbf{R}.$$

If $\hat{A}$ is nonnegative and, in addition, there is more than one nonnegative selfadjoint extension of A_0, then we have either $\lambda = 0$ or $\lambda = \infty$. If $\hat{A}$ is not nonnegative, there exist nonzero neutral elements $g_0 \in \mathcal{N}(A_0)$, $g_\infty \in A_0(0)$, such that

$$\mathcal{R}(A_0-z)^{[\perp]} \subset \operatorname{span}\{g_0,g_\infty\}, \qquad z \in \mathbf{C}\setminus\mathbf{R}.$$

Proof. It remains only to verify that (6.7) implies the nonnegativity of $\hat{A}$. The other assertions are consequences of Theorem 6.1 and its proof. The relation (6.7) implies $g_\lambda \in \mathcal{D}(A_0)^{[\perp]} \cap \mathcal{R}(A_0)^{[\perp]}$ from which together with $[g_\lambda,g_\lambda] = 0$ it follows that $\hat{A} = A_0 \dot{+} \operatorname{span}\{\binom{g_\lambda}{0},\binom{0}{g_\lambda}\}$ is nonnegative. The Corollary 6.2 is proved.

If the assumptions of Corollary 6.2 are fulfilled, the holomorphic defect vectors $g(z)$, $z \in \mathbf{C} \setminus \mathbf{R}$, of A_0 satisfying (3.5) have the form

$$g(z) = (\lambda - i)(\lambda - z)^{-1} g_\lambda, \qquad z \in \mathbf{C} \setminus \mathbf{R}, \tag{6.8}$$

if $\hat{A}$ is nonnegative and

$$g(z) = iz^{-1} g_0 + g_\infty, \qquad z \in \mathbf{C} \setminus \mathbf{R}, \tag{6.9}$$

if $\hat{A}$ is not nonnegative. Here g_λ, g_0 and g_∞ are as in Corollary 6.2.

7. The resolvents of the selfadjoint extensions

Let A_0 be a nonnegative c.l.r. of regular defect one and let A be a selfadjoint extension of A_0 with $\rho(A) \neq \emptyset$. We assume in the following that i, $-i \in \rho(A)$. This is no restriction as it can always be achieved replacing A_0 and A by αA_0 and αA with some $\alpha > 0$ (see Theorem 1.1). We set $R(z) := (A - z)^{-1}$, $z \in \rho(A)$. Let, as in Section 3, $z \longmapsto g(z)$ denote a non-zero $\mathcal{H}$-valued locally holomorphic function on $\rho(A)$ with

$$g(z) - g(\zeta) = (z - \zeta) R(z) g(\zeta), \qquad z,\ \zeta \in \rho(A), \tag{7.1}$$

and

$$\mathcal{R}(A_0 - \bar{z})^{[\perp]} = \operatorname{span}\{g(z)\}, \qquad z \in \rho(A). \tag{7.2}$$

In M. G. Krein's description of the generalized resolvents of a hermitian operator the so-called Q-function plays an essential role (comp. also (2.5)). For the operator A_0 we define the Q-function (more exactly, the Q-function of the pair A_0, A) (comp. [10]):

$$Q(z) := i[g(i), g(i)] + (z - i)[g(z), g(-i)], \qquad z \in \rho(A). \tag{7.3}$$

It has the following properties which can be checked easily:

$$\operatorname{Re} Q(i) = 0, \qquad Q(\bar{z}) = \overline{Q(z)}, \qquad z \in \rho(A). \tag{7.4}$$

$$Q(z) - Q(\zeta) = (z - \zeta)[g(z), g(\bar{\zeta})], \qquad z,\ \zeta \in \rho(A). \tag{7.5}$$

The situation where the nonnegative c.l.r. A_0 has a selfadjoint extension $\hat{A}$ with $\rho(\hat{A}) = \emptyset$ (see Section 6) can be characterized by the Q-function.

Lemma 7.1. *Let g and Q be locally holomorphic functions on $\rho(A)$ such that the relations (7.1-3) hold. Then the following assertions are equivalent:*

(i) A_0 has a selfadjoint extension $\hat{A}$ with $\rho(\hat{A}) = \emptyset$.

(ii) $Q(z) = 0$ if $z \in \rho(A)$.

(iii) $Q(z) = c$ if $z \in \mathbf{C}^+ \cap \rho(A)$ for some $c \in \mathbf{C}$.

Proof. Suppose that (i) holds. Then according to (4.10), (6.8) and (6.9) we have $[g(z), g(\zeta)] = 0$ for all $z, \zeta \in \rho(A) \setminus \mathbf{R}$. Hence by (7.3) we see that (ii) holds. The implication (ii)$\Longrightarrow$(iii) is trivial.

Assume (iii). In order to prove (i) we can suppose (see Corollary 1.2) that there exists a nonnegative selfadjoint extension with real spectrum. Let $\widetilde{A}$ be such an extension, $\widetilde{R}(z) := (\widetilde{A} - z)^{-1}$, $z \in \rho(\widetilde{A})$, and let $z \longmapsto \widetilde{g}(z)$ be a locally holomorphic function on $\rho(\widetilde{A})$ which has the properties (7.1) and (7.2) with A replaced by $\widetilde{A}$. From (7.3) it follows that $[g(z), g(-i)] = 0$ for $z \in \rho(A) \cap \mathbf{C}^+$. This implies

$$[\widetilde{g}(z), \widetilde{g}(-i)] = 0, \qquad z \in \rho(\widetilde{A}) \cap \mathbf{C}^+, \tag{7.6}$$

and, in particular, $[\widetilde{g}(i), \widetilde{g}(-i)] = 0$. Put $\widetilde{g} := \widetilde{g}(i)$. Then

$$[(\widetilde{R}(i) + \widetilde{R}(-i))\widetilde{g}, \widetilde{g}] = -\mathrm{Re}\{i[\widetilde{g}, \widetilde{g}(-i)]\} = 0$$

and as in the proof of Theorem 6.1 it follows that

$$\widetilde{g}(z) = iz^{-1} g_0 + g_\infty, \qquad z \in \rho(\widetilde{A}), \tag{7.7}$$

with some $g_0 \in \mathcal{N}(\widetilde{A})$, $g_\infty \in \widetilde{A}(0)$. Inserting (7.7) into (7.6) we obtain

$$[g_0, g_0] = [g_\infty, g_\infty] = 0.$$

This implies $\Delta_0 = \mathbf{C} \setminus \mathbf{R}$, and by Theorem 6.1 the assertion (i) holds.

In the following theorem the resolvents of the selfadjoint extensions of A_0 are described. For the definition of the gap metric between subspaces we refer to [11, Section IV.2.1].

Theorem 7.2. *Let A_0 be a nonnegative c.l.r. of regular defect one in the Krein space $\mathcal{H}$ and let A be a selfadjoint extension of A_0 with $i, -i \in \rho(A)$. If g and Q are as in (7.1-3), the following holds.*

(a) If all the selfadjoint extensions of A_0 have a nonempty resolvent set, then the formula

$$R(z; A_{(\gamma)}) = R(z) - (\gamma + Q(z))^{-1} [\cdot, g(\bar{z})] g(z), \; z \in \rho(A), \; \gamma + Q(z) \neq 0, \tag{7.8}$$

establishes a bijective correspondence between all $\gamma \in \bar{\mathbf{R}}$ and all selfadjoint extensions $A_{(\gamma)}$ of A_0.

(b) If A_0 has a nonnegative selfadjoint extension $\hat{A}$ with $\rho(\hat{A}) = \emptyset$ there exists a $\lambda \in \widetilde{\sigma}_p(A_0)$ (the extended point spectrum of A_0, see Section 1) and a corresponding eigenvector g_λ of A_0 such that the formula

$$R(z; A_{(\gamma)}) = R(z) - \gamma^{-1} |\lambda - i|^2 (\lambda - z)^{-1} (\bar{\lambda} - z)^{-1} [\cdot, g_\lambda] g_\lambda, \; z \in \rho(A), \tag{7.9}$$

establishes a bijection between all selfadjoint extensions $A_{(\gamma)}$ of A_0 with $\rho(A_{(\gamma)}) \neq \emptyset$ and all $\gamma \in \bar{\mathbf{R}} \setminus \{0\}$. In this case $\lim_{\gamma \to 0} A_{(\gamma)}$ exists with respect to the gap metric and is equal to

$$A_0 \dotplus \mathrm{span}\{ \binom{g_\lambda}{0}, \binom{0}{g_\lambda} \} = \hat{A}.$$

(c) *If A_0 has a selfadjoint extension $\hat{A}$ which is not nonnegative and such that $\rho(\hat{A}) = \emptyset$, then there exist nonzero neutral elements $g_0 \in \mathcal{N}(A_0)$, $g_\infty \in A_0(0)$, such that the formula*

$$R(z; A_{(\gamma)}) = R(z) - \gamma^{-1}[\cdot, i\bar{z}^{-1}g_0 + g_\infty](iz^{-1}g_0 + g_\infty),\ z \in \rho(A), \tag{7.10}$$

establishes a bijective correspondence between the selfadjoint extensions $A_{(\gamma)}$ of A_0 with $\rho(A_{(\gamma)}) \neq \emptyset$ and all $\gamma \in \bar{\mathbf{R}} \setminus \{0\}$. In this case $\lim_{\gamma \to 0} A_{(\gamma)}$ exists with respect to the gap metric and is equal to

$$A_0 \dotplus \operatorname{span}\{\binom{g_\infty}{ig_0}\} = \hat{A}.$$

Remark 1. If the assumption of statement (a) of Theorem 7.2 is fulfilled, then it follows from Lemma 7.1, (7.1) and (7.3) that the right hand side of (7.8) defines a meromorphic function on $\mathbf{C}^+$ and on $\mathbf{C}^-$ for every $\gamma \in \mathbf{R}$. It is easy to see that, in addition to (7.8), we have

$$\rho(A_{(\gamma)}) \cap \rho(A) = \{z \in \rho(A) :\ \gamma + Q(z) \neq 0\}$$

for every real γ.

Remark 2. The relation (7.8) does also hold in cases (b) and (c). In fact, (7.9) and (7.10) are the specific forms of (7.8) under the assumptions of (b) or (c), respectively (observe (4.10), (6.8), (6.9) and Lemma 7.1).

Before proving Theorem 7.3 we observe that it is easy to pass from the formula for the resolvent $R(z; A_{(\gamma)})$, e.g. from (7.8), to a formula for the linear relation $A_{(\gamma)}$ itself. Indeed, if $\gamma + Q(z) \neq 0$, it holds

$$A_{(\gamma)} = \{\begin{pmatrix} R(z)u - (\gamma+Q(z))^{-1}[u, g(\bar{z})]g(z) \\ u + zR(z)u - z(\gamma+Q(z))^{-1}[u, g(\bar{z})]g(z) \end{pmatrix} :\ u \in \mathcal{H}\}.$$

This equality and the fact that u runs through $\mathcal{H}$ if and only if $\begin{pmatrix} R(z)u \\ u+zR(z)u \end{pmatrix} =: \binom{x}{y}$ runs through A shows that Theorem 7.2 implies the following

Corollary 7.3. *Let A_0, A, g, Q and $A_{(\gamma)}$ and cases (a), (b), (c) be as in Theorem 7.2. Then, for every $\gamma \in \mathbf{R}$ in case (a), or for every $\gamma \in \mathbf{R} \setminus \{0\}$ in the cases (b) and (c) the selfadjoint extension $A_{(\gamma)}$ of A_0 can be written as*

$$A_{(\gamma)} = \{(\gamma + Q(z))\binom{x}{y} - [y - zx, g(\bar{z})]\binom{g(z)}{zg(z)} :\ \binom{x}{y} \in A\}; \tag{7.11}$$

here $z \in \rho(A)$ such that $\gamma + Q(z) \neq 0$.

If we choose $\binom{x'}{y'} \in A \setminus A_0$ then the relation (7.11) takes the form

$$A_{(\gamma)} = A_0 \dotplus \operatorname{span}\{(\gamma + Q(z))\binom{x'}{y'} - [y' - zx', g(\bar{z})]\binom{g(z)}{zg(z)}\}, \tag{7.12}$$

where $z \in \rho(A)$. In (7.11) and (7.12) the expressions $[y - zx, g(\bar{z})]$ and $[y' - zx', g(\bar{z})]$ are independent of z, hence, in particular, they can be replaced by $[y - \bar{z}x, g(z)]$ and $[y' - \bar{z}x', g(z)]$, respectively.

Proof of Theorem 7.2. It is not hard to check that the expression on the right hand side of (7.8) (without the additional assumption of (a)) satisfies the resolvent equation for all $z \in \rho(A)$ such that $\gamma + Q(z) \neq 0$, and that this expression defines the resolvent of a selfadjoint c.l.r. which extends A_0. On the other hand, given a selfadjoint extension of A_0 with a nonempty resolvent set, as in the proofs of Theorem 3.1 and Proposition 4.1 of [10] it can be verified that its resolvent is of the form (7.8).

Now we use the representation of $A_{(\gamma)}$, $\gamma \neq 0$, in Corollary 7.3 and conclude from (7.12) that in cases (b) and (c) the limits for $\gamma \to 0$ exist. Making use of (4.10), (6.8) and (6.9) it can be seen that these limits coincide with $\hat{A}$ (see Lemma 4.3 and Theorem 6.1).

Let A_0 be again a nonnegative c.l.r. of regular defect one which has a selfadjoint extension $\hat{A}$ with $\rho(\hat{A}) = \emptyset$. If A_0 has a nonreal eigenvalue λ_0 it was shown in Theorem 4.1 that all the selfadjoint extensions of A_0 coincide with the exception of their restrictions to the two-dimensional subspace $\mathcal{L}_0 = \mathcal{R}(A_0 - \lambda_0)^{[\perp]}$. We shall show in the following Theorem 7.4, that in the cases (b) and (c) of Theorem 7.2 all the selfadjoint extensions of A_0 with nonempty resolvent set coincide outside of the "spectral subspaces" at λ, $\bar{\lambda}$ and $0, \infty$, respectively. Because of Theorem 4.1 we can restrict ourselves to the case that λ in (b) is real. By E ($E_{(\gamma)}$) we denote the spectral function of A ($A_{(\gamma)}$, respectively).

Theorem 7.4. *Let A_0 be a nonnegative c.l.r. of regular defect one with $\sigma_p(A_0) \subset$ $\mathbf{R}$ such that there exists a selfadjoint extension $\hat{A}$ of A_0 with $\rho(\hat{A}) = \emptyset$, and let A be a selfadjoint extension of A_0 with $\sigma(A) \subset \mathbf{R}$. Then, in case (b), if $\lambda \in \bar{\mathbf{R}}$ we have*

$$E = E_{(\gamma)},$$

and in case (c),

$$E \,|\, \mathbf{R} \setminus \{0\} = E_{(\gamma)} \,|\, \mathbf{R} \setminus \{0\}.$$

If Δ is an open subset of $\bar{\mathbf{R}}$ such that the boundary points of Δ are no critical points of E, $\lambda \notin \Delta$ in case (b) and $0, \infty \notin \Delta$ in case (c), then we have

$$A_{(\gamma)} \cap (E(\Delta)\mathcal{H})^2 = A \cap (E(\Delta)\mathcal{H})^2 \subset A_0 \tag{7.13}$$

for all $\gamma \in \mathbf{R} \setminus \{0\}$.

Proof. Let I be a bounded closed interval of the real axis which does not contain the point $z = 0$ in case (c). If the boundary points of I are not critical points of A and $A_{(\gamma)}$, the representation of the spectral projection as an integral over the resolvent (see [15]) yields

$$E_{(\gamma)}(I) = E(I),$$

as the second terms on the right hand sides of (7.9) and (7.10) are holomorphic on I or have residue zero at $z = \lambda$ $(= \bar{\lambda})$. It remains to prove the relation (7.13). According to (7.9) and (7.10) we have

$$\begin{aligned} A_{(\gamma)} \cap (E(\Delta)\mathcal{H})^2 &= \left\{ \begin{pmatrix} R(z;A_{(\gamma)})x \\ x + zR(z;A_{(\gamma)})x \end{pmatrix} : x \in E(\Delta)\mathcal{H} \right\} \\ &= \left\{ \begin{pmatrix} R(z)x \\ x + zR(z)x \end{pmatrix} : x \in E(\Delta)\mathcal{H} \right\} = A \cap (E(\Delta)\mathcal{H})^2. \end{aligned}$$

The last inclusion of (7.13) follows from the fact that the intersection of two different selfadjoint extensions of A_0 is equal to A_0.

We conclude this section with some remarks on rank one perturbations of a nonnegative selfadjoint c.l.r.

Let A be a nonnegative selfadjoint c.l.r. with $\rho(A) \neq \emptyset$, $R(z) := (A-z)^{-1}$, $z \in \rho(A)$, and let $e \in \mathcal{H} \setminus \mathcal{D}(A)^{[\perp]} = \mathcal{H} \setminus A(0)$. If $\alpha \in \mathbf{R}$, then

$$A_{[\alpha]} := A + \alpha[\cdot, e]e \tag{7.14}$$

is a selfadjoint c.l.r. with $\rho(A_{[\alpha]}) \neq \emptyset$. An easy computation yields

$$R(z; A_{[\alpha]}) = R(z) - (\alpha^{-1} + [R(z)e, e])^{-1}[\cdot, R(\bar{z})e]R(z)e$$

for all $z \in \rho(A)$ with $\alpha^{-1} + [R(z)e, e] \neq 0$. As in the proof of Theorem 6.1 one verifies that the function $z \longmapsto [R(z)e, e]$ is constant in $\mathbf{C}^{+}$ if and only if $e = e_0 + e_\infty$ with $e_0 \in \mathcal{N}(A)$, $e_\infty \in A(0)$, $[e_0, e_0] = [e_\infty, e_\infty] = 0$. In this case it is even identically equal to zero.

Define a nonnegative c.l.r. A_0 by

$$A_0 := \{\tbinom{x}{y} \in A : \ [x, e] = 0\}.$$

Evidently, A_0 is of regular defect one. Then with the help of Theorem 7.2 we see that $\lim_{\alpha \to \infty} A_{[\alpha]} =: A_{[\infty]}$ exists with respect to the gap metric, and the family $A_{[\beta]}$, $\beta \in \bar{\mathbf{R}}$, contains all the selfadjoint relation extensions of A_0.

More generally, let A be as above, and fix $z \in \rho(A)$, $g \in \mathcal{H}$, $g \neq 0$. Define the linear relations

$$A_{\{\delta\}} := \{\tbinom{x}{y} - (\delta + i\mathrm{Im}\, z\, [g, g])^{-1}[y - \bar{z}x, g]\tbinom{g}{zg} : \ \tbinom{x}{y} \in A\} \tag{7.15}$$

for all $\delta \in \mathbf{R}$ such that $\delta + i\mathrm{Im}\, z\, [g, g] \neq 0$. Making use of Theorem 7.2 and Corollary 7.3 with

$$A_0 := \{\tbinom{x}{y} \in A : \ [y - \bar{z}x, g] = 0\}$$

it is easy to see that the linear relations $A_{\{\delta\}}$ are selfadjoint and that in the case $(\mathrm{Im}\, z)[g, g] = 0$ the limit $\lim_{\delta \to 0} A_{\{\delta\}} =: A_{\{0\}}$ exists with respect to the gap metric. The family $A_{\{\delta\}}$, $\delta \in \bar{\mathbf{R}}$, contains all the selfadjoint extensions of A_0. We have $\rho(A_{\{\delta\}}) \neq \emptyset$ if $\delta \neq 0$ and $\rho(A_{\{0\}}) = \emptyset$ if and only if g has the form $g = g_0 + g_\infty$, $g_0 \in \mathcal{N}(A)$, $g_\infty \in A(0)$, $[g_0, g_0] = [g_\infty, g_\infty] = 0$. Assume, additionally, that $g \in \mathcal{D}(A)$ and let $A_{[\alpha]}$ be defined as in (7.14) with some $e \in \mathcal{H}$ such that $R(z)e = g$. Then it is not hard to verify that

$$A_{\{\delta\}} = A_{[\alpha]} \quad \text{with} \quad \alpha = (\delta - [Ag, g] + \mathrm{Re}\, z\, [g, g])^{-1}. \tag{7.16}$$

Here the expression $[Ag, g]$ makes sense as $g \in \mathcal{D}(A) \subset A(0)^{[\perp]}$. The relation (7.16) will be used in Section 8.

8. The nonnegative selfadjoint extensions

In this section we describe the *nonnegative* selfadjoint extensions of a nonnegative c.l.r. A_0 of regular defect one under the assumption that A_0 admits a *nonnegative* selfadjoint extension A with $\rho(A) \neq \emptyset$. According to Theorem 6.1 this holds e.g. if A_0 has more than one nonnegative selfadjoint extension (that is $A_F \neq A_K$). If A_0 has only one nonnegative selfadjoint extension A (that is $A = A_F = A_K$) both cases $\rho(A) = \emptyset$ or $\rho(A) \neq \emptyset$ are possible (see Lemma 4.3 and Theorem 8.2 below). A characterization of the fact that A_0 has a nonnegative selfadjoint extension A with $\rho(A) \neq \emptyset$ is given in Theorem 8.1.

If $\mathcal{M}$ is a linear manifold of the Krein space $\mathcal{H}$, by $\mathcal{M}^0$ we denote its isotropic part: $\mathcal{M}^0 = \mathcal{M} \cap \mathcal{M}^{[\perp]}$.

Theorem 8.1. *The nonnegative c.l.r. A_0 of regular defect one has a nonnegative selfadjoint extension A with $\rho(A) \neq \emptyset$ if and only if*

$$\mathcal{D}[A_0]^0 \cap \mathcal{D}[A_0^{-1}]^0 = \{0\}. \tag{8.1}$$

Proof. As $A_0^+(0) = \mathcal{D}(A_0)^{[\perp]}$ and $\mathcal{D}(A_0)^{[\perp]} \supset \mathcal{D}[A_0]^{[\perp]}$ we have $\mathcal{D}[A_0]^0 \subset \mathcal{D}[A_0] \cap A_0^+(0)$. If $x \in \mathcal{D}[A_0] \cap A_0^+(0)$, then x is orthogonal to $\mathcal{D}(A_0)$ and hence also orthogonal to the closure of $\mathcal{D}(A_0)$ with respect to the norm $(\|x\|^2 + [A_0x, x])^{\frac{1}{2}}$. Here $\|\cdot\|$ is an arbitrary Hilbert majorant of $[\cdot,\cdot]$. It follows that $\mathcal{D}[A_0]^0 = \mathcal{D}[A_0] \cap A_0^+(0)$ and, analogously, $\mathcal{D}[A_0^{-1}]^0 = \mathcal{D}[A_0^{-1}] \cap \mathcal{N}(A_0^+)$. Thus the relation (8.1) is equivalent to

$$\mathcal{D}[A_0]^0 \cap \mathcal{N}(A_0^+) \cap \mathcal{D}[A_0^{-1}]^0 \cap A_0^+(0) = \{0\}. \tag{8.2}$$

Further, according to the definitions of A_F and A_K in (2.3) and (2.4)

$$\mathcal{D}[A_0]^0 \cap \mathcal{N}(A_0^+) = \mathcal{D}[A_0] \cap \mathcal{N}(A_0^+) \cap A_0^+(0) = \mathcal{N}(A_F) \cap A_F(0),$$

$$\mathcal{D}[A_0^{-1}]^0 \cap A_0^+(0) = \mathcal{D}[A_0^{-1}] \cap A_0^+(0) \cap \mathcal{N}(A_0^+) = \mathcal{N}(A_K) \cap A_K(0).$$

The Theorem 1.3 implies that (8.2) is equivalent to

$$\mathcal{R}(A_F - z)^{[\perp]} \cap \mathcal{R}(A_K - z)^{[\perp]} = \{0\}, \qquad z \neq \bar{z}.$$

This relation holds if and only if either $A_F = A_K$ and $\mathcal{R}(A_F - z) = \mathcal{R}(A_K - z) = \mathcal{H}$ for all $z \neq \bar{z}$, or if $A_F \neq A_K$ (since A_F or A_K must have a nonempty resolvent set). Hence (8.1) holds if and only if either $\rho(A_F) \neq \emptyset$ or $\rho(A_K) \neq \emptyset$.

Let in the following A be a fixed nonnegative selfadjoint extension of A_0 with $\rho(A) \neq \emptyset$, $g(z)$ be a non-zero $\mathcal{H}$-valued locally holomorphic function on $\rho(A)$ with the properties (7.1) and (7.2) and let Q be defined as in (7.3). Consider the selfadjoint extensions $A_{(\gamma)}$ of A_0 parametrized by $\gamma \in \bar{\mathbf{R}}$ through the resolvent formula in Theorem 7.2 with $A_{(\infty)} = A$. We characterize those parameters γ for which $A_{(\gamma)}$ is nonnegative.

Let J be a fundamental symmetry of $(\mathcal{H}, [\cdot,\cdot])$. Then the relations

$$H := JA, \qquad K := AJ$$

are nonnegative and selfadjoint in $(\mathcal{H}, (\cdot,\cdot)_J)$. We denote by P_0^H and P_∞^K the $(\cdot,\cdot)_J$-orthogonal projections onto $\overline{\mathcal{D}(H)}$ and $\overline{\mathcal{R}(K)}$, respectively, and define

$$H_{op} := H \cap (P_0^H\mathcal{H})^2, \qquad (K^{-1})_{op} := K^{-1} \cap (P_\infty^K\mathcal{H})^2$$

and, for $x \in \mathcal{H}$,

$$(8.3)\qquad \begin{aligned} |x|_+^{(A)} := |x|_+^{(H)} &= \begin{cases} \|H_{op}^{\frac{1}{2}}x\|_J & \text{if} \quad x \in \mathcal{D}(H_{op}^{\frac{1}{2}}) \\ \infty & \text{otherwise} \end{cases} \\ |x|_-^{(A)} := |x|_-^{(K)} &= \begin{cases} \|(K^{-1})_{op}^{\frac{1}{2}}x\|_J & \text{if} \quad x \in \mathcal{D}((K^{-1})_{op}^{\frac{1}{2}}) \\ \infty & \text{otherwise} \end{cases} \end{aligned}$$

(see (2.6)). It is easy to see that $\mathcal{D}(H_{op}^{\frac{1}{2}}) = \mathcal{D}[A]$ and $\mathcal{D}((K^{-1})_{op}^{\frac{1}{2}}) = \mathcal{D}[A^{-1}]$ (see Section 2). Moreover, the functions $x \longmapsto |x|_\pm$ on $\mathcal{D}[A^{\pm 1}]$ are the extensions by continuity of

$$\mathcal{D}(A^{\pm 1}) \ni x \longmapsto [y, x]^{\frac{1}{2}}, \qquad \binom{x}{y} \in A^{\pm 1},$$

to $\mathcal{D}[A^{\pm 1}]$ with respect to the norm

$$x \longmapsto (\|x\|_J^2 + [y, x])^{\frac{1}{2}}, \qquad \binom{x}{y} \in A^{\pm 1}.$$

Therefore the expressions $|\cdot|_\pm$ do not depend on the choice of the fundamental symmetry J. In the following, the defect element $g(i)$ plays a special role (according to the fact that the Q-function was normalized by the first relation of (7.4)). We write $g := g(i)$ and define

$$\gamma_+ := |g|_+^2, \qquad \gamma_- := -|g|_-^2.$$

Theorem 8.2. *Let A_0 be a nonnegative c.l.r. of regular defect one which has a nonnegative selfadjoint extension A with $\rho(A) \neq \emptyset$. Then the c.l.r. $A_{(\gamma)}$ is nonnegative if and only if $\gamma \notin (\gamma_-, \gamma_+)$; $A_{(\gamma_-)}$ is the von Neumann-Krein extension and $A_{(\gamma_+)}$ is the Friedrichs extension of A_0.*

Proof. 1. Assume first that for all selfadjoint extensions A of A_0 we have $\rho(A) \neq \emptyset$ and that $[g, g] \neq 0$, say $\delta := \operatorname{sign}[g, g]$. We choose a fundamental symmetry J in $(\mathcal{H}, [\cdot, \cdot])$ such that $Jg = \delta g$, that is $g \in P_\pm \mathcal{H} = \frac{1}{2}(I \pm J)\mathcal{H}$ if $\delta = \pm 1$. Let $\binom{x'}{y'} \in A$ be such that $[y' + ix', g] \neq 0$.

If $\delta = 1$, then $0 = [y + ix, g] = (Jy + ix, g)_J$ for $\binom{x}{y} \in A_0$, hence $g \in \mathcal{R}(JA_0 + i)^\perp$. Further,

$$(8.4)\qquad (Jy' + ix', g)_J = [y' + ix', g] \neq 0.$$

The relation (7.12) with $z = i$ implies

$$A_{(\gamma)} = A_0 \dot{+} \operatorname{span}\{(\gamma + i[g, g])\binom{x'}{y'} - [y' + ix', g]\binom{g}{ig}\}.$$

We parametrize the selfadjoint extensions of JA_0 in $(\mathcal{H}, (\cdot, \cdot)_J)$ also by M. G. Krein's formula (see (2.5)):

$$R(z; H_{(\gamma)}) = R(z; H) - (\gamma + Q_1(z))^{-1}(\cdot, g_1(\bar{z}))g_1(z), \quad z \neq \bar{z},$$

where $H = JA$, $g_1(z) := g + (z - i)R(z; H)g$ and

$$Q_1(z) := i(g_1(i), g_1(i))_J + (z - i)(g_1(z), g_1(-i))_J.$$

Making use of (8.4) we conclude from Corollary 7.3 with $z = i$ that

$$H_{(\gamma)} = JA_0 \dotplus \operatorname{span}\{(\gamma + i(g,g)_J)\binom{x'}{Jy'} - (Jy' + ix', g)_J\binom{g}{ig}\}.$$

Hence, in view of the relation $Jg = g$,

$$A_{(\gamma)} = JH_{(\gamma)}, \qquad \gamma \in \mathbf{R}.$$

By Lemma 2.3 $H_{(\gamma)}$ is nonnegative in $(\mathcal{H}, (\cdot,\cdot)_J)$ if and only if

$$(8.5) \qquad \begin{aligned} \gamma \notin (-|g|_-^{(H)2}, |g|_+^{(H)2}) &= (-|Jg|_-^{(H)2}, |g|_+^{(H)2}) = \\ = (-|g|_-^{(K)2}, |g|_+^{(H)2}) &= (\gamma_-, \gamma_+). \end{aligned}$$

It follows that $A_{(\gamma)}$ is nonnegative if and only if $\gamma \notin (\gamma_-, \gamma_+)$.

Assume now that $\delta = -1$. Then $\binom{x}{y} \in A_0$ implies $0 = [y + ix, g] = (Jy - ix, g)_J$, hence $g \in \mathcal{R}(JA_0 - i)^\perp$. Further,

$$(8.6) \qquad (Jy' - ix', g)_J = [y' + ix', g] \neq 0.$$

In this case we parametrize the selfadjoint extensions of JA_0 in $(\mathcal{H}, (\cdot,\cdot)_J)$ as follows:

$$R(z; H_{(\gamma)}) = R(z; H) - (\gamma + Q_{-1}(z))^{-1}(\cdot, g_{-1}(\bar{z}))g_{-1}(z), \quad z \neq \bar{z},$$

where $g_{-1}(z) := g + (z + i)R(z; H)g$ and

$$Q_{-1}(z) := i(g_{-1}(i), g_{-1}(i))_J + (z - i)(g_{-1}(z), g_{-1}(-i))_J.$$

Using (8.6) we conclude from Corollary 7.3 with $z = -i$ that

$$\begin{aligned} H_{(\gamma)} &= JA_0 \dotplus \operatorname{span}\{(\gamma + Q_{-1}(-i))\binom{x'}{Jy'} - (Jy' - ix', g)_J\binom{g}{-ig}\} = \\ &= JA_0 \dotplus \operatorname{span}\{(\gamma - i(g,g)_J)\binom{x'}{Jy'} - (Jy' - ix', g)_J\binom{g}{-ig}\}. \end{aligned}$$

Hence, in view of the relation $Jg = -g$,

$$A_{(\gamma)} = JH_{(\gamma)}, \qquad \gamma \in \mathbf{R}.$$

By Lemma 2.3 and (2.8), $H_{(\gamma)}$ is nonnegative in $(\mathcal{H}, (\cdot,\cdot)_J)$ if and only if

$$\gamma \notin (-|g|_-^{(H)2}, |g|_+^{(H)2}) = (\gamma_-, \gamma_+)$$

(cf. (8.5)). Hence $A_{(\gamma)}$ is nonnegative if and only if $\gamma \notin (\gamma_-, \gamma_+)$.

That $A_{(\gamma_-)}$ is the von Neumann-Krein extension of A_0 and $A_{(\gamma_+)}$ the Friedrichs extension follows from Lemma 2.3. This proves Theorem 8.2 if $[g,g] \neq 0$.

2. Assume now that $[g,g] = 0$. This holds e.g. if there exists a selfadjoint extension $\hat{A}$ with $\rho(\hat{A}) = \emptyset$ as in this case all the defect vectors are neutral. Then $Q(i) = 0$ and, by Corollary 7.3, we have

$$A_{(\gamma)} = \{\binom{x}{y} - \gamma^{-1}[y + ix, g]\binom{g}{ig} : \binom{x}{y} \in A\}.$$

Hence $A_{(\gamma)}$ is nonnegative if and only if

$$\begin{aligned} 0 &\leq [x-\gamma^{-1}[y+ix,g]g, y-i\gamma^{-1}[y+ix,g]g] = \\ &= [x,y]-\gamma^{-1}[y+ix,g][g,y]+i\gamma^{-1}[g,y+ix][x,g] = \\ &= [x,y]-\gamma^{-1}|[y,g]|^2+\gamma^{-1}|[x,g]|^2 \end{aligned}$$

for all $\binom{x}{y} \in A$.

Consider first the case $\gamma > 0$. Let J be an arbitrary fundamental symmetry of $(\mathcal{H}, [\cdot,\cdot])$. Then $A_{(\gamma)}$ is nonnegative if and only if

$$\gamma(u,v)_J + |(u,Jg)_J|^2 \geq |(v,g)_J|^2 \tag{8.7}$$

for all $\binom{u}{v} \in H$. If $g \notin \overline{\mathcal{D}(H)}$ we have $|g|_+^{(A)} = \infty$ and, since there exists a $v' \in H(0)$ with $(v',g)_J \neq 0$, the relation (8.7) cannot hold for any γ. Therefore, it remains to consider the case $g \in \overline{\mathcal{D}(H)}$. Assume that (8.7) holds. If $F(\cdot)$ is the spectral function of H_{op} and $\Delta_n := (-n,n)$, $n = 1, 2, \ldots$, then (8.7) implies

$$\gamma(F(\Delta_n)g, H_{op}F(\Delta_n)g)_J + |[F(\Delta_n)g,g]|^2 \geq |(H_{op}F(\Delta_n)g,g)_J|^2. \tag{8.8}$$

Since by assumption we have $\lim_{n\to\infty}[F(\Delta_n)g,g] = 0$ it follows that

$$\gamma \liminf_{n\to\infty}(H_{op}F(\Delta_n)g,g)_J \geq \liminf_{n\to\infty}(H_{op}F(\Delta_n)g,g)_J^2. \tag{8.9}$$

If $g \in \overline{\mathcal{D}(H)} \setminus \mathcal{D}(H_{op}^{\frac{1}{2}})$, then $|g|_+^{(A)} = \infty$ and $\liminf_{n\to\infty}(H_{op}F(\Delta_n)g,g)_J = \infty$. The latter relation contradicts (8.9). Hence, in this case, all selfadjoint extensions $A_{(\gamma)}$, $\gamma > 0$, are not nonnegative. Let now $g \in \mathcal{D}(H_{op}^{\frac{1}{2}})$. Then we have $|g|_+^{(A)2} = \liminf_{n\to\infty}(H_{op}F(\Delta_n)g,g)_J$, and if $A_{(\gamma)}$ is nonnegative it follows from (8.9) that $\gamma \geq |g|_+^{(A)2}$. Conversely, $\gamma \geq |g|_+^{(A)2}$ implies for $\binom{u}{v} \in H$

$$\begin{aligned} |(v,g)_J|^2 &= |(H_{op}u,g)_J|^2 = |(H_{op}^{\frac{1}{2}}u, H_{op}^{\frac{1}{2}}g)_J|^2 \leq \\ &\leq (H_{op}^{\frac{1}{2}}u, H_{op}^{\frac{1}{2}}u)_J(H_{op}^{\frac{1}{2}}g, H_{op}^{\frac{1}{2}}g)_J = \\ &= (u,v)_J|g|_+^{(A)2} \leq \gamma(u,v)_J + |(u,Jg)_J|^2, \end{aligned}$$

hence $A_{(\gamma)}$ is nonnegative and the first assertion of Theorem 8.2 is proved for $\gamma > 0$.

Let now $\gamma < 0$. Since in this case

$$0 \leq [x,y]-\gamma^{-1}|[y,g]|^2+\gamma^{-1}|[x,g]|^2 \quad \text{for all} \quad \binom{x}{y} \in A$$

is equivalent to

$$|\gamma|(u,v)_J + |(u,Jg)_J|^2 \geq |(v,g)_J|^2 \quad \text{for all} \quad \binom{u}{v} \in K^{-1}$$

we get the result in a similar way.

Consider now the extension $A_{(0)}$. If $\gamma_- = 0$ or $\gamma_+ = 0$ the c.l.r. $A_{(0)}$ is nonnegative since a gap metric limit of a sequence of nonnegative c.l.r.'s is nonnegative.

Assume that $A_{(0)}$ is nonnegative. Then by Corollary 7.3 we have

$$A_{(0)} = A_0 \dot{+} \mathrm{sp}\{\binom{g}{ig}\}. \tag{8.10}$$

Then, for $\binom{x_0}{y_0} \in A_0$ and $\alpha \in \mathbf{C}$

$$\begin{aligned} 0 &\leq [x_0 + \alpha g, y_0 + \alpha i g] = [x_0, y_0] + \alpha[g, y_0] - \bar{\alpha} i [x_0, g] = \\ &= [x_0, y_0] + 2\mathrm{Re}\{\alpha[g, y_0]\}, \end{aligned}$$

since $[y_0 + ix_0, g] = 0$. This implies $[g, y_0] = 0$ for every $y_0 \in \mathcal{R}(A_0)$ and $[g, x_0] = 0$ for every $x_0 \in \mathcal{D}(A_0)$. Then, by (8.10), $g[\perp]\mathcal{R}(A_{(0)})$ and $g[\perp]\mathcal{D}(A_{(0)})$, i.e. $g \in \mathcal{N}(A_{(0)}) \cap A_{(0)}(0)$. It follows that $\rho(A_{(0)}) = \emptyset$. Since $\mathcal{R}(A_0 - z)^{[\perp]} = \mathrm{span}\{g\}$ for every $z \in \mathbf{C} \setminus \mathbf{R}$, on account of Corollary 6.2, we have either $\binom{0}{g} \in A_0$ or $\binom{g}{\lambda g} \in A_0$ for some $\lambda \in \mathbf{C}$. This implies either $\gamma_- = 0$ or $\gamma_+ = 0$, which proves the first assertion of Theorem 8.2.

3. It remains to prove that, in the case when $[g, g] = 0$, $A_{(\gamma_-)}$ is the von Neumann-Krein extension and $A_{(\gamma_+)}$ the Friedrichs extension. We assume that either $\gamma_- \neq -\infty$ or $\gamma_+ \neq \infty$. In view of [4, Theorem 5] it is sufficient to show that

$$(JA_{(\gamma_+)} + 1)^{-1} \leq (JA_{(\gamma)} + 1)^{-1} \leq (JA_{(\gamma_-)} + 1)^{-1} \tag{8.11}$$

for every $\gamma \in \bar{\mathbf{R}} \setminus (\gamma_-, \gamma_+)$.

We shall prove that

$$(JA_{(\gamma_2)} + 1)^{-1} \leq (JA_{(\gamma_1)} + 1)^{-1} \tag{8.12}$$

for every pair $\gamma_1, \gamma_2 \in \bar{\mathbf{R}} \setminus (\gamma_-, \gamma_+)$ with

$$\gamma_-^{-1} < \gamma_1^{-1} < \gamma_2^{-1} < \gamma_+^{-1}, \tag{8.13}$$

where we set $\gamma_-^{-1} = -\infty$ if $\gamma_- = 0$ and $\gamma_+^{-1} = \infty$ if $\gamma_+ = 0$. Since

$$\lim_{\gamma \to \gamma_-} (JA_{(\gamma)} + 1)^{-1} = (JA_{(\gamma_-)} + 1)^{-1},$$

$$\lim_{\gamma \to \gamma_+} (JA_{(\gamma)} + 1)^{-1} = (JA_{(\gamma_+)} + 1)^{-1}, \qquad \gamma \in \bar{\mathbf{R}} \setminus (\gamma_-, \gamma_+),$$

with respect to the operator norm, the relation (8.12) implies (8.11).

In order to prove (8.12) we define operators H_n, $n = 1, 2, \ldots$:

$$H_n := n^{-1}E([0, n^{-1}); H) + HE([n^{-1}, n]; H) + nE((n, \infty) \cup \{\infty\}; H).$$

The sequence (H_n) converges to H with respect to the gap metric. Let $A_n := JH_n$, $K_n := A_nJ = JH_nJ$, $n = 1, 2, \ldots$, and choose γ such that $\gamma_-^{-1} < \gamma^{-1} < \gamma_+^{-1}$. Then

$$A_{(\gamma)} = \{\binom{u}{v} - \gamma^{-1}[v + iu, g]\binom{g}{ig} : \binom{u}{v} \in A\}.$$

Consider the selfadjoint c.l.r.'s

$$A_{n,\{\gamma\}} := \{\binom{u}{v} - \gamma^{-1}[v + iu, g]\binom{g}{ig} : \binom{u}{v} \in A_n\}, \quad n = 1, 2, \ldots$$

(see (7.15)). It is not difficult to verify that $\lim_{n\to\infty} A_{n,\{\gamma\}} = A_{(\gamma)}$ with respect to the gap metric. As we have proved above, $A_{n,\{\gamma\}}$ is nonnegative if and only if

$$\gamma \notin (-|g|_-^{(A_n)2}, |g|_+^{(A_n)2}),$$

where $|g|_-^{(A_n)} = \|K_n^{-\frac{1}{2}} g\|_J$, $|g|_+^{(A_n)} = \|H_n^{\frac{1}{2}} g\|_J$. Evidently,

$$\lim_{n\to\infty} (-\|K_n^{-\frac{1}{2}} g\|_J^2) = \gamma_-, \qquad \lim_{n\to\infty} \|H_n^{\frac{1}{2}}\|_J^2 = \gamma_+.$$

Hence, for fixed γ_1, γ_2 as in (8.13) we may choose an n_0 such that for all $n \geq n_0$ we have

$$(-|g|_-^{(A_n)})^{-2} < \gamma_1^{-1} < \gamma_2^{-1} < (|g|_+^{(A_n)})^{-2} = \|H_n^{\frac{1}{2}} g\|_J^{-2}.$$

By (7.16) $A_{n,\{\gamma_1\}}$ and $A_{n,\{\gamma_2\}}$, $n \geq n_0$, are operators:

$$A_{n,\{\gamma\}} = A_n + (\gamma - \|H_n^{\frac{1}{2}} g\|_J^2)^{-1} [\cdot, (A_n - i)g](A_n - i)g, \quad \gamma = \gamma_1, \gamma_2.$$

It follows that for $n \geq n_0$ we have

$$JA_{n,\{\gamma_1\}} \leq JA_{n,\{\gamma_2\}}$$

and, hence,

$$(JA_{n,\{\gamma_1\}} + 1)^{-1} \geq (JA_{n,\{\gamma_2\}} + 1)^{-1}.$$

For $n \to \infty$ we obtain the relation (8.12) which completes the proof of Theorem 8.2.

Let the assumptions of Theorem 8.2 be fulfilled. If $\gamma \in (\gamma_-, \gamma_+)$, the selfadjoint relation $JA_{(\gamma)}$ in the Hilbert space $(\mathcal{H}, (\cdot,\cdot)_J)$ has one (including the multiplicity) negative eigenvalue. It can be shown that it depends nonincreasingly and continuously on γ with

$$\lim_{\gamma\downarrow\gamma_-} \lambda_\gamma = 0, \qquad \lim_{\gamma\uparrow\gamma_+} \lambda_\gamma = -\infty.$$

In a similar way, the selfadjoint relation $A_{(\gamma)}$ in the Krein space $(\mathcal{H}, [\cdot,\cdot])$, $\gamma \in (\gamma_-, \gamma_+)$, has one eigenvalue λ_γ ($\in \tilde{\sigma}_p(A_{(\gamma)})$) which is "exceptional" in the sense that it is either in $\mathbf{C}^+$ or, if e.g. it is real and $\neq 0$, there exists an eigenvector $x_\gamma \neq 0$ corresponding to λ_γ such that $[x_\gamma, x_\gamma] \operatorname{sign} \lambda_\gamma \leq 0$. As we have mentioned in the introduction, this eigenvalue will be considered elsewhere.

References

[1] Azizov, T. Ya.; Iohvidov, I. S.: Foundations of the Theory of Linear Operators in Spaces with Indefinite Metric, Moscow, 1986; English transl.: Linear Operators in Spaces with Indefinite Metric, Wiley, New York, 1989.

[2] Bognár, J.: Indefinite Inner Product Spaces, Springer-Verlag, Berlin Heidelberg New York, 1974.

[3] Coddington, E. A.: Extension theory of formally normal and symmetric subspaces, Mem. Amer. Math. Soc. **134** (1973).

[4] Coddington, E. A.; de Snoo, H. S. V.: Positive selfadjoint extensions of positive symmetric subspaces, Math. Z. **159** (1978), 203-214.

[5] Derkach, V. A.: Generalized resolvents of hermitian operators in Krein space, Preprint, Donetsk 1992.

[6] Dijksma, A.; Langer, H.; de Snoo, H. S. V.: Unitary colligations in Krein spaces and their role in the extension theory of isometric and symmetric linear relations in Hilbert spaces, Functional Analysis II, Proceedings Dubrovnik 1985, Lecture Notes in Mathematics 1242 (1986), 1-42.

[7] Dijksma, A.; de Snoo, H. S. V.: Selfadjoint extensions of symmetric subspaces, Pac. J. Math. **54** (1974), 71-100.

[8] Dijksma, A.; de Snoo, H. S. V.: Symmetric and selfadjoint relations in Krein spaces I, Operator Theory: Advances and Applications Vol. 24 (1987), Birkhäuser Verlag Basel, 145-166.

[9] Dijksma, A.; de Snoo, H. S. V.: Symmetric and selfadjoint relations in Krein spaces II, Ann. Acad. Sci. Fenn., Ser. A.I. Mathematica **12** (1987), 199-216.

[10] Jonas, P.; Langer, H.: Some questions in the perturbation theory of J-nonnegative operators in Krein spaces, Math. Nachr. **114** (1983), 205-226.

[11] Kato, T.: Perturbation Theory for Linear Operators, Springer-Verlag New York, 1966.

[12] Krein, M. G.; Langer, H.: On defect subspaces and generalized resolvents of a Hermitian operator in the space Π_κ, Funktsional.Anal. i Prilozhen. **5**, n.2 (1971), 59-71; **5**, n.3 (1971), 54-69 (Russian); English transl.: Functional Anal. Appl. **5** (1971/1972), 139-146, 217-228.

[13] Krein, M. G.; Shmul'yan, Yu. L.: Plus-operators in a space with indefinite metric, Mat. Issled. **1**, n. 1 (1966), 131-161; English transl.: Amer. Math. Soc. Transl.(2) **85** (1969), 93-113.

[14] Langer, H.: Verallgemeinerte Resolventen eines J-nichtnegativen Operators mit endlichem Defekt, J. Functional Analysis **8** (1971), 287-320.

[15] Langer, H.: Spectral functions of definitizable operators in Krein spaces, Functional Analysis, Proceedings Dubrovnik, 1981, Lecture Notes in Mathematics **948** (1982), 1-46.

[16] Langer, H.; Textorius, B.: On generalized resolvents and Q-functions of symmetric linear relations (subspaces) in Hilbert space, Pac. J. Math. **72** (1977), 135-165.

[17] Shmul'yan, Yu. L.: On a class of holomorphic operator functions, Mat. Zametki **5** (1969), 351-359 (Russian).

[18] Shmul'yan, Yu. L.: Extension theory for operators and spaces with indefinite metric, Izv. Akad. Nauk SSSR, Ser.Mat. **38** (1974), 896-908; English transl.: Math. USSR Izvestiya **8**, n.4 (1974), 895-907.

Peter Jonas
Institut für Mathematik
Universität Potsdam
Postfach 601553
D-14415 Potsdam, Germany

Heinz Langer
Institut für Analysis, Technische Mathematik
und Versicherungsmathematik
Technische Universität Wien
Wiedner Hauptstrasse 8-10
A-1040 Wien, Austria

1991 Mathematics Subject Classification: 47 B 50, 47 A 20

Operator Theory:
Advances and Applications, Vol. 80

DIFFERENTIAL GEOMETRY OF GENERALIZED GRASSMANN MANIFOLDS IN C^*-ALGEBRAS

by

Mircea Martin and **Norberto Salinas**

The main goal of this article is to set off and to study some genuine differential geometric objects that naturally occur in the framework of C^*-algebras. The intent was to develop a unified treatment of a few specific situations that were considered by the authors in previous articles (cf. [M3–4], [MS1–2], [S]), as well as by many others (cf. [ARS], [AS], [A], [CPR1–4], [LM], [Ma], [MR], [PR1–2], [W1–3]). The advantage of our present approach seems to be that that instead of certain, more or less, ad-hoc methods, we tried to find an appropiate setting and suitable simple tools which facilitate the introduction of techniques from differential geometry into operator algebras. It should be mentioned that our contribution in this paper is strongly motivated by the program initiated by M. J. Cowen and R. G. Douglas (cf. [CD1–2]). More evidence for these aspects can be found in [S], and also in [M2–3] and [MS1–2].

The geometric objects we are going to discuss reflect in a specific way the underlying structure of the algebras involved in their construction, and are, in many respects, related to the well known Grassmann manifolds. The finite dimensional Grassmann manifolds, as well as their even more interesting relatives, the flag manifolds, enjoy a lot of nice topological and geometric properties and can be described in many different ways. To illustrate the point we first recall the classical setting.

Let n and N be two positive integers, with $n \leqslant N$. An n-flag in the Hilbert space $\mathbb{C}^N$ is a filtration of length n of $\mathbb{C}^N$, that is, an increasing chain

$$0 = V_0 \subseteq V_1 \subseteq \cdots \subseteq V_n = \mathbb{C}^N,$$

where V_k $(0 \leqslant k \leqslant n)$ are vector spaces. We let $\mathfrak{F}_n(\mathbb{C}^N)$ denote the set of all n-flags in $\mathbb{C}^N$. The simplest case $n = 2$ corresponds to the Grassmann manifold of $\mathbb{C}^N$.

There are at least two other very elementary realizations of the flag manifold $\mathfrak{F}_n(\mathbb{C}^N)$, each of them providing specific new information about $\mathfrak{F}_n(\mathbb{C}^N)$. Since these alternative descriptions also provide a motivation for our subsequent approach, we explain them briefly.

1. Systems of orthogonal projections. Let $A = \mathcal{L}(\mathbb{C}^N)$ be the C^*-algebra of linear operators on $\mathbb{C}^N$. Any n-flag in $\mathbb{C}^N$ determines, and is uniquely determined by, an n-tuple $(e_1, e_2, \ldots, e_n)$ of mutually orthogonal projections in A, such that

$$e_1 + e_2 + \cdots + e_n = 1.$$

More specifically, e_k is the orthogonal projection onto the subspace $V_k \ominus V_{k-1}$, $1 \leqslant k \leqslant n$. Consequently, $\mathfrak{F}_n(\mathbb{C}^N)$ can be consider as a closed subset of the algebra $A^n = A \oplus A \oplus \cdots \oplus A$.

2. Cyclic group representations. Let C_n be the cyclic group of order n. Any unitary representation of C_n on $\mathbb{C}^N$ is uniquely defined by a unitary element $u \in A$, subject to the condition $u^n = 1$. In its turn, each u as above is related to an n-tuple $(e_1, e_2, \ldots, e_n)$ of mutually orthogonal projections in A that decomposes the identity. We simply let e_k be the spectral projection of u corresponding to $\exp(2\pi(k-1)\mathrm{i}/n)$, $1 \leqslant k \leqslant n$ ($e_k = 0$ is allowed for some k).

In this new realization, the space $\mathfrak{F}_n(\mathbb{C}^N)$ is a closed subset of the algebra $\mathbb{C}[C_n] \otimes A$, where $\mathbb{C}[C_n]$ denotes the complex group algebra of C_n.

From both of these descriptions one easily gets the well known realization of $\mathfrak{F}_n(\mathbb{C}^N)$ as a disjoint union of the reductive homogeneous spaces

$$U(N)/U(N_1) \times U(N_2) \times \cdots \times U(N_n),$$

where $U(N)$ is the unitary group of $\mathcal{L}(\mathbb{C}^N)$ and $(N_1, N_2, \ldots, N_n)$ is an n-tuple of non-negative integers, such that $N_1 + N_2 + \cdots + N_n = N$.

Another important feature illustrated by the remarks above is that the flag manifolds can be embedded in, and described in terms of, some algebras which interact with the algebra A. It is this very specific feature that prompts us to introduce the notion of environments over algebras as the basic starting point of our approach to a generalization of Grassmann manifolds.

The organization of the paper is as follows. In Section 1 we start with the definition of environments and present a few examples. We also discuss some algebraic geometric properties of the Grassmannians associated to environments. Section 2 is devoted to the topological properties of the Grassmannians related to Banach environments. A specific aspect, the lifting problem for continuous curves in a Grassmannian, is discussed in Section 3. In Section 4 we prove the existence of natural differentiable structures on our general Grassmannians and show that each Grassmannian is a discrete disjoint union of reductive homogeneous spaces. This specific property implies the existence of invariant linear connections on the generalized Grassmann manifolds, an aspect developed in Section 5. We next exhibit a canonical linear connection and then, in Section 6, we find an explicit formula of the associated exponential map. We conclude Section 6 with some results on the geodesics of the canonical linear connection.

Our presentation is somewhat expository and it can be partly regarded as a complement of the general study of infinite dimensional reductive homogeneous spaces made in [MR] and [W2].

We are grateful to Professor R. D. Wilkins for his helpful comments on a preliminary version of our paper. We are also grateful to Professor G. Corach for having drawn our attention to [MR].

1. ENVIRONMENTS AND THEIR GRASSMANNIANS

Our approach to a generalization of the classical Grassmann and flag manifolds is based on the notion of *environments*. At this early stage we will be primarily concerned with a few specific algebraic properties of these objects.

1.1. To start with, let A be a fixed unital complex algebra.

DEFINITION. By an *environment* over A we will mean a pair $\mathcal{E} = (E, \Pi)$, where

(i) E is a complex algebra equipped with a compatible A-bimodule structure, and

(ii) $\Pi : E \to A$ is a left and right A-linear map.

To be more specific, we always assume that

$$(x \cdot \varphi) \cdot y = x \cdot (\varphi \cdot y) \quad \text{and} \quad 1 \cdot \varphi = \varphi \cdot 1 = \varphi,$$

$$x \cdot (\varphi \times \psi) = (x \cdot \varphi) \times \psi \quad \text{and} \quad (\varphi \times \psi) \cdot x = \varphi \times (\psi \cdot x),$$

$$(\varphi \cdot x) \times \psi = \varphi \times (x \cdot \psi),$$

for all $x, y \in A$ and $\varphi, \psi \in E$, where 1 is the identity of A and $\times$ stands for the multiplication in E.

1.2. In the case when A is an involutive algebra we define a particular kind of environments as follows.

DEFINITION. An environment $\mathcal{E} = (E, \Pi)$ over an involutive algebra A is called an *involutive environment* if E is an involutive algebra and Π satisfies the condition

$$\Pi(\varphi^\sharp) = \Pi(\varphi)^*, \quad \varphi \in E,$$

where $*$ and $\sharp$ denote the involution on A and E, respectively.

For involutive environments we assume that

$$(x \cdot \varphi)^\sharp = \varphi^\sharp \cdot x^*, \quad x \in A, \varphi \in E.$$

1.3. DEFINITION. We introduce the Grassmannian $\mathfrak{G} = \mathfrak{G}(\mathcal{E})$ of an environment $\mathcal{E} = (E, \Pi)$ as the set of all $\alpha \in E$ satisfying

(i) $\Pi(\alpha \times \alpha) = 1$,

(ii) $\alpha \cdot \Pi(\alpha \times \varphi) = \alpha \times \varphi$,

(iii) $\Pi(\varphi \times \alpha) \cdot \alpha = \varphi \times \alpha$,

for every $\varphi \in E$.

If $\mathcal{E}$ is an involutive environment we will also consider the set $\mathfrak{U} = \mathfrak{U}(\mathcal{E})$, defined by

$$\mathfrak{U} = \{\alpha \in \mathfrak{G} : \alpha^\sharp = \alpha\}, \tag{1.1}$$

and referred to as the self-adjoint Grassmannian of $\mathcal{E}$.

One of the goals of the present paper is to study the geometry of the spaces $\mathfrak{G}(\mathcal{E})$ and $\mathfrak{U}(\mathcal{E})$.

As a first simple remark we notice that any α in $\mathfrak{G}$ is an idempotent of E, that is,

$$\alpha \times \alpha = \alpha \tag{1.2}$$

and, consequently,

$$\Pi(\alpha) = 1. \tag{1.3}$$

On the other hand, we easily observe that given φ in E with $\Pi(\varphi) = 1$ and such that either $\alpha \times \varphi = \varphi$, or $\varphi \times \alpha = \varphi$, for some α in $\mathfrak{G}$, then $\varphi = \alpha$. In particular, this shows that if α and β are in $\mathfrak{G}$ and $\Pi(\alpha \times \beta) = 1$, then $\alpha = \beta$. Finally, the above definitions obviously imply that in the case when $\mathcal{E}$ is an involutive environment, the conditions $\alpha \in \mathfrak{G}$ and $\alpha^\sharp \in \mathfrak{G}$ are equivalent and condition (iii) in Definition 1.3 is a consequence of condition (ii).

1.4. Before continuing the general study of Grassmannians associated with environments, we consider some simple examples. Throughout all the next examples A is a fixed unital algebra.

EXAMPLE 1. We explain first how the general framework described above includes, as specific cases, the classical Grassmann and flag manifolds. Let $n \geqslant 2$ be a given integer and let $A^n = A \oplus A \oplus \cdots \oplus A$ be the direct sum of n copies of A. The addition, multiplication and both left and right scalar multiplications in A^n are defined componentwise. Further, let $\Pi : A^n \to A$ be the map defined by

$$\Pi(\varphi) = x_1 + x_2 + \cdots + x_n, \quad \varphi = (x_1, x_2, \ldots, x_n) \in A^n.$$

The pair (A^n, Π) is an environment over A and a straightforward computation shows that its Grassmannian $\mathfrak{G}(A^n, \Pi)$ consists of all n-tuples $\alpha = (e_1, e_2, \cdots, e_n)$ of mutually orthogonal idempotents of A satisfying the condition

$$e_1 + e_2 + \cdots + e_n = 1.$$

If A is an involutive algebra, then A^n carries a natural involution given by

$$\varphi^\sharp = (x_1^*, x_2^*, \ldots, x_n^*), \quad \varphi = (x_1, x_2, \ldots, x_n) \in A^n.$$

Accordingly, it makes sense to consider the self-adjoint Grassmannian $\mathfrak{U}(A^n, \Pi)$. In the particular case when A is the algebra $\mathcal{L}(\mathcal{H})$ of bounded linear operators on a complex Hillbert space $\mathcal{H}$, the preceding space $\mathfrak{U}(A^n, \Pi)$ is nothing else than the space of all n-flags on $\mathcal{H}$. It is exactly this reason that prompts us to call $\mathfrak{U}(A^n, \Pi)$ the space of n-flags of A, for any involutive algebra A.

The next example shows that spaces of group representations can be also realized as Grassmannians. This example includes as a particular case the second alternative description of flag manifolds.

EXAMPLE 2. Let G be a finite group and let $A[G]$ be the set of all A-valued functions on G. The addition and scalar multiplication in $A[G]$ are defined pointwise. For $\varphi, \psi \in A[G]$ we define their convolution product $\varphi \times \psi \in A[G]$ by

$$\varphi \times \psi(g) = |G|^{-1} \sum_{h \in G} \varphi(h)\psi(h^{-1}g), \quad g \in G,$$

where $|G|$ denotes the order of G. Under this operation as multiplication, $A[G]$ becomes an algebra. We let $\Pi : A[G] \to A$ denote the map given by

$$\Pi(\varphi) = \varphi(e), \quad \varphi \in A[G],$$

where e is the identity of G.

One easily checks that $(A[G], \Pi)$ is an environment over A, and its Grassmannian $\mathfrak{G}(A[G], \Pi)$ consists of those elements $\alpha \in A[G]$ that satisfy the conditions

$$\alpha(gh) = \alpha(g)\alpha(h), \quad g, h \in G, \text{ and } \alpha(e) = 1.$$

In other words, $\mathfrak{G}(A[G], \Pi)$ coincides with the set of all group homomorphisms from G into the group of invertible elements of A.

If A is an involutive algebra, then $A[G]$ inherits an involution by a standard device, namely,

$$\varphi^\sharp(g) = \varphi(g^{-1})^*, \quad g \in G.$$

The self-adjoint Grassmannian $\mathfrak{U}(A[G], \Pi)$ is precisely the set of group homomorphisms from G into the group of unitary elements of A.

EXAMPLE 3. As a last elementary example, we consider the set $M_n(A)$ of all $n \times n$ matrices over A, where $n \geqslant 2$ is fixed. The addition and scalar multiplication are the usual ones, and the product $\varphi \times \psi$ of two matrices φ and ψ in $M_n(A)$ is defined by

$$\varphi \times \psi = n^{-1} \varphi \cdot \psi,$$

where $\cdot$ stands for the usual matrix multiplication. Let $\Pi : M_n(A) \to A$ be the A-valued trace, defined by

$$\Pi(\varphi) = x_{11} + x_{22} + \cdots + x_{nn}, \quad \varphi = [x_{ij}]_{i,j=1}^n \in M_n(A).$$

It is easy to see that an element $\alpha = [a_{ij}]_{i,j=1}^n \in M_n(A)$ belongs to the Grassmannian of the environment $(M_n(A), \Pi)$ if and only if its entries form a matrix unit in A, that is,

$$a_{ij}a_{jk} = a_{ik}, \quad 1 \leqslant i, j, k \leqslant n,$$

$$a_{ij}a_{kl} = 0, \quad 1 \leqslant i, j, k, l \leqslant n, \ j \neq k,$$

$$a_{11} + a_{22} + \cdots + a_{nn} = 1.$$

Consequently, we may identify $\mathfrak{G}(M_n(A), \Pi)$ with the set of all algebra homomorphisms from $M_n(\mathbb{C})$ into A. If A is an involutive algebra, then $M_n(A)$ has a natural involution and $\mathfrak{U}(M_n(A), \Pi)$ can be described as the set of all involutive algebra homomorphisms from $M_n(\mathbb{C})$ into A.

1.5. As these few examples suggest, there is a great variety of natural environments over algebras. Moreover, a common feature of the examples above indicates a quite general method of producing environments. Specifically, any complex algebra C with a trace $\pi : C \to \mathbb{C}$ yields the environment (E, Π) given by $E = A \otimes C$ and $\Pi = \mathrm{id} \otimes \pi$. In our previous examples C was $\mathbb{C}^n$, $\mathbb{C}[G]$ and $M_n(\mathbb{C})$, respectively.

It should be pointed out also that in an appropriate topological setting all the previous examples have more general analogues. Of course in such cases, some restrictions concerning continuity are needed. To ilustrate this point we mention the far reaching generalization of Example 1 discussed in [ARS], as well as the extension of Example 2 for compact groups considered in [M4].

1.6. We now return to the general case of an environment $\mathcal{E} = (E, \Pi)$ over an arbitrary algebra A. As before, we let $\mathfrak{G}$ denote the Grassmannian of $\mathcal{E}$.

Given α in $\mathfrak{G}$ we define the subalgebra A^α of A by

$$A^\alpha = \{x \in A : x \cdot \alpha = \alpha \cdot x\}, \tag{1.4}$$

and the similarity orbit $\mathfrak{G}(\alpha)$ of α by

$$\mathfrak{G}(\alpha) = \{a \cdot \alpha \cdot a^{-1} : a \in G(A)\}, \tag{1.5}$$

where $G(A)$ is the group of all invertible elements of A.

If one considers the action $\rho : G(A) \times \mathfrak{G} \to \mathfrak{G}$ of the group $G(A)$ on $\mathfrak{G}$ given by

$$\rho(a, \alpha) = a \cdot \alpha \cdot a^{-1}, \quad a \in G(A),\ \alpha \in \mathfrak{G}, \tag{1.6}$$

then the isotropy subgroup of a fixed element $\alpha \in \mathfrak{G}$ coincides with $G(A^\alpha)$, and, consequently, we can identify the similarity orbit $\mathfrak{G}(\alpha)$ with the quotient space $G(A)/G(A^\alpha)$, in the usual manner.

A nice and important property of the subagebras A^α is the existence of some ready-made conditional expectations onto them (see, for instance, [FD, XI.4.13]. More specifically, for each $\alpha \in \mathfrak{G}$ let $P^\alpha : A \to A$ be the linear map defined by

$$P^\alpha(x) = \Pi(\alpha \cdot x \times \alpha), \quad a \in A. \tag{1.7}$$

A straightforward computation, based on equations (i), (ii), (iii) and the first subsequent remark in 1.3, shows that $P^\alpha \circ P^\alpha = P^\alpha$, $P^\alpha(x) = x$ if and only if $x \in A^\alpha$, and $P^\alpha(x'xx'') = x'P^\alpha(x)x''$, for every $x', x'' \in A^\alpha$ and $x \in A$. This construction also provides a vector subspace A_α of A that is a complement of A^α in A, namely, $A_\alpha = \{x \in A : P^\alpha(x) = 0\}$. Obviously, A_α is an invariant subspace under left and right multiplications by elements of A^α. The space A_α is the range of the complementary projection map $P_\alpha = \mathrm{id} - P^\alpha$ from

A into A. Using equations (ii) and (iii) from 1.3, and the basic equality $\alpha \times \alpha = \alpha$, one checks easily that

$$\alpha \cdot P^\alpha(x) \times \alpha = \alpha \cdot x \times \alpha, \quad x \in A.$$

This relation in turn implies that the subspace A_α is characterized by

$$A_\alpha = \{x \in A : \alpha \cdot x \times \alpha = 0\}. \tag{1.8}$$

1.7. Analogous remarks can be done for the self-adjoint Grassmannian $\mathfrak{U}$ of an involutive environment. Given α in $\mathfrak{U}$ one considers its unitary orbit $\mathfrak{U}(\alpha)$ defined by

$$\mathfrak{U}(\alpha) = \{u \cdot \alpha \cdot u^* : u \in U(A)\},$$

where $U(A)$ is the group of unitary elements of A. There is a natural action of the group $U(A)$ on $\mathfrak{U}$ obtained as a restriction of the action ρ defined above. If $\alpha \in \mathfrak{U}$ is fixed, then the isotropy subgroup of α coincides with $U(A^\alpha)$, and the unitary orbit $\mathfrak{U}(\alpha)$ could be identified with the quotient space $U(A)/U(A^\alpha)$.

1.8. The rest of this section will be concerned with the study of a Grassmannian $\mathfrak{G}$ as the "closed algebraic subset" defined by equations (i), (ii) and (iii) in 1.3. Our specific purpose is to introduce the Zariski tangent space to $\mathfrak{G}$ at a given point α and then to find simpler descriptions of this space.

In algebraic geometry one uses various descriptions of the Zariski tangent space to an algebraic set at a point (see, for instance, [Mu]). An elementary construction, relying on a trick with dual numbers, is easily transferable in our setting. This construction was successfully used in the framework of algebraic geometry, for the study of spaces of finite dimensional representations of finitely generated groups as algebraic varieties, in [LM] and [Ma].

From now on $\mathcal{E} = (E, \Pi)$ will be a fixed environment over a fixed algebra A and $\mathfrak{G}$ will denote the Grassmannian of $\mathcal{E}$.

Let $D = \mathbb{C}[\delta] = \mathbb{C} + \mathbb{C}\delta$, with $\delta^2 = 0$, be the complex algebra of dual numbers and consider the tensor product $\tilde{A} = A \otimes D$. Any element of $A \otimes D$ can be represented uniquely as $\tilde{a} = a_0 + a_1\delta$, where $a_0, a_1 \in A$.

For $\tilde{a} = a_0 + a_1\delta$ and $\tilde{b} = b_0 + b_1\delta$ in $A \otimes D$ one defines

$$\tilde{a} + \tilde{b} = (a_0 + b_0) + (a_1 + b_1)\delta,$$

$$\tilde{a}\tilde{b} = a_0 b_0 + (a_0 b_1 + a_1 b_0)\delta.$$

With these operations $A \otimes D$ is a unital complex algebra.

Consider next the algebra $\tilde{E} = E \otimes D$, with a compatible $\tilde{A}$-bimodule structure defined in an obvious way. Finally, let $\tilde{\Pi} : \tilde{E} \to \tilde{A}$ be the map given by

$$\tilde{\Pi}(\varphi_0 + \varphi_1)\delta) = \Pi(\varphi_0) + \Pi(\varphi_1)\delta.$$

It follows easily that $\tilde{\mathcal{E}} = (\tilde{\mathcal{E}}, \tilde{\Pi})$ is an environment over $\tilde{A}$ and, therefore, it makes sense to introduce its Grassmannian $\tilde{\mathfrak{G}} = \mathfrak{G}(\tilde{\mathcal{E}})$. A short computation yields the following result. The details of the proof are left to the reader.

LEMMA. *An element $\alpha + \theta\delta \in \tilde{\mathcal{E}}$ is a point in the Grassmannian $\tilde{\mathfrak{G}}$ if and only if $\alpha \in \mathfrak{G}$ and*

$$\Pi(\theta \times \alpha + \alpha \times \theta) = 0, \tag{1.9}$$

$$\theta \cdot \Pi(\alpha \times \varphi) + \alpha \cdot \Pi(\theta \times \varphi) = \theta \times \varphi, \tag{1.10}$$

$$\Pi(\varphi \times \theta) \cdot \alpha + \Pi(\varphi \times \alpha) \cdot \theta = \varphi \times \theta, \tag{1.11}$$

for every $\varphi \in E$.

1.9. The next definition is motivated by the already mentioned characterization of tangent spaces to algebraic sets.

DEFINITION. An element $\theta \in E$ is said to be a *Zariski tangent vector* to $\mathfrak{G}$ at the point $\alpha \in \mathfrak{G}$ provided that $\alpha + \theta\delta$ belongs to $\tilde{\mathfrak{G}}$.

The definition requires that θ satisfies equations (1.9), (1,10) and (1.11). The set $\mathrm{T}_\alpha^{\mathrm{alg}}\mathfrak{G}$ of all tangent vectors θ will be called the *Zariski tangent space* to $\mathfrak{G}$ at α.

It may be worth mentioning that for $\varphi = \alpha$ relations (1.10) and (1.11) become

$$\theta \times \alpha = \theta + \alpha \cdot \Pi(\theta \times \alpha),$$

$$\alpha \times \theta = \Pi(\alpha \times \theta) \cdot \alpha + \theta.$$

In particular, each of the these equalities shows that

$$\Pi(\theta) = 0. \tag{1.12}$$

The same conclusion follows from $\tilde{\Pi}(\alpha + \theta\delta) = 1$. Since, on the other hand, $\alpha + \theta\delta$ must be an idempotent in $\tilde{E}$, we get

$$\alpha \times \theta + \theta \times \alpha = \theta. \tag{1.13}$$

Actually, as we will show below, the last two conditions characterize completely a tangent vector.

1.10. We now introduce two maps that will enable us to find alternative simpler descriptions of tangent vectors.

Assume that $\alpha \in \mathfrak{G}$ is fixed and consider the next linear maps

$$\partial_\alpha : A \to E, \quad \partial_\alpha(x) = x \cdot \alpha - \alpha \cdot x \qquad (x \in A); \tag{1.14}$$

$$\varepsilon_\alpha : E \to A, \quad \varepsilon_\alpha(\varphi) = \frac{1}{2}\Pi(\varphi \times \alpha - \alpha \times \varphi) \qquad (\varphi \in E). \tag{1.15}$$

PROPOSITION. *With the above notations we have*

(i) $\partial_\alpha(x) \in \mathrm{T}^{\mathrm{alg}}_\alpha \mathfrak{G}$ *for all* $x \in A$;

(ii) *for any* $\varphi \in E$ *which satisfies the conditions*

$$\Pi(\varphi) = 0 \text{ and } \varphi \times \alpha + \alpha \times \varphi = \varphi, \tag{1.16}$$

we have

$$\partial_\alpha \varepsilon_\alpha(\varphi) = \varphi. \tag{1.17}$$

Proof. Assertion (i) follows by a straightforward computation based on equations (i), (ii) and (iii) in Definition 1.3. All one has to do is to check that $\theta = x \cdot \alpha - \alpha \cdot x$ satisfies conditions (1.9), (1.10) and (1.11). For the second assertion, we first notice that

$$\Pi(\varphi \times \alpha + \alpha \times \varphi) = 0,$$

whence

$$\varepsilon_\alpha(\varphi) = \Pi(\varphi \times \alpha) = -\Pi(\alpha \times \varphi).$$

According to (1.14) and Definition 1.3 it follows that

$$\partial_\alpha \varepsilon_\alpha(\varphi) = \varepsilon_\alpha(\varphi) \cdot \alpha - \alpha \cdot \varepsilon_\alpha(\varphi) =$$

$$= \Pi(\varphi \times \alpha) \cdot \alpha + \alpha \cdot \Pi(\alpha \times \varphi) = \varphi \times \alpha + \alpha \times \varphi = \varphi,$$

as asserted. ∎

1.11. COROLLARY. *Let* $\alpha \in \mathfrak{G}$ *be fixed.*

(i) *If* $\theta \in \mathrm{T}^{\mathrm{alg}}_\alpha \mathfrak{G}$ *then*

$$\partial_\alpha \varepsilon_\alpha(\theta) = \theta. \tag{1.18}$$

(ii) *The linear map*

$$\pi_\alpha : E \to E, \quad \pi_\alpha = \partial_\alpha \varepsilon_\alpha, \tag{1.19}$$

is a projection of E *onto* $\mathrm{T}_\alpha^{\mathrm{alg}}\mathfrak{G}$.

Proof. If $\theta \in \mathrm{T}_\alpha^{\mathrm{alg}}\mathfrak{G}$ then, by (1.12) and (1.13), we find that θ satisfies (1.16), thus (1.18) follows from (1.17).

In order to prove (ii), by view of assertion (i) in Proposition 1.10, we first observe that $\pi_\alpha(\varphi) \in \mathrm{T}_\alpha^{\mathrm{alg}}\mathfrak{G}$ for any $\varphi \in E$. On the other hand, if $\theta \in \mathrm{T}_\alpha^{\mathrm{alg}}\mathfrak{G}$ then (1.18) gives $\pi_\alpha(\theta) = \theta$. By combining these two remarks we conclude that $\pi_\alpha \circ \pi_\alpha = \pi_\alpha$ and $\operatorname{ran}\pi_\alpha = \mathrm{T}_\alpha^{\mathrm{alg}}\mathfrak{G}$. ■

1.12. The existence of the projection map π_α from E onto $\mathrm{T}_\alpha^{\mathrm{alg}}\mathfrak{G}$ shows that $\mathrm{T}_\alpha^{\mathrm{alg}}\mathfrak{G}$ has a complement in E, namely, the subspace $\mathrm{T}_\alpha^{\perp}\mathfrak{G} = \ker \pi_\alpha$.

The spaces $\mathrm{T}_\alpha^{\mathrm{alg}}\mathfrak{G}$ and $\mathrm{T}_\alpha^{\perp}\mathfrak{G}$ have the following simple descriptions.

PROPOSITION. *Let* $\varphi \in E$ *be fixed. Then*

(i) $\varphi \in \mathrm{T}_\alpha^{\mathrm{alg}}\mathfrak{G}$ *if and only if* $\Pi(\varphi) = 0$ *and* $\varphi \times \alpha + \alpha \times \varphi = \varphi$;

(ii) $\varphi \in \mathrm{T}_\alpha^{\perp}\mathfrak{G}$ *if and only if* $\varepsilon_\alpha(\varphi) = 0$, *that is,* $\Pi(\varphi \times \alpha) = \Pi(\alpha \times \varphi)$.

Proof. Assertion (i) summarizes a few previous remarks. For (ii) it is enough to recall first that $\pi_\alpha = \partial_\alpha \varepsilon_\alpha$ and then to observe that $\varepsilon_\alpha = \varepsilon_\alpha \pi_\alpha$. Indeed, a direct computation based on Definition 1.3 gives

$$\alpha \cdot \varepsilon_\alpha(\varphi) \times \alpha = \frac{1}{2}\alpha \cdot \Pi(\varphi \times \alpha - \alpha \times \varphi) \times \alpha =$$

$$= \frac{1}{2}(\alpha \times \varphi \times \alpha - \alpha \times \varphi \times \alpha) = 0,$$

for every $\varphi \in E$. Therefore, by using also the equalities (1.2) and (1.3), we find that

$$\varepsilon_\alpha \pi_\alpha(\varphi) = \frac{1}{2}\Pi(\pi_\alpha(\varphi) \times \alpha - \alpha \times \pi_\alpha(\varphi)) =$$

$$= \frac{1}{2}\Pi((\varepsilon_\alpha(\varphi) \cdot \alpha - \alpha \cdot \varepsilon_\alpha(\varphi)) \times \alpha - \alpha \times (\varepsilon_\alpha(\varphi) \cdot \alpha - \alpha \cdot \varepsilon_\alpha(\varphi))) =$$

$$= \frac{1}{2}\Pi(\varepsilon_\alpha(\varphi) \cdot \alpha - 2\alpha \cdot \varepsilon_\alpha(\varphi) \times \alpha + \alpha \cdot \varepsilon_\alpha(\varphi)) = \varepsilon_\alpha(\varphi).$$

The proof is complete. ■

1.13. The computations above, combined with formula (1.7) that defines the projection $P^\alpha : A \to A$ of A onto A^α, show that

$$P^\alpha \circ \varepsilon_\alpha = 0.$$

On the other hand clearly

$$\partial_\alpha \circ P^\alpha = 0.$$

Actually, for any $x \in A$, we have

$$\varepsilon_\alpha \partial_\alpha(x) = \frac{1}{2}\Pi((x \cdot \alpha - \alpha \cdot x) \times \alpha - \alpha \times (x \cdot \alpha - \alpha \cdot x)) =$$

$$= \frac{1}{2}\Pi(x \cdot \alpha - 2\alpha \cdot x \times \alpha + \alpha \cdot x) =$$

$$= x - \Pi(\alpha \cdot x \times \alpha) = x - P^\alpha(x),$$

that is (for notation see 1.6)

$$\varepsilon_\alpha \partial_\alpha = \mathrm{id} - P^\alpha = P_\alpha.$$

We summarize this discussion in the following proposition.

PROPOSITION. *The short sequences of vector spaces*

$$0 \to A^\alpha \hookrightarrow A \xrightarrow{\partial_\alpha} \mathrm{T}_\alpha \mathfrak{G} \to 0,$$

$$0 \to \mathrm{T}_\alpha \mathfrak{G} \xrightarrow{\varepsilon_\alpha} A \xrightarrow{P^\alpha} A^\alpha \to 0$$

are exact.

1.14. All previous definitions, constructions and results have natural counterparts if $\mathcal{E}$ is an involutive environment and one replaces $\mathfrak{G}$ by the self-adjoint Grassmannian $\mathfrak{U} = \mathfrak{U}(\mathcal{E})$. For instance, the definition of the Zariski tangent space to $\mathfrak{U}$ at a point $\alpha \in \mathfrak{U}$ will be (compare with Definition 1.9)

$$\mathrm{T}_\alpha^{\mathrm{alg}} \mathfrak{U} = \{\theta \in E_{\mathrm{h}} : \alpha + \theta\delta \in \mathfrak{U}(\tilde{\mathcal{E}})\},$$

where E_{h} denotes the set of all hermitian elements of E, and the involution in the algebra $\tilde{\mathcal{E}}$ is defined componentwise, that is,

$$(\varphi_0 + \varphi_1\delta)^\sharp = \varphi_0^\sharp + \varphi_1^\sharp \delta.$$

This implies that $\mathrm{T}_\alpha^{\mathrm{alg}}\mathfrak{U} = \mathrm{T}_\alpha^{\mathrm{alg}}\mathfrak{G} \cap E_{\mathrm{h}}$. In addition, given $\alpha \in \mathfrak{U}$, instead of the maps ∂_α and ε_α defined by (1.14) and (1.15), respectively, we have to use some suitable restrictions, namely

$$\partial_\alpha | A_{\mathrm{sh}} : A_{\mathrm{sh}} \to E_{\mathrm{h}},$$

$$\varepsilon_\alpha | E_{\mathrm{h}} : E_{\mathrm{h}} \to A_{\mathrm{sh}},$$

where $A_{\rm sh}$ is the real subspace of all skew-hermitian elements of A. The projection of $E_{\rm h}$ onto $\mathrm{T}^{\rm alg}_\alpha \mathfrak{U}$ is the map $\pi_\alpha | \mathrm{T}^{\rm alg}_\alpha \mathfrak{U}$.

We omit the rest of the details.

2. TOPOLOGICAL PROPERTIES

Starting with this section we will be concerned with Banach environments. However, it should be mentioned that some of the subsequent results remain true under more general assumptions.

2.1. An environment $\mathcal{E} = (E, \Pi)$ over an algebra A will be called a *Banach environment*, whenever

(i) both A and E are Banach algebras,

(ii) E is a Banach A-bimodule, and

(iii) Π is a continuous map.

From time to time, in the sequel, we will assume that A is a C^*-algebra. We also notice that for Banach environments all the maps defined and used in Section 1 are continuous.

From now on $\mathcal{E} = (E, \Pi)$ will denote a fixed Banach environment. Definition 1.3 obviously implies that the Grassmannian $\mathfrak{G} = \mathfrak{G}(\mathcal{E})$ is a closed subspace of E. The same conclusion is true, in the involutive case, for the self-adjoint Grassmannian $\mathfrak{U} = \mathfrak{U}(\mathcal{E})$.

In what follows we consider $\mathfrak{G}$ and $\mathfrak{U}$ as topological spaces with the induced topologies.

2.2. Our next purpose is to continue the study of similarity and unitary orbits in $\mathfrak{G}$ and $\mathfrak{U}$, respectively. The notations will be as in Section 1.

Let $\alpha, \beta \in \mathfrak{G}$ be given and let $a = \Pi(\beta \times \alpha)$. Then $a \cdot \alpha = \beta \times \alpha$ and $\beta \cdot a = \beta \times \alpha$, hence

$$a \cdot \alpha = \beta \cdot a. \tag{2.1}$$

Assume in addition that $\|\alpha - \beta\| < (\|\Pi\| \, \|\alpha\|)^{-1}$. Then (1.2) and (1.3) yield $1 - a = \Pi((\alpha - \beta) \times \alpha)$, and, consequently, we have

$$\|1 - a\| \leq \|\Pi\| \, \|\alpha - \beta\| \, \|\alpha\| < 1.$$

Therefore, a is invertible, and, according to (2.1), β and α are in the same similarity orbit.

Suppose now that $\mathcal{E}$ is an involutive environment and $\alpha, \beta \in \mathfrak{U}$, such that $\|\alpha - \beta\| < \|\Pi\|^{-1} \|\alpha\|^{-2}$. As before, let $a = \Pi(\beta \times \alpha)$. Since $\alpha^\sharp = \alpha$ and $\beta^\sharp = \beta$ (cf. (1.1)), from (2.1) we get $\alpha \cdot a^* = a^* \cdot \beta$, and hence

$$\alpha \cdot (a^* a) = a^* \cdot \beta \cdot a = (a^* a) \cdot \alpha. \tag{2.2}$$

From Definition 1.3 and (1.2) we have

$$a^*a = \Pi(\alpha \times \beta)\Pi(\beta \times \alpha) = \Pi(\alpha \times \beta \times \alpha),$$

and also,

$$1 - a^*a = \Pi(\alpha \times (\alpha - \beta) \times \alpha).$$

It follows that

$$\|1 - a^*a\| \leq \|\Pi\| \, \|\alpha\| \, \|\alpha - \beta\| \, \|\alpha\| < 1.$$

Now, by using the holomorphic functional calculus, we define the absolute value of a as

$$|a| = \exp\left(\frac{1}{2}\log(a^*a)\right),$$

where exp is the usual exponential map and log denotes the principal branch of the logarithm function. Equation (2.2) clearly implies $\alpha \cdot |a| = |a| \cdot \alpha$, hence $\alpha \cdot |a|^{-1} = |a|^{-1} \cdot \alpha$. Finally, let $u = a|a|^{-1}$. Then u is a unitary element of A and

$$\beta \cdot u = \beta \cdot a|a|^{-1} = a \cdot \alpha \cdot |a|^{-1} = a|a|^{-1} \cdot \alpha = u \cdot \alpha.$$

In conclusion, β and α are unitarily equivalent.

2.3. The next result is a simple consequence of the previous remarks, with an obvious proof.

PROPOSITION. (i) *Given $\alpha \in \mathfrak{G}$ let $\mathfrak{G}(\alpha) = \{a \cdot \alpha \cdot a^{-1} : a \in \mathrm{G}(A)\}$ be its similarity orbit. Then $\mathfrak{G}(\alpha)$ is an open and closed subset of $\mathfrak{G}$.*

(ii) *Given $\alpha \in \mathfrak{U}$ let $\mathfrak{U}(\alpha) = \{u \cdot \alpha \cdot u^* : u \in U(A)\}$ be its unitary orbit. Then $\mathfrak{U}(\alpha)$ is an open and closed subset of $\mathfrak{U}$.*

2.4. Assume, as before, that $\alpha \in \mathfrak{G}$ is fixed and let ρ_α be the continuous map from $\mathrm{G}(A)$ onto $\mathfrak{G}(\alpha)$ defined by

$$\rho_\alpha(a) = a \cdot \alpha \cdot a^{-1} \quad (a \in \mathrm{G}(A)). \tag{2.3}$$

Recall that $A^\alpha = \{x \in A : x \cdot \alpha = \alpha \cdot x\}$. Clearly A^α is a closed unital subalgebra of A, therefore, $\mathrm{G}(A^\alpha)$ is a closed subgroup of $\mathrm{G}(A)$.

The next result is a generalization of [CPR1, Theorem 2.1].

THEOREM. *If $\alpha \in \mathfrak{G}$ then $\rho_\alpha : \mathrm{G}(A) \to \mathfrak{G}(\alpha)$ is a principal fiber bundle with the structural group $\mathrm{G}(A^\alpha)$.*

In particular, $\mathfrak{G}(\alpha)$ is homeomorphic to the coset space $\mathrm{G}(A)/\mathrm{G}(A^\alpha)$ with the usual quotient topology.

Proof. The proof reduces essentially to the construction of some local cross-sections of ρ_α. Fix an arbitrary point $\beta \in \mathfrak{G}(\alpha)$ and let

$$\mathcal{D}(\beta) = \left\{\gamma \in \mathfrak{G} : \|\gamma - \beta\| < (\|\Pi\|\,\|\beta\|)^{-1}\right\}.$$

Choose an element $a \in \mathrm{G}(A)$ such that $\beta = a \cdot \alpha \cdot a^{-1}$ and let $\sigma^a_\beta : \mathcal{D}(\beta) \to \mathrm{G}(A)$ be the map defined by

$$\sigma^a_\beta(\gamma) = \Pi(\gamma \times \beta)a \quad (\gamma \in \mathcal{D}(\beta)).$$

According to 2.2 we know that $\Pi(\gamma \times \beta)$ is invertible and

$$\gamma = \Pi(\gamma \times \beta) \cdot \beta \cdot \Pi(\gamma \times \beta)^{-1},$$

for any $\gamma \in \mathcal{D}(\beta)$. This implies that σ^a_β is a well-defined map and furthermore, by the preceding equality we obtain easily that $\rho_\alpha \circ \sigma^a_\beta = \mathrm{id}_{\mathcal{D}(\beta)}$. Thus σ^a_β is a local cross-section of ρ_α over $\mathcal{D}(\beta)$, a remark that concludes the proof. ■

2.5. An analogous result can be proved for the unitary orbit of a point $\alpha \in \mathfrak{U}$. Let $\mathfrak{U}(\alpha)$ be the unitary orbit of α and define

$$\tilde{\rho}_\alpha : \mathrm{U}(A) \to \mathfrak{U}(\alpha), \quad \tilde{\rho}_\alpha = u \cdot \alpha \cdot u^* \qquad (u \in \mathrm{U}(A)). \tag{2.4}$$

Since $\alpha^\sharp = \alpha$, it follows that A^α is a $*$-subalgebra of A, so it makes sense to consider its unitary group $\mathrm{U}(A^\alpha)$, a closed subgroup of $\mathrm{U}(A)$.

THEOREM. *If $\alpha \in \mathfrak{U}$ then $\tilde{\rho}_\alpha : \mathrm{U}(A) \to \mathfrak{U}(\alpha)$ is a principal fiber bundle with the structural group $\mathrm{U}(A^\alpha)$.*

In particular, $\mathfrak{U}(\alpha)$ is homeomorphic to the coset space $\mathrm{U}(A)/\mathrm{U}(A^\alpha)$ with the usual quotient topology.

Proof. As in the previous case it will be enough to construct explicitly local cross-sections of $\tilde{\rho}_\alpha$. Start with a fixed point $\beta \in \mathfrak{U}(\alpha)$ and let

$$\tilde{\mathcal{D}}(\beta) = \{\gamma \in \mathfrak{U} : \|\gamma - \beta\| < \|\Pi\|^{-1}\|\beta\|^{-2}\}.$$

Choose an element $u \in \mathrm{U}(A)$ such that $\beta = u \cdot a \cdot u^*$ and let $\tilde{\sigma}^u_\beta : \tilde{\mathcal{D}}(\beta) \to \mathrm{U}(A)$ be the map defined by

$$\tilde{\sigma}^u_\beta(\gamma) = \Pi(\gamma \times \beta)|\Pi(\gamma \times \beta)|^{-1}u \quad (\gamma \in \tilde{\mathcal{D}}(\beta)).$$

The fact that $\tilde{\sigma}^u_\beta$ is well-defined and the equality $\tilde{\rho}_\alpha \circ \tilde{\sigma}^u_\beta = \mathrm{id}_{\tilde{\mathcal{D}}(\beta)}$ are simple consequences of the remark contained in 2.2. ■

3. THE STANDARD LIFT

This section is concerned with the lifting problem for continuous curves in $\mathfrak{G} = \mathfrak{G}(\mathcal{E})$ or $\mathfrak{U} = \mathfrak{U}(\mathcal{E})$. The construction of standard lifts presented in the sequel uses multiplicative integrals (see [P]). A similar but rather concise discussion may be found in [CPR1], [CPR3], and [PR1].

3.1. Let us begin with the case of $\mathfrak{G}$. Start with an open interval $I = (-\varepsilon, \varepsilon)$ of the real line, where $\varepsilon \in (0, \infty]$, and let $\gamma : I \to \mathfrak{G}$, be a continuous curve in $\mathfrak{G}$. The point $\alpha = \gamma(0)$ will be called the *origin of* γ and clearly Proposition 2.3 implies that γ is a curve in $\mathfrak{G}(\alpha)$. Consider also the principal fiber bundle $\rho_\alpha : \mathrm{G}(A) \to \mathfrak{G}(\alpha)$.

DEFINITION. By a *normalized* lift of the continuous curve $\gamma : I \to \mathfrak{G}$ with the origin α we mean any continuous curve $\Gamma : I \to \mathrm{G}(A)$ such that

$$\Gamma(0) = 1 \tag{3.1}$$

and

$$\gamma = \rho_\alpha \circ \Gamma. \tag{3.2}$$

The last equation has the explicit form

$$\gamma(t) = \Gamma(t) \cdot \alpha \cdot \Gamma(t)^{-1} \quad (t \in I). \tag{3.3}$$

3.2. Of course, as a consequence of the local triviality of the bundles $\rho_\alpha : \mathrm{G}(A) \to \mathfrak{G}(\alpha)$ ($\alpha \in \mathfrak{G}$), given a curve γ in $\mathfrak{G}$ it has a lot of normalized lifts. The next construction provides a special lift under the additional assumption that γ, as a function from I into the Banach space E, is locally of bounded variation. This last condition means that for each $J = [s, t]$, a compact subinterval of I, there exists a positive number $V(J)$ satisfying the condition

$$\mathrm{var}(\Delta) \leq V(J), \tag{3.4}$$

where

$$\Delta : s = t_0 < t_1 < \ldots < t_{n-1} < t_n = t \tag{3.5}$$

is an arbitrary partition of J, and, by definition,

$$\mathrm{var}(\Delta) = \sum_{i=0}^{n-1} \|\gamma(t_{i+1}) - \gamma(t_i)\|. \tag{3.6}$$

Suppose that γ and J are fixed. Let us introduce also the number $M(J) = \sup_{r\in J} \|\gamma(r)\|$ and choose $\delta > 0$ such that

$$\|\gamma(r) - \gamma(r')\| \leq \|\Pi\|^{-1}(2M(J))^{-1} \tag{3.7}$$

for any $r, r' \in J$ with $|r - r'| \leq \delta$. Of course the existence of δ follows from the continuity of γ and the compactness of J.

If Δ is a partition as in (3.5) then $|\Delta| = \max_{0\leq i\leq n-1}(t_{i+1} - t_i)$ will denote its mesh.

Assume now that Δ is a partition of J with $|\Delta| \leq \delta$. For any $0 \leq i \leq n-1$ let $a_i = \Pi(\gamma(t_{i+1}) \times \gamma(t_i))$. By a previous computation (see 2.2), from (3.7) we obtain the next estimates

$$\|1 - a_i\| \leq \|\Pi\|\, \|\gamma(t_{i+1}) - \gamma(t_i)\|\|\gamma(t_i)\| \leq \|\Pi\|\, \|\gamma(t_{i+1}) - \gamma(t_i)\|M(J) \leq 2^{-1}, \tag{3.8}$$

hence a_i is an invertible element of A. In addition we know that (see (2.1))

$$\gamma(t_{i+1}) \cdot a_i = a_i \cdot \gamma(t_i) \quad (0 \leq i \leq n-1). \tag{3.9}$$

Now let $a(\Delta) = a_{n-1}a_{n-2}\cdots a_1 a_0$. Then $a(\Delta)$ is invertible and (3.9) yields

$$\gamma(t) \cdot a(\Delta) = a(\Delta) \cdot \gamma(s). \tag{3.10}$$

By Definition 1.3 and formula (1.2) we obtain a different expression for $a(\Delta)$, namely

$$a(\Delta) = \Pi(\gamma(t_n) \times \gamma(t_{n-1}) \times \cdots \times \gamma(t_1) \times \gamma(t_0)). \tag{3.11}$$

From (3.8) we find, for any invertible element a_i, that the following estimates hold:

$$\|a_i\| \leq 1 + \|1 - a_i\| \leq \exp(\|1 - a_i\|) \leq \exp(\|\Pi\|\, \|\gamma(t_{i+1}) - \gamma(t_i)\|M(J)) \tag{3.12}$$

The inequalities (3.8) and (3.7) imply also that

$$\begin{aligned} \|a_i^{-1}\| &\leq (1 - \|1 - a_i\|)^{-1} \leq 1 + 2\|1 - a_i\| \leq \\ &\leq \exp(2\|1 - a_i\|) \leq \exp(2\|\Pi\|\, \|\gamma(t_{i+1} - \gamma(t_i)\|M(J)). \end{aligned} \tag{3.13}$$

3.3. LEMMA. *There are two positive constants $K(J)$ and $L(J)$ such that*

(i) *for any partition Δ with $|\Delta| \leq \delta$ and any refinement Δ' of Δ we have*

$$\|a(\Delta') - a(\Delta)\| \leq K(J)(\mathrm{var}(\Delta') - \mathrm{var}(\Delta)); \tag{3.14}$$

(ii) *for any partition* Δ *with* $|\Delta| \leq \delta$ *we have*

$$\|a(\Delta)^{-1}\| \leq L(J). \tag{3.15}$$

Proof. (i) It is enough to show that (3.14), with a suitable constant $K(J)$, holds in the particular case when Δ' is obtained from Δ by adding only a new point r. Fix $0 \leq k \leq n-1$ such that $t_k < r < t_{k+1}$ and let $b = \Pi(\gamma(t_{k+1}) \times \gamma(r))$, $c = \Pi(\gamma(r) \times \gamma(t_k))$. Then

$$a(\Delta') = a_{n-1} \cdots a_{k+1} bc a_{k-1} \cdots a_0,$$

hence

$$a(\Delta') - a(\Delta) = a_{n-1} \cdots a_{k+1}(bc - a_k)a_{k-1} \cdots a_0. \tag{3.16}$$

But $bc - a_k = \Pi(\gamma(t_{k+1}) \times (\gamma(r) - \gamma(t_k)) \times \gamma(t_k))$. Therefore

$$\|bc - a_k\| \leq M(J)^2 \|\Pi\| \, \|\gamma(r) - \gamma(t_k)\| \leq M(J)^2 \|\Pi\| (\mathrm{var}(\Delta') - \mathrm{var}(\Delta)).$$

This estimate, combined with (3.16), (3.12) and (3.4), leads to

$$\|a(\Delta') - a(\Delta)\| \leq \exp(\|\Pi\| V(J) M(J)) M(J)^2 \|\Pi\| (\mathrm{var}(\Delta') - \mathrm{var}(\Delta)),$$

and (3.14) is proved.

(ii) For (3.15) we observe first that $a(\Delta)^{-1} = a_0^{-1} a_1^{-1} \cdots a_{n-1}^{-1}$. Using (3.13) one obtains

$$\|a(\Delta)^{-1}\| \leq \|a_0^{-1}\| \, \|a_1^{-1}\| \cdots \|a_{n-1}^{-1}\| \leq \exp(2\|\Pi\| \, \mathrm{var}(\Delta) M(J)) \leq \exp(2\|\Pi\| V(J) M(J)),$$

and the proof is complete. ∎

3.4. The inequality (3.14) clearly implies the existence of the limit

$$\Gamma_{t,s} = \lim_{|\Delta| \to 0} a(\Delta).$$

From (3.10) one finds that

$$\gamma(t) \cdot \Gamma_{t,s} = \Gamma_{t,s} \cdot \gamma(s). \tag{3.17}$$

Now let us show that $\Gamma_{t,s}$ is invertible. Fix a partition Δ with a sufficiently small mesh, such that

$$\mathrm{var}(\Delta') - \mathrm{var}(\Delta) \leq 2^{-1} K(J)^{-1} L(J)^{-1}$$

for any refinement Δ' of Δ. Using (3.14) and (3.15) we have

$$\|a(\Delta') - a(\Delta)\| \leq 2^{-1}L(J)^{-1} \leq 2^{-1}\|a(\Delta)^{-1}\|^{-1},$$

and, consequently,

$$\|\Gamma_{t,s} - a(\Delta)\| \leq 2^{-1}\|a(\Delta)^{-1}\|^{-1}.$$

Since $a(\Delta)$ is invertible, using a well-known device, from the last inequality we conclude that $\Gamma_{t,s}$ is invertible.

Finally, let us define the elements $\Gamma_{t,s}$ for arbitrary points $t, s \in I$ by the rule

$$\Gamma_{t,s} = \begin{cases} 1 & \text{if } t = s \\ (\Gamma_{s,t})^{-1} & \text{if } t < s. \end{cases}$$

Then we have

$$\Gamma_{t,s}\Gamma_{s,r}\Gamma_{r,t} = 1 \tag{3.18}$$

and

$$\Gamma_{t,t} = 1 \tag{3.19}$$

for any $r, s, t \in I$.

Now let $\Gamma : I \to \mathrm{G}(A)$ be the map defined by

$$\Gamma(t) = \Gamma_{t,0} \quad (t \in I). \tag{3.20}$$

It is not difficult to show, by using the previous estimates, that Γ is a continuous map. Actually, as an A-valued map, Γ is also of locally bounded variation. We leave the proof of these two assertions to the reader.

From (3.17) one concludes easily that Γ satisfies the conditions (3.1) and (3.2). In conclusion Γ is a normalized lift of the curve γ.

We also observe that Γ satisfies the condition

$$\Gamma(s) = \Gamma_{s,t}\Gamma(t) \quad (t, s \in I). \tag{3.21}$$

For a later use we insert the next definition.

DEFINITION. The map Γ associated, as in the previous construction, to a continuous curve γ in $\mathfrak{G}$ of locally bounded variation, will be called *the standard lift of* γ.

3.5. The last result of this section is concerned with a basic property of the standard lift of a C^∞ curve.

PROPOSITION. *Let $\gamma : I \to \mathfrak{G}$ be a curve with origin α and assume that γ, as an E-valued function, is smooth. Then the standard lift $\Gamma : I \to \mathrm{G}(A)$ of γ is a smooth map from I to A and, moreover, it is the solution of the first-order initial value problem*

$$\dot{\Gamma}(t) = \Pi(\dot{\gamma}(t) \times \gamma(t))\Gamma(t) \tag{3.22}$$

$$\Gamma(0) = 1, \tag{3.23}$$

where $\dot{\gamma}(t)$ and $\dot{\Gamma}(t)$ denote the derivatives with respect to t of γ and Γ, respectively.

Proof. Fix a point $t \in I$ and let h be a sufficiently small positive number such that $t+h \in I$. From (3.21) we find

$$\Gamma(t+h) = \Gamma_{t+h,t}\Gamma(t). \tag{3.24}$$

Since γ is smooth one obtains

$$\gamma(t+h) = \gamma(t) + h\dot{\gamma}(t) + \mathrm{o}(h)$$

and, consequently,

$$\Pi(\gamma(t+h) \times \gamma(t)) = 1 + h\Pi(\dot{\gamma}(t) \times \gamma(t)) + \mathrm{o}(h).$$

From the construction of the standard lift it follows that $\Pi(\gamma(t+h) \times \gamma(t))$ approximates $\Gamma_{t+h,t}$. Actually we have an estimate of the form

$$\Gamma_{t+h,h} = 1 + h \cdot \Pi(\dot{\gamma}(t) \times \gamma(t)) + \mathrm{o}(h).$$

By substituting this equality in (3.24) we get

$$\Gamma(t+h) = \Gamma(t) + h \cdot \Pi(\dot{\gamma}(t) \times \gamma(t))\Gamma(t) + \mathrm{o}(h),$$

a relation that proves the differentiability of Γ and also the fact that Γ satisfies the equation (3.22). ■

3.6. Since we purposely omitted some details of the previous proof, let us show, by using a straightforward argument, that the solution Γ of the problem (3.22)–(3.23) is indeed a normalized lift of γ.

The fact that Γ takes values in $\mathrm{G}(A)$ follows from some general arguments, due to the special form of the equation (3.22). The problem is to prove that γ and Γ satisfy the condition

$$\gamma(t) = \Gamma(t) \cdot \alpha \cdot \Gamma(t)^{-1} \quad (t \in I).$$

To this end, let $\delta : I \to E$ be the map defined by $\delta(t) = \Gamma(t)^{-1} \cdot \gamma(t) \cdot \Gamma(t)$. Then, according to (3.22), we have

$$\dot{\delta}(t) = -\Gamma(t)^{-1}(\Pi(\dot{\gamma}(t) \times \gamma(t)) \cdot \gamma(t) - \dot{\gamma}(t) - \gamma(t) \cdot \Pi(\dot{\gamma}(t) \times \gamma(t)))\Gamma(t). \tag{3.25}$$

But γ satisfies the condition (cf. (1.2))

$$\gamma(t) \times \gamma(t) = \gamma(t),$$

therefore

$$\dot{\gamma}(t) \times \gamma(t) + \gamma(t) \times \dot{\gamma}(t) = \dot{\gamma}(t). \tag{3.26}$$

Since, in addition, $\Pi(\gamma(t)) = 1$, we obtain also $\Pi(\dot{\gamma}(t)) = 0$, whence

$$\Pi(\dot{\gamma}(t) \times \gamma(t)) + \Pi(\gamma(t) \times \dot{\gamma}(t)) = 0.$$

Using this last equality, formula (3.25) becomes

$$\dot{\delta}(t) = -\Gamma(t)^{-1}(\Pi(\dot{\gamma}(t) \times \gamma(t)) \cdot \gamma(t) + \gamma(t) \cdot \Pi(\gamma(t) \times \dot{\gamma}(t)) - \dot{\gamma}(t))\Gamma(t). \tag{3.27}$$

On the other hand, by Definition 1.3 we have $\Pi(\dot{\gamma}(t) \times \gamma(t)) \cdot \gamma(t) = \dot{\gamma}(t) \times \gamma(t)$, and $\gamma(t)\Pi(\gamma(t) \times \dot{\gamma}(t)) = \gamma(t) \times \dot{\gamma}(t)$. By substituting these two equations in (3.27) and using (3.26) we conclude that $\dot{\delta}(t) \equiv 0$. Since, on the other hand, $\delta(0) = \gamma(0) = \alpha$ we get $\delta(t) \equiv \alpha$, a condition equivalent to $\gamma(t) = \Gamma(t) \cdot \alpha \cdot \Gamma(t)^{-1}$ for all $t \in I$.

The proof is complete. ■

3.7. Finally, let us mention that a completely analogous construction with similar conclusions is possible for a continuous curve $\gamma : I \to \mathfrak{U}$ that is of locally bounded variation, or smooth. The standard lift will be a continuous curve $\Gamma : I \to \mathrm{U}(A)$, that in the case when γ is smooth represents also the solution of the initial value problem (3.22)–(3.23). The only essential difference is that instead of the elements a_i considered in 3.2 we have to use the unitary elements $u_i = a_i|a_i|^{-1}$ $(0 \le i \le n-1)$.

Remark. The construction of the standard lift was sketched in the particular case when $\mathcal{E} = (A^n, \Pi)$ is the environment discussed in Example 1 (see 1.4) in [CPR1], where an analog of Proposition 3.5 was obtained. In Section 5 we will prove that the standard lift of a smooth curve is, actually, the horizontal lift of that curve with respect to a canonical connection, a fact also established in [CPR1] for (A^n, Π).

4. DIFFERENTIABLE STRUCTURES

This section deals with the natural differentiable structures on the spaces $\mathfrak{G}(\mathcal{E})$ and $\mathfrak{U}(\mathcal{E})$. We will prove that the space $\mathfrak{G}(\mathcal{E})$ is a complex analytic submanifold of E, and $\mathfrak{U}(\mathcal{E})$ is a real analytic submanifold of the real Banach space E_{h}. Recall that E_{h} denotes the set of hermitian elements of E.

For terminology and general results concerning Banach differentiable manifolds we refer in what follows to [B] and [U].

4.1. We first consider $\mathfrak{G}(\mathcal{E})$. By Proposition 2.3 above, the space $\mathfrak{G}(\mathcal{E})$ is a discrete union of open and closed subspaces of the form $\mathfrak{G}(\alpha)$. Therefore, in order to introduce a generalized differentiable structure on $\mathfrak{G}$ it will be enough to provide each $\mathfrak{G}(\alpha)$ with a differentiable structure. This can be done simply by one of the conclusions of Theorem 2.4, namely, the fact that $\mathfrak{G}(\alpha)$ is homeomorphic with the coset space $\mathrm{G}(A)/\mathrm{G}(A^{\alpha})$. Consequently, our first task will be to prove that

$$\mathrm{G}(A)/\mathrm{G}(A^{\alpha}) \textit{ is a complex analytic manifold.} \tag{4.1}$$

At the same time, assuming that $\mathfrak{G}(\alpha)$ inherits the differentiable structure from $\mathrm{G}(A)/\mathrm{G}(A^{\alpha})$, it is desirable to show that

$$\mathfrak{G}(\alpha) \textit{ is a submanifold of } E. \tag{4.2}$$

It turns out that it is possible to prove simultaneously both (4.1) and (4.2), as well as that $\mathfrak{G}(\alpha)$ and $\mathrm{G}(A)/\mathrm{G}(A^{\alpha})$ are diffeomorphic, using techniques from [B, Section 5.12.5]. More precisely, let $\rho : \mathrm{G}(A) \times E \to E$ be the map defined by

$$\rho(a, \varphi) = a \cdot \varphi \cdot a^{-1} \quad (a \in \mathrm{G}(A),\ \varphi \in E). \tag{4.3}$$

The group $\mathrm{G}(A)$ is a complex Lie group and the map ρ is complex analytic.

Fix an element $\alpha \in \mathfrak{G}$. Then $\mathfrak{G}(\alpha)$ is exactly the orbit of α under the action of $\mathrm{G}(A)$ on E induced by ρ. Let $\rho_\alpha : \mathrm{G}(A) \to E$ be the map

$$\rho_\alpha(a) = \rho(a, \alpha) = a \cdot \alpha \cdot a^{-1} \quad (a \in \mathrm{G}(A)). \tag{4.4}$$

According to [B, 5.12.5], in order to prove (4.1) and (4.2) it will be necessary to show that ρ_α is a subimmersion at any point $a \in \mathrm{G}(A)$ (the rest of the conditions appearing in [B, 5.12.5] are, more or less, obvious). To be more specific, we have to show that the subspaces $\ker(\mathrm{d}\rho_\alpha)(a) \subset \mathrm{T}_a\mathrm{G}(A)$ and $\mathrm{ran}(\mathrm{d}\rho_\alpha)(a) \subset \mathrm{T}_{\rho_\alpha(a)}E = E$ are closed and have closed complements, where $(\mathrm{d}\rho_\alpha)(a) : \mathrm{T}_a\mathrm{G}(A) \to E$ is the differential of the function ρ_α at a point $a \in \mathrm{G}(A)$.

First, let us notice that the tangent space to $\mathrm{G}(A)$ at a point a is $\mathrm{T}_a\mathrm{G}(A) = aA = \{ax : x \in A\}$. From (4.4) we obtain

$$(\mathrm{d}\rho_\alpha)(a)(ax) = a \cdot (x \cdot \alpha - \alpha \cdot x) \cdot a^{-1} \quad (a \in \mathrm{G}(A),\ x \in A) \tag{4.5}$$

It follows now that $\ker(\mathrm{d}\rho_\alpha)(a) = aA^\alpha$, where A^α has exactly the same meaning as in 1.6, namely, $A^\alpha = \{x \in A : x \cdot \alpha = \alpha \cdot x\}$. On the other hand, by (4.5) we find that

$$(\mathrm{d}\rho_\alpha)(a)(ax) = a \cdot \mathrm{d}\rho_\alpha(1)(x) \cdot a^{-1}.$$

Consequently, it will be enough to show that

$$A^\alpha \textit{ has a closed complement in } A, \tag{4.6}$$

and also that

$$\begin{gathered}\operatorname{ran}(\mathrm{d}\rho_\alpha)(1) = \{x \cdot \alpha - \alpha \cdot x : x \in A\} \\ \textit{is a closed subspace with a closed complement in } E.\end{gathered} \tag{4.7}$$

The proof of (4.1) and (4.2) will be completed after proving (4.6) and (4.7). Moreover, assuming that (4.6) and (4.7) are true, it will follow that the tangent space to $\mathfrak{G}$ at α, which actually coincides with the tangent space to $\mathfrak{G}(\alpha)$ at α, is precisely the space $\operatorname{ran}(\mathrm{d}\rho_\alpha)(1)$. Actually, both assertions (4.6) and (4.7) are simple consequences of some results discussed in Section 1. Indeed, A^α has in A the complementary subspace A_α defined by (1.8), and abviously A_α is closed. On the other hand, Proposition 1.10 shows that $\operatorname{ran}(\mathrm{d}\rho_\alpha)(1)$ coincides with the tangent space $\mathrm{T}_\alpha^{\mathrm{alg}}\mathfrak{G}$, hence, it has a complement in E, namely, the space $\mathrm{T}_\alpha^\perp\mathfrak{G}$. Proposition 1.12 clearly implies that both these subspaces are closed in E.

4.2. The existence, in the involutive case, of a natural differentiable structure on the space $\mathfrak{U} = \mathfrak{U}(\mathcal{E})$ follows from similar arguments. As counterparts of (4.1) and (4.2) we have to prove that given $\alpha \in \mathfrak{U}$ then

$$\mathrm{U}(A)/\mathrm{U}(A^\alpha) \textit{ is a real analytic manifold;} \tag{4.8}$$

$$\mathfrak{U}(\alpha) \textit{ is a submanifold of } E_{\mathrm{h}}. \tag{4.9}$$

The unitary group $\mathrm{U}(A)$ is a real analytic Lie group and the tangent space to $\mathrm{U}(A)$ at a point $u \in \mathrm{U}(A)$ is given by $\mathrm{T}_u\mathrm{U}(A) = uA_{\mathrm{sh}}$, where A_{sh} denotes the real space of all skew-hermitian elements of A. Instead of ρ we consider now the real analytic map $\tilde{\rho} : \mathrm{U}(A) \times E_{\mathrm{h}} \to E_{\mathrm{h}}$, defined by

$$\tilde{\rho}(u, \varphi) = u \cdot \varphi \cdot u^* \quad (u \in \mathrm{U}(A),\ \varphi \in E_{\mathrm{h}}), \tag{4.10}$$

and let $\tilde{\rho}_\alpha : \mathrm{U}(A) \to E_{\mathrm{h}}$, $\tilde{\rho}_\alpha(u) = \tilde{\rho}(u, \alpha)$ $(u \in \mathrm{U}(A))$. We have to show that $\ker(\mathrm{d}\tilde{\rho}_\alpha)(u) \subset \mathrm{T}_u\mathrm{U}(A)$ and $\mathrm{ran}(\mathrm{d}\tilde{\rho}_\alpha)(u) \subset \mathrm{T}_{\tilde{\rho}_\alpha(u)}E_{\mathrm{h}} = E_{\mathrm{h}}$ are closed and have closed complements, for any $u \in \mathrm{U}(A)$. As before, after some simple computations, the problem reduces to proving that

$$A^\alpha_{\mathrm{sh}} = A^\alpha \cap A_{\mathrm{sh}} \textit{ has a closed complement in } A_{\mathrm{sh}}; \tag{4.11}$$

$$\begin{gathered}\mathrm{ran}(\mathrm{d}\tilde{\rho}_\alpha)(1) = \{x \cdot \alpha - \alpha \cdot x : x \in A_{\mathrm{sh}}\} = \mathrm{ran}(\mathrm{d}\rho_\alpha)(1) \cap E_{\mathrm{h}} \\ \textit{is a closed subspace with a closed complement in } E_{\mathrm{h}}.\end{gathered} \tag{4.12}$$

But, clearly, both (4.11) and (4.12) are consequences of (4.6) and (4.7), respectively.

4.3. At this moment the fact that $\mathfrak{G}$ is a complex analytic manifold and that $\mathfrak{U}$ is a real analytic manifold, is completely proved.

In addition, as we already observed at the end of 4.1, the tangent space to $\mathfrak{G}$ at α in the sense of differential geometry, denoted as usual by $\mathrm{T}_\alpha\mathfrak{G}$, coincides with $\mathrm{ran}(\mathrm{d}\rho_\alpha)(1)$, hence $\mathrm{T}_\alpha\mathfrak{G} = \mathrm{T}^{\mathrm{alg}}_\alpha\mathfrak{G}$. From now on we will use only the notation $\mathrm{T}_\alpha\mathfrak{G}$.

4.4. We conclude this section with some brief remarks concerning tangential vector fields on the manifold $\mathfrak{G}$ or $\mathfrak{U}$.

By a tangential vector field on $\mathfrak{G}$ (resp. $\mathfrak{U}$) we will mean any C^∞ map X from $\mathfrak{G}$ (resp. $\mathfrak{U}$) into E (resp. E_{h}) such that $X(\alpha) \in \mathrm{T}_\alpha\mathfrak{G}$ (resp. $\mathrm{T}_\alpha\mathfrak{U}$), for any $\alpha \in \mathfrak{G}$ (resp. $\mathfrak{U}$). Clearly the tangential vector fields can be identified as C^∞ cross-sections of the tangent bundles $\mathrm{T}\mathfrak{G}$ (resp. $\mathrm{T}\mathfrak{U}$). Sometimes $X(\alpha)$ will be denoted by X_α. For the spaces of all tangential vector fields on $\mathfrak{G}$ or $\mathfrak{U}$, we shall use the notation $\mathcal{X}(\mathfrak{G})$ or $\mathcal{X}(\mathfrak{U})$, respectively. The description of the tangent spaces to $\mathfrak{G}$ or $\mathfrak{U}$ shows that any element $x \in A$ (resp. A_{sh}) defines a tangential vector field $X = \tilde{x}$ on $\mathfrak{G}$ (resp. $\mathfrak{U}$) by the next formula

$$X_\alpha = \partial_\alpha(x) = x \cdot \alpha - \alpha \cdot x \quad (\alpha \in \mathfrak{G} \text{ (resp. } \mathfrak{U})). \tag{4.13}$$

In some of our subsequent proofs it will be enough to work only with tangential vector fields of this type.

For a later use, we point out also the action of the group $\mathrm{G}(A)$ (resp. $\mathrm{U}(A)$) on $\mathcal{X}(\mathfrak{G})$ (resp. $\mathcal{X}(\mathfrak{U})$).

If $a \in \mathrm{G}(A)$ and $X \in \mathcal{X}(\mathfrak{G})$ are given, let X^a be the map from $\mathfrak{G}$ into E defined by

$$X^a(\alpha) = a^{-1} \cdot X(a \cdot \alpha \cdot a^{-1}) \cdot a \quad (\alpha \in \mathfrak{G}). \tag{4.14}$$

Then $X^a \in \mathcal{X}(\mathfrak{G})$, and $(X^a)^b = X^{ab}$ for any $b \in \mathrm{G}(A)$. Similarly, if $u \in \mathrm{U}(A)$ and $X \in \mathcal{X}(\mathfrak{U})$, then the map X^u from $\mathfrak{U}$ into E_{h} defined by

$$X^u(\alpha) = u^* \cdot X(u \cdot \alpha \cdot u^*) \cdot u \quad (\alpha \in \mathfrak{U}) \tag{4.15}$$

is a vector field on $\mathfrak{U}$, and $(X^u)^v = X^{uv}$ for any $v \in \mathrm{U}(A)$.

In the particular case when $X = \bar{x}$ for a fixed element $x \in A$ (resp. A_{sh}), then $X^a = \overline{axa^{-1}}$ (resp. $X^u = \overline{uxu^*}$) for each $a \in \mathrm{G}(A)$ (resp. $u \in \mathrm{U}(A)$).

In order to derive an expression for the Lie bracket of two tangential vector fields on $\mathfrak{G}$ or $\mathfrak{U}$, we proceed as follows. We begin with tangential vector fields X and Y on $\mathfrak{G}$ (resp. $\mathfrak{U}$), and, for any $\alpha \in \mathfrak{G}$ (resp. $\mathfrak{U}$), let $(\partial_X Y)_\alpha = (\mathrm{d}Y)(\alpha)(X_\alpha)$ be the directional derivative of Y along X at α, where $(\mathrm{d}Y)(\alpha) : \mathrm{T}_\alpha\mathfrak{G} \to E$ (resp. $(\mathrm{d}Y)(\alpha) : \mathrm{T}_\alpha\mathfrak{U} \to E_{\mathrm{h}}$) denotes the differential of Y at α. Alternatively, we choose an integral curve of X around α, that is a C^∞ map $\gamma : (-\varepsilon, \varepsilon) \to \mathfrak{G}$ (resp. $\mathfrak{U}$) such that $\gamma(0) = \alpha$, $\dot{\gamma}(0) = X_\alpha$, and

$$(\partial_X Y)_\alpha = \lim_{t \to 0} \frac{1}{t}(Y_{\gamma(t)} - Y_\alpha). \tag{4.16}$$

In general $(\partial_X Y)_\alpha$ is no longer a tangent vector to $\mathfrak{G}$ (resp. $\mathfrak{U}$) at α. However, the Lie bracket $[X, Y]$, which is a tangential vector field, can be computed using directional derivatives. More precisely, we have

$$[X, Y]_\alpha = (\partial_X Y)_\alpha - (\partial_Y X)_\alpha \quad (\alpha \in \mathfrak{G} \text{ (resp. } \mathfrak{U})). \tag{4.17}$$

Assume now that $X = \bar{x}$ and $Y = \bar{y}$ for some x and y in A (resp. A_{sh}). Then, according to (4.13) and using (4.16), we find

$$(\partial_X Y)_\alpha = y \cdot (x \cdot \alpha - \alpha \cdot x) - (x \cdot \alpha - \alpha \cdot x) \cdot y. \tag{4.18}$$

Therefore,

$$(\partial_X Y)_\alpha - (\partial_Y X)_\alpha = -([x, y] \cdot \alpha - \alpha \cdot [x, y]),$$

where $[x, y] = xy - yx$ is the usual commutator bracket of x and y in A, hence

$$[X, Y] = -\overline{[x, y]}. \tag{4.19}$$

5. INVARIANT LINEAR CONNECTIONS

The existence of linear connections on the manifold $\mathfrak{G} = \mathfrak{G}(\mathcal{E})$ (resp. $\mathfrak{U}(\mathcal{E})$) that are invariant with respect to the action of $\mathrm{G}(A)$ (resp. $\mathrm{U}(A)$) is essentially a consequence of the fact that the spaces $\mathrm{G}(A)/\mathrm{G}(A^\alpha)$ for $\alpha \in \mathfrak{G}$ (resp. $\mathrm{U}(A)/\mathrm{U}(A^\alpha)$ for $\alpha \in \mathfrak{U}$) are reductive coset spaces (see [N]).

In this section we choose a canonical linear connection on $\mathfrak{G}$. A similar construction is possible also for $\mathfrak{U}$. Since the case of the space $\mathfrak{U}$ follows, by performing appropriate modifications, from the case of the space $\mathfrak{G}$, we restrict ourselves to discuss only about $\mathfrak{G}$.

5.1. Recall that the space $\mathfrak{G}$ is a submanifold of the Banach space E and for any point $\alpha \in \mathfrak{G}$ there is a projection map π_α from E onto $\mathrm{T}_\alpha\mathfrak{G}$ (see Corollary 1.11 and formula (1.19)). This map induces, by a simple standard construction, a linear connection on $\mathfrak{G}$,

$$\nabla : \mathcal{X}(\mathfrak{G}) \times \mathcal{X}(\mathfrak{G}) \to \mathcal{X}(\mathfrak{G}) : (X, Y) \mapsto \nabla_X Y,$$

defined as follows

$$(\nabla_X Y)_\alpha = \pi_\alpha((\partial_X Y)_\alpha) \quad (X, Y \in \mathcal{X}(\mathfrak{G}),\ \alpha \in \mathfrak{G}), \tag{5.1}$$

where $\partial_X Y$ denotes the directional derivative of Y along X (see (4.16) above). One obtains easily that the map ∇ is well-defined and it satisfies the usual axioms of a linear connection. Moreover, ∇ is invariant with respect to the action of $\mathrm{G}(A)$ on $\mathfrak{G}$, namely,

$$\nabla_{X^a} Y^a = (\nabla_X Y)^a \quad (X, Y \in \mathcal{X}(\mathfrak{G}),\ a \in \mathrm{G}(A)), \tag{5.2}$$

where X^a and Y^a are defined as in (4.14). It follows directly from (4.16) and (5.1) that ∇ is a torsionless linear connection.

DEFINITION. The linear connection ∇ defined above will be called *the canonical linear connection on* $\mathfrak{G}$.

5.2. We postpone for the moment the study of the canonical linear connection in order to discuss about another invariant linear connection on $\mathfrak{G}$, that arises naturally because $\mathfrak{G}$ is a discrete union of reductive homogeneous spaces (for more details concerning this subject see, for instance, [BC], [GHV], [H], [KN] and [N]).

Assume that $\alpha \in \mathfrak{G}$ is fixed and consider the principal fiber bundle $\rho_\alpha : \mathrm{G}(A) \to \mathfrak{G}(\alpha)$. We already know that $\mathfrak{G}(\alpha)$ is diffeomorphic with the coset space $\mathrm{G}(A)/\mathrm{G}(A^\alpha)$. The Lie algebra of $\mathrm{G}(A)$, denoted in what follows by $\mathrm{gl}(A)$, is the vector space $\mathrm{T}_1\mathrm{G}(A) = A$ with the

Lie algebra structure given by the usual commutator $[x, y] = xy - yx$ $(x, y \in A)$. Similarly, we have the Lie algebra $\mathrm{gl}(A^\alpha)$, which as a vector space coincides with $\mathrm{T}_1\mathrm{G}(A^\alpha) = A^\alpha$, and is a Lie subalgebra of $\mathrm{gl}(A)$. Recall that A^α has a closed complement in A, denoted by A_α (see Section 1), therefore

$$\mathrm{gl}(A) = \mathrm{gl}(A^\alpha) + A_\alpha \quad \text{(direct sum of vector spaces).} \tag{5.3}$$

Moreover, a straightforward computation shows that if $a \in \mathrm{G}(A^\alpha)$ and $x \in A_\alpha$ then $axa^{-1} \in A_\alpha$. According to the usual terminology (see [KN] or [H]) one says that $(\mathrm{G}(A), \mathrm{G}(A^\alpha))$ is a reductive pair, or, in other words, $\mathfrak{G}(\alpha)$ is a reductive homogeneous space.

For any $a \in \mathrm{G}(A)$ we denote by V_a and H_a the complementary subspaces of $\mathrm{T}_a\mathrm{G}(A) = aA$ defined by

$$V_a = aA^\alpha \quad H_a = aA_\alpha. \tag{5.4}$$

If $b \in \mathrm{G}(A^\alpha)$ then, since $A_\alpha b = bA_\alpha$, one obtains that $H_a b = H_{ab}$.

The projections of $\mathrm{T}_a\mathrm{G}(A)$ onto V_a and H_a are given, respectively, by (for notations see 1.6)

$$v_a(ax) = aP^\alpha(x), \quad h_a(ax) = aP_\alpha(x) \qquad (x \in A),$$

and clearly $a \mapsto v_a$ and $a \mapsto h_a$ are differentiable maps from $\mathrm{G}(A)$ into the space of bounded linear operators on A.

As a conclusion of all these remarks one obtains that the distribution $a \mapsto H_a$ defines a connection on the principal fiber bundle $\rho_\alpha : \mathrm{G}(A) \to \mathfrak{G}(\alpha)$ (see, for instance, [BC]).

5.3. We recall now the notion of the horizontal lift of a curve with respect to this connection.

DEFINITION. Let $\gamma : (-\varepsilon, \varepsilon) \to \mathfrak{G}$ be a C^∞ curve with the origin $\gamma(0) = \alpha$. By *a horizontal lift of* γ one means any C^∞ curve $\Gamma : (-\varepsilon, \varepsilon) \to \mathrm{G}(A)$ such that

(i) Γ is horizontal, that is,

$$\dot{\Gamma}(t) \in H_{\Gamma(t)} = \Gamma(t)A_\alpha, \tag{5.5}$$

(ii) $\gamma(t) = \Gamma(t) \cdot \alpha \cdot \Gamma(t)^{-1}$ for all $t \in (-\varepsilon, \varepsilon)$.

It is well-known that for any $a \in \mathrm{G}(A^\alpha)$ there exists a unique horizontal lift Γ of γ such that $\Gamma(0) = a$. In particular, if $a = 1$, we obtain a unique normalized horizontal lift. As we will see below, this is not a new object.

5.4. THEOREM. *Let* $\gamma : (-\varepsilon, \varepsilon) \to \mathfrak{G}$ *be a* C^∞ *curve with* $\gamma(0) = \alpha$. *Then the standard lift of* γ *coincides with the normalized horizontal lift.*

Proof. Recall that the standard lift of γ is the unique solution of the initial value problem

$$\dot{\Gamma}(t) = \Pi(\dot{\gamma}(t) \times \gamma(t))\Gamma(t), \tag{5.6}$$

$$\Gamma(0) = 1. \tag{5.7}$$

Since the normalized horizontal lift is also unique it suffices to show that the standard lift satisfies the condition (5.5), that is,

$$P^{\alpha}(\Gamma(t)^{-1} \cdot \dot{\Gamma}(t)) = 0 \quad (t \in (-\varepsilon, \varepsilon)). \tag{5.8}$$

Since $\dot{\gamma}(t) \in \mathrm{T}_{\gamma(t)}\mathfrak{G}$ we observe first that (see 3.6)

$$\Pi(\dot{\gamma}(t) \times \gamma(t)) = \varepsilon_{\gamma(t)}(\dot{\gamma}(t)).$$

According to (5.6) we have

$$\begin{aligned} P^{\alpha}(\Gamma(t)^{-1}\dot{\Gamma}(t)) &= \Pi(\alpha \cdot \Gamma(t)^{-1}\dot{\Gamma}(t) \times \alpha) = \\ &= \Pi(\alpha \cdot \Gamma(t)^{-1}\Pi(\dot{\gamma}(t) \times \gamma(t))\Gamma(t) \times \alpha) = \\ &= \Gamma(t)^{-1}\Pi(\gamma(t) \cdot \Pi(\dot{\gamma}(t) \times \gamma(t)) \times \gamma(t))\Gamma(t) = \\ &= \Gamma(t)^{-1}P^{\gamma(t)}\varepsilon_{\gamma(t)}(\dot{\gamma}(t))\Gamma(t) = 0, \end{aligned}$$

the last equality being a consequence of a remark in 1.13.

Thus (5.8) is established and the proof is complete. ∎

5.5. The normalized horizontal lift of a C^{∞} curve γ with origin α induces a parallel translation along that curve in the principal bundle $\rho_{\alpha} : \mathrm{G}(A) \to \mathfrak{G}(\alpha)$ (see [BC]). At the same time there exists an associated parallel translation along γ in the tangent bundle of $\mathfrak{G}(\alpha)$. Its definition goes as follows:

DEFINITION. Let $\Gamma : (-\varepsilon, \varepsilon) \to \mathrm{G}(A)$ be the normalized horizontal lift of the curve $\gamma : (-\varepsilon, \varepsilon) \to \mathfrak{G}$ with the origin α. The *parallel translation along γ of tangent vectors from $\gamma(0) = \alpha$ to $\gamma(t)$* ($t \in (-\varepsilon, \varepsilon)$) is the map

$$\tau_t : \mathrm{T}_{\alpha}\mathfrak{G} \to \mathrm{T}_{\gamma(t)}\mathfrak{G}, \quad \tau_t(\theta) = \Gamma(t) \cdot \theta \cdot \Gamma(t)^{-1} \qquad (\theta \in \mathrm{T}_{\alpha}\mathfrak{G}). \tag{5.9}$$

The definition makes sense because, in general, $\mathrm{T}_{a \cdot \alpha \cdot a^{-1}}\mathfrak{G} = a \cdot \mathrm{T}_{\alpha}\mathfrak{G} \cdot a^{-1}$, for any $a \in \mathrm{G}(A)$.

5.6. Now we can use the parallel translation of tangent vectors to $\mathfrak{G}$ along C^{∞} curves to introduce a new invariant linear connection on $\mathfrak{G}$ (compare with [H]).

Specifically, we define a map

$$\nabla^+ : \mathcal{X}(\mathfrak{G}) \times \mathcal{X}(\mathfrak{G}) \to \mathcal{X}(\mathfrak{G}) : (X,Y) \mapsto \nabla^+_X Y \quad (X,Y \in \mathcal{X}(\mathfrak{G}))$$

by the formula

$$(\nabla^+_X Y)_\alpha = \lim_{t\to 0} \frac{1}{t}\left(\tau_t^{-1}(Y_{\gamma(t)}) - Y_\alpha\right), \tag{5.10}$$

for any $X,Y \in \mathcal{X}(\mathfrak{G})$ and $\alpha \in \mathfrak{G}$, where $\gamma : (-\varepsilon, \varepsilon) \to \mathfrak{G}$ satisfies the conditions $\gamma(0) = \alpha$, $\dot\gamma(0) = X_\alpha$, and τ_t denotes the parallel translation along this curve γ. Observe that from now on $\alpha \in \mathfrak{G}$ is no longer fixed.

It is not difficult to prove that ∇^+ is indeed an invariant linear connection on $\mathfrak{G}$. It will be referred to as the direct linear connection on $\mathfrak{G}$. From Theorem 5.4 we obtain the next explicit description of ∇^+.

PROPOSITION. *The direct linear connection ∇^+ acts via the formula*

$$(\nabla^+_X Y)_\alpha = (\partial_X Y)_\alpha + Y_\alpha \cdot \Pi(X_\alpha \times \alpha) - \Pi(X_\alpha \times \alpha)\cdot Y_\alpha. \tag{5.11}$$

Proof. Given $X,Y \in \mathcal{X}(\mathfrak{G})$ and $\alpha \in \mathfrak{G}$ choose a C^∞ curve $\gamma : (-\varepsilon, \varepsilon) \to \mathfrak{G}$ with $\gamma(0) = \alpha$ and $\dot\gamma(0) = X_\alpha$. Let $\Gamma : (-\varepsilon, \varepsilon) \to \mathfrak{G}$ be the standard lift of γ. From (5.10) and (5.9) we have

$$(\nabla^+_X Y)_\alpha = \lim_{t\to 0} \frac{1}{t}\left(\Gamma(t)^{-1}\cdot Y_{\gamma(t)} \cdot \Gamma(t) - Y_\alpha\right). \tag{5.12}$$

On the other hand

$$\begin{gathered}\Gamma(t)^{-1}\cdot Y_{\gamma(t)}\cdot\Gamma(t) - Y_\alpha = \\ \Gamma(t)^{-1}\cdot(Y_{\gamma(t)} - Y_\alpha)\cdot\Gamma(t) + \Gamma(t)^{-1}\left(Y_\alpha\cdot(\Gamma(t)-1) - (\Gamma(t)-1)\cdot Y_\alpha\right)\end{gathered} \tag{5.13}$$

and, moreover,

$$\lim_{t\to 0}\frac{1}{t}\left(Y_{\gamma(t)} - Y_\alpha\right) = (\partial_X Y)_\alpha, \tag{5.14}$$

$$\lim_{t\to 0}\frac{1}{t}(\Gamma(t) - 1) = \dot\Gamma(0) = \Pi(\dot\gamma(0)\times\gamma(0))\Gamma(0) = \Pi(X_\alpha\times\alpha). \tag{5.15}$$

Clearly (5.11) follows from (5.12)–(5.15). ■

5.7. Let $T^+ : \mathcal{X}(\mathfrak{G}) \times \mathcal{X}(\mathfrak{G}) \to \mathcal{X}(\mathfrak{G})$ be the torsion tensor field of ∇^+, that is

$$T^+(X,Y) = \nabla^+_X Y - \nabla^+_Y X - [X,Y] \quad (X,Y\in\mathcal{X}(\mathfrak{G})). \tag{5.16}$$

From (5.11) above and by using also (4.17) one obtains the next simple formula:

$$\begin{gathered}T^+(X,Y)_\alpha = Y_\alpha\cdot\Pi(X_\alpha\times\alpha) - \Pi(X_\alpha\times\alpha)\cdot Y_\alpha - \\ -X_\alpha\cdot\Pi(Y_\alpha\times\alpha) + \Pi(Y_\alpha\times\alpha)\cdot X_\alpha,\end{gathered} \tag{5.17}$$

for any $X, Y \in \mathcal{X}(\mathfrak{G})$ and $\alpha \in \mathfrak{G}$.

It follows that, in general, ∇^+ is not a symmetric linear connection, and, consequently, it is different from the canonical linear connection ∇ introduced in 5.1.

5.8. However, due to the natural way in which both linear connections ∇ and ∇^+ were defined, they are related in the simplest possible manner that could be expected, namely, ∇ coincides with the symmetrical part of ∇^+. To be more specific, the symmetrical part of ∇^+ is a new linear connection ∇^0 on $\mathfrak{G}$ defined by

$$\nabla^0_X Y = \nabla^+_X Y - \frac{1}{2} T^+(X, Y) \quad (X, Y \in \mathcal{X}(\mathfrak{G})). \tag{5.18}$$

In order to prove that $\nabla^0 = \nabla$ let us consider their difference $D = \nabla^0 - \nabla$. It is well-known that such a difference is a tensor field. Therefore, D induces a bilinear map $D_\alpha : \mathrm{T}_\alpha\mathfrak{G} \times \mathrm{T}_\alpha\mathfrak{G} \to \mathrm{T}_\alpha\mathfrak{G}$, for any $\alpha \in \mathfrak{G}$, and D is uniquely determined by these maps. Assume now that $\alpha \in \mathfrak{G}$ is fixed. Recall that any tangent vector $\theta \in \mathrm{T}_\alpha\mathfrak{G}$ is given by $\theta = \partial_\alpha(x) = x \cdot \alpha - \alpha \cdot x$, where $x = \varepsilon_\alpha(\theta) \in A_\alpha$ (for notations see Section 1). Consequently, in order to describe D_α completely it is enough to compute $D_\alpha(\partial_\alpha(x), \partial_\alpha(y))$, where $x, y \in A_\alpha$, that is $\alpha \cdot x \times \alpha = \alpha \cdot y \times \alpha = 0$. Let $X = \bar{x}$ and $Y = \bar{y}$, with $x, y \in A_\alpha$. Then, according to formula (4.18) one finds

$$(\partial_X Y)_\alpha = yx \cdot \alpha - y \cdot \alpha \cdot x - x \cdot \alpha \cdot y + \alpha \cdot xy. \tag{5.19}$$

Since $\Pi(X_\alpha \times \alpha) = \varepsilon_\alpha(X_\alpha) = x$, from (5.11) and the previous equality it follows that

$$\begin{aligned}(\nabla^+_X Y)_\alpha &= yx \cdot \alpha - y \cdot \alpha \cdot x - x \cdot \alpha \cdot y + \alpha \cdot xy + \\ &+(y \cdot \alpha - \alpha \cdot y) \cdot x - x \cdot (y \cdot \alpha - \alpha \cdot y) = \\ &= -([x, y] \cdot \alpha - \alpha \cdot [x, y]) = \\ &= -\partial_\alpha([x, y]).\end{aligned}$$

After some similar computations, (5.17) gives

$$T^+(X, Y)_\alpha = -\partial_\alpha([x, y]).$$

Therefore, according to (5.18) we have

$$(\nabla^0_X Y)_\alpha = -\frac{1}{2}\partial_\alpha([x, y]). \tag{5.20}$$

On the other hand, from (5.1) and (5.19) one obtains

$$(\nabla_X Y)_\alpha = \partial_\alpha \varepsilon_\alpha(yx \cdot \alpha - y \cdot \alpha \cdot x - x \cdot \alpha \cdot y + \alpha \cdot xy).$$

But $\alpha \cdot x \times \alpha = \alpha \cdot y \times \alpha = 0$. Using these equalities and, of course, the definitions of ε_α and ∂_α, we have

$$(\nabla_X Y)_\alpha = -\frac{1}{2}\partial_\alpha([x,y]). \tag{5.21}$$

Finally, (5.20) and (5.21) yield $D_\alpha(\partial_\alpha(x), \partial_\alpha(y))$, thus $D_\alpha \equiv 0$, hence $\nabla^0 = \nabla$.

5.9. REMARK. The previous results lead easily to explicit formulae for the curvature tensors of both linear connections ∇ and ∇^+. We do not present these computations here. The interested reader may find a similar calculation in the already quoted papers [MR] and [W2], where the canonical connection on a (possibly infinite-dimensional) reductive homogeneous Banach manifold is studied.

6. GEODESICS

The goal of this section is to discuss about the geodesics of the canonical linear connection and to obtain an explicit formula for the corresponding exponential map. In the particular case when $\mathcal{E} = (E, \Pi)$ is a Banach environment over a C^*-algebra A we will prove that short geodesics in $\mathfrak{U} = \mathfrak{U}(\mathcal{E})$ have minimal length with respect to a suitable metric structure on $\mathfrak{U}$.

6.1. As before, let ∇ and ∇^+ be the canonical, resp. direct, linear connection on $\mathfrak{G}$. Since ∇ is the symmetrical part of ∇^+, both connections have the same geodesics.

Following the standard terminology and notation, recall that a C^∞ curve $\gamma : (-\varepsilon, \varepsilon) \to \mathfrak{G}$ is said to be a *geodesic with respect to* ∇ if

$$\left(\nabla_{\dot\gamma(t)}\dot\gamma(t)\right)_{\gamma(t)} = 0 \quad (t \in (-\varepsilon, \varepsilon)). \tag{6.1}$$

By (5.1) one obtains the equation of a geodesic in a simpler form

$$\pi_{\gamma(t)}(\ddot\gamma(t)) = 0 \quad (t \in (-\varepsilon, \varepsilon)). \tag{6.2}$$

The next result gives other alternative descriptions.

6.2. PROPOSITION. *Let $\gamma : (-\varepsilon, \varepsilon) \to \mathfrak{G}$ be a C^∞-curve with origin $\gamma(0) = \alpha$, and denote $\dot\gamma(0) = \theta \in \mathrm{T}_\alpha\mathfrak{G}$. Consider the standard lift $\Gamma : (-\varepsilon, \varepsilon) \to \mathrm{G}(A)$ of γ, and let $x = \varepsilon_\alpha(\theta) \in A_\alpha$. The following conditions are equivalent:*

(i) *γ is a geodesic;*

(ii) *$\varepsilon_{\gamma(t)}(\dot\gamma(t)) = x$ for any $t \in (-\varepsilon, \varepsilon)$;*

(iii) *$\Gamma(t) = \exp(tx)$ for any $t \in (-\varepsilon, \varepsilon)$;*

(iv) *for any* $t, s \in (-\varepsilon, \varepsilon)$ *with* $t + s \in (-\varepsilon, \varepsilon)$ *we have*

$$\Gamma(s+t) = \Gamma(s)\Gamma(t). \tag{6.3}$$

Proof. Equation (6.2) means exactly that $\ddot{\gamma}(t)$ is a normal vector to $\mathfrak{G}$ at $\gamma(t)$ for any $t \in (-\varepsilon, \varepsilon)$. From Proposition 1.12 one obtains that (i) is equivalent to

$$\varepsilon_{\gamma(t)}(\ddot{\gamma}(t)) = 0 \quad \text{for any } t \in (-\varepsilon, \varepsilon). \tag{6.4}$$

But

$$\varepsilon_{\gamma(t)}(\ddot{\gamma}(t)) = \frac{1}{2}\Pi(\ddot{\gamma}(t) \times \gamma(t) - \gamma(t) \times \ddot{\gamma}(t)).$$

Since

$$\varepsilon_{\gamma(t)}(\dot{\gamma}(t)) = \frac{1}{2}\Pi(\dot{\gamma}(t) \times \gamma(t) - \gamma(t) \times \dot{\gamma}(t)).$$

one finds easily that

$$\frac{\mathrm{d}}{\mathrm{d}t}\left(\varepsilon_{\gamma(t)}(\dot{\gamma}(t))\right) = \varepsilon_{\gamma(t)}(\ddot{\gamma}(t)). \tag{6.5}$$

Now clearly (6.5) and (6.4) imply

$$\varepsilon(t)(\dot{\gamma}(t)) = \varepsilon_{\gamma(0)}(\dot{\gamma}(0)) = \varepsilon_{\alpha}(\theta) = x,$$

and conversely, assertion (ii) combined with (6.5) yield (6.4). Therefore (i) and (ii) are equivalent. On the other hand, recall that Γ is the solution of the initial value problem

$$\dot{\Gamma}(t) = \Pi(\dot{\gamma}(t) \times \gamma(t))\Gamma(t); \quad \Gamma(0) = 1. \tag{6.6}$$

Since $\dot{\gamma}(t) \in \mathrm{T}_{\gamma(t)}\mathfrak{G}$ we have

$$\Pi(\dot{\gamma}(t) \times \gamma(t)) = \varepsilon_{\gamma(t)}(\dot{\gamma}(t))$$

hence, according to (ii), (6.6) becomes

$$\dot{\Gamma}(t) = x\Gamma(t); \quad \Gamma(0) = 1, \tag{6.7}$$

and (iii) follows now easily.

Thus (ii) implies (iii). Conversely, assume that $\Gamma(t) = \exp(tx)$. Then

$$\gamma(t) = \Gamma(t) \cdot \alpha \cdot \Gamma(t)^{-1} = \exp(tx) \cdot \alpha \cdot \exp(-tx),$$

therefore

$$\dot\gamma(t) = \exp(tx)\cdot(x\cdot\alpha - \alpha\cdot x)\cdot\exp(-tx) = \exp(tx)\cdot\theta\cdot\exp(-tx).$$

It follows that

$$\varepsilon_{\gamma(t)}(\dot\gamma(t)) = \Gamma(t)\varepsilon_\alpha(\theta)\Gamma(t)^{-1} = \Gamma(t)x\Gamma(t)^{-1}.$$

But $\Gamma(t)$ and x are commuting, hence $\varepsilon_{\gamma(t)}(\dot\gamma(t)) = x$.

Finally, the equivalence between (iii) and (iv) is a general fact. The proof is complete. ■

6.3. COROLLARY. *Given $\alpha \in \mathfrak{G}$ and $\theta \in \mathrm{T}_\alpha\mathfrak{G}$ there exists a unique geodesic $\gamma : (-\infty, \infty) \to \mathfrak{G}$ such that $\gamma(0) = \alpha$ and $\dot\gamma(0) = \theta$. It is given by*

$$\gamma(t) = \exp\varepsilon_\alpha(t\theta)\cdot\alpha\cdot\exp\varepsilon_\alpha(-t\theta) \quad (t \in (-\infty, \infty)). \tag{6.8}$$

In particular, all geodesics on $\mathfrak{G}$ are total.

6.4. COROLLARY. *The exponential map $E_\alpha : \mathrm{T}_\alpha\mathfrak{G} \to \mathfrak{G}$ of the canonical linear connection ∇ at a point $\alpha \in \mathfrak{G}$ has the explicit equation*

$$E_\alpha(\theta) = \exp\varepsilon_\alpha(\theta)\cdot\alpha\cdot\exp\varepsilon_\alpha(-\theta) \quad (\theta \in T_\alpha\mathfrak{G}). \tag{6.9}$$

6.5. Formula (6.9) above enables us to verify simply that E_α is a complex analytic diffeomorphism from a neighborhood of 0 in $\mathrm{T}_\alpha\mathfrak{G}$ onto a neighborhood of α in $\mathfrak{G}$. Indeed, $E_\alpha(0) = \alpha$, and, for any $\theta \in \mathrm{T}_\alpha\mathfrak{G}$ we have also

$$(\mathrm{d}E_\alpha)(0)(\theta) = \varepsilon_\alpha(\theta)\cdot\alpha - \alpha\cdot\varepsilon_\alpha(\theta) = \partial_\alpha\varepsilon_\alpha(\theta) = \pi_\alpha(\theta) = \theta.$$

On the other hand $\mathrm{T}_\alpha\mathfrak{G}$ is isomorphic with the space A_α. These observations allow us to exhibit a total system of charts for the complexe analytic manifold $\mathfrak{G}(\alpha)$, with the model space A_α.

6.6. All the constructions and results above concerned with the canonical linear connection and its geodesics have natural and simple counterparts for the manifold $\mathfrak{U} = \mathfrak{U}(\mathcal{E})$. We will conclude this section with some remarks about the minimality of geodesics in $\mathfrak{U}$ in the case when A is a C^*-algebra.

First of all we have to define the length of a smooth path in $\mathfrak{U}$. This goal requires a suitable norm in each tangent spaces to $\mathfrak{U}$. As a matter of fact, given $\alpha \in \mathfrak{U}$, there are at least two possibilities to introduce a norm in the tangent space $\mathrm{T}_\alpha\mathfrak{U}$, using either the norm of E, or identifying isometrically $\mathrm{T}_\alpha\mathfrak{U}$ with $(A_\alpha)_{\mathrm{sh}}$ and, consequently, using the norm of A. More precisely, we obtain two norms in $\mathrm{T}_\alpha\mathfrak{U}$, denoted respectively by $\|\cdot\|_{\alpha,E}$ and $\|\cdot\|_\alpha$, and defined as follows:

$$\|\theta\|_{\alpha,E} = \|\theta\|, \quad \theta \in \mathrm{T}_\alpha\mathfrak{U}, \tag{6.10}$$

$$\|\theta\|_\alpha = \|\varepsilon_\alpha(\theta)\|, \quad \theta \in \mathrm{T}_\alpha\mathfrak{U}. \tag{6.11}$$

Recall that ε_α induces a linear isomorphism from $\mathrm{T}_\alpha\mathfrak{U}$ onto $(A_\alpha)_{\mathrm{sh}}$.

Actually, if we assume that $\|x \cdot \varphi\| = \|x\|\,\|\varphi\|$ for every $x \in A$ and $\varphi \in E$, then these two norms are equivalent. Indeed, given $\theta \in \mathrm{T}_\alpha\mathfrak{U}$, let $x = \varepsilon_\alpha(\theta) \in (A_\alpha)_{\mathrm{sh}}$. Then $\theta = x \cdot \alpha - \alpha \cdot x$, hence

$$\|\theta\|_{\alpha,E} = \|\theta\| \leqslant 2\|\alpha\|\,\|x\| = 2\|\alpha\|\,\|\theta\|_\alpha.$$

On the other hand, since $\alpha \cdot x \times \alpha = 0$, we get $\theta \times \alpha = x \cdot \alpha$, and therefore

$$\|\theta\|_\alpha = \|x\| = \|x \cdot \alpha\|\,\|\alpha\|^{-1} = \|\theta \times \alpha\|\,\|\alpha\|^{-1} \leqslant \|\theta\|\,\|\alpha\|\,\|\alpha\|^{-1} = \|\theta\|_{\alpha,E}.$$

Moreover, if $\gamma : (-\varepsilon, \varepsilon) \to \mathfrak{U}$ is a smooth curve, then the associated parallel translation along γ from $\alpha = \gamma(0)$ to $\gamma(t)$ is an isometry (compare with Definition 5.5) from $\mathrm{T}_\alpha\mathfrak{U}$ onto $\mathrm{T}_{\gamma(t)}\mathfrak{U}$, with respect to both types of norms.

However, the norm $\|\cdot\|_\alpha$ seems to be more appropriate. It is intrinsically related to the norm of the C^*-algebra A, while the norm $\|\cdot\|_{\alpha,E}$ depends on the choice of a norm in the ambient space E. Due to this reason, in what follows we will use the norm $\|\cdot\|_\alpha$ in the computation of the length of a curve.

DEFINITION. The length of a smooth path $\gamma : [0, \tau] \to \mathfrak{U}$ is defined as

$$\operatorname{length}(\gamma) = \int_0^\tau \|\dot\gamma(t)\|_{\gamma(t)} \mathrm{d}t. \tag{6.12}$$

A first advantage of this definition is offered by the next simple remark.

6.7. LEMMA. *Let $\gamma : [0, \tau] \to \mathfrak{U}$ be a smoth path and let $\Gamma : [0, \tau] \to \mathrm{U}(A)$ be its standard lift. Then*

$$\operatorname{length}(\gamma) = \int_0^\tau \|\dot\Gamma(t)\| \mathrm{d}t. \tag{6.13}$$

Proof. The standard lift Γ satisfies the equation $\dot\Gamma(t) = \varepsilon_{\gamma(t)}(\dot\gamma(t))\Gamma(t)$. Since $\Gamma(t)$ is unitary we get

$$\|\dot\Gamma(t)\| = \|\varepsilon_{\gamma(t)}(\dot\gamma(t))\| = \|\dot\gamma(t)\|_{\gamma(t)}. \qquad \blacksquare$$

We are now ready to prove that a short geodesic in $\mathfrak{U}$ has minimal length in a class of suitably related smooth paths.

6.8. THEOREM. *Let* $\gamma_0 : [0, \tau_0] \to \mathfrak{U}$ *and* $\gamma : [0, \tau] \to \mathfrak{U}$ *be two smooth paths and let* $\Gamma_0 : [0, \tau_0] \to \mathrm{U}(A)$ *and* $\Gamma : [0, \tau] \to \mathrm{U}(A)$ *be the standard lift of* γ_0 *and* γ*, respectively. Assume that*

(i) $\Gamma(\tau) = u\Gamma_0(\tau_0)u^*$ *for some unitary* $u \in \mathrm{U}(A)$*, and*

(ii) γ_0 *is a geodesic arc such that* $\mathrm{length}(\gamma_0) < \pi$.

Then

$$\mathrm{length}(\gamma_0) \leqslant \mathrm{length}(\gamma).$$

Proof. According to Proposition 6.2 the standard lift Γ_0 is given by $\Gamma_0(t) = \exp(tx)$, where $x = \varepsilon_{\gamma_0(0)}(\dot\gamma_0(0)) \in (A_\alpha)_{\mathrm{sh}}$.

Let ϕ be a state of A that satisfies the condition $\phi(-x^2) = \|x\|^2$. The GNS construction yields a cyclic representation $(\pi_\phi, \mathcal{H}_\phi)$ of A with the cyclic vector ξ_ϕ, such that

$$\phi(x) = \langle \pi_\phi(x)\xi_\alpha \mid \xi_\alpha \rangle, \quad x \in A. \tag{6.14}$$

We consider the smooth paths $c_0 : [0, \tau_0] \to \mathcal{H}_\phi$ and $c : [0, \tau] \to \mathcal{H}_\phi$ defined, respectively, by

$$c_0(t) = \pi_\phi(\Gamma_0(t))\xi_\phi, \quad t \in [0, \tau_0], \tag{6.15}$$

$$c(t) = \pi_\phi(u^*\Gamma(t)u)\xi_\phi, \quad t \in [0, \tau]. \tag{6.16}$$

Both c_0 and c are curves on the unit sphere $\mathcal{S}_\phi$ of $\mathcal{H}_\phi$, with the same endpoints. Since $\|\dot c(t)\| \leqslant \|\dot\Gamma(t)\|$ for any $t \in [0, \tau]$, by Lemma 6.7 we get

$$\mathrm{length}(c) \leqslant \mathrm{length}(\gamma). \tag{6.17}$$

On the other hand for c_0 we have

$$\begin{aligned}\|\dot c_0(t)\|^2 &= \left\langle \pi_\phi\left(\dot\Gamma_0(t)^*\dot\Gamma_0(t)\right)\xi_\phi \mid \xi_\phi \right\rangle = \\ &= \langle \pi_\phi(-x^2)\xi_\phi \mid \xi_\phi \rangle = \phi(-x^2) = \|x\|^2,\end{aligned}$$

hence $\|\dot c_0(t)\| = \|x\| = \|\dot\Gamma_0(t)\|$, therefore

$$\mathrm{length}(c_0) = \mathrm{length}(\gamma_0). \tag{6.18}$$

We show now that c_0 is a geodesic with respect to the usual Riemann structure of the sphere $\mathcal{S}_\phi$. Since the geodesic arcs on the unit sphere of a Hilbert space of length less than

π are minimal, the conclusion of our theorem will follow from (6.18) and (6.17). In order to obtain that c_0 is a geodesic we have to prove that $\ddot{c}_0(t)$ is collinear with $c_0(t)$. To this end we observe that

$$|\langle \ddot{c}_0(t) \mid c_0(t)\rangle| = \left|\left\langle \pi_\phi \left(\ddot{\Gamma}_0(t)\right) \xi_\phi \mid \pi_\phi(\Gamma_0(t))\xi_\phi \right\rangle\right| = \\ = |\langle \pi_\phi(x^2)\xi_\phi \mid \xi_\phi\rangle| = |\phi(x^2)| = \|x\|^2.$$

and, also,

$$\|\ddot{c}_0(t)\| \; \|c_0(t)\| = \|\ddot{c}_0(t)\| = \left\|\pi_\phi \left(\ddot{\Gamma}_0(t)\right) \xi_\phi\right\| = \\ = \|\pi_\phi(x^2)\xi_\phi\| \leqslant \|x^2\| = \|x\|^2,$$

hence

$$\|\ddot{c}_0(t)\| \; \|c_0(t)\| \leqslant |\langle \ddot{c}_0(t) \mid c_0(t)\rangle| . \tag{6.19}$$

Combined with the Cauchy-Schwartz inequality, (6.19) becomes an equality, a relation equivalent to our claim. The proof is complete. ■

References

[ARS] Andruchow, E.; Recht, L.; Stojanoff, D., The space of spectral measures is a homogeneous reductive space, *Integral Equations Operator Theory*, **16**(1993), 1–14.

[AS] Andruchow, E.; Stojanoff, D., Geometry of unitary orbits, *J. Operator Theory*, **26**(1991), 25–41.

[A] Aupetit, B., Projections in real Banach algebras, *Bull. London Math. Soc.*, **13**(1981), 412–414.

[BC] Bishop, R. L.; Crittenden, R. J., *Geometry of Manifolds*, Academic Press, New York, 1964.

[B] Bourbaki, N., *Variété Différentielles et Analytiques, Fascicule de results*, Hermann, Paris, 1967.

[CPR1] Corach, G.; Porta, H.; Recht, L., Differential geometry of systems of projections in Banach algebras, *Pacific J. Math.*, **143**(1990), 209–228.

[CPR2] Corach, G.; Porta, H.; Recht, L., Differential geometry of spaces of relatively regular operators, *Integral Equations Operator Theory*, **13**(1990), 771–794.

[CPR3] Corach, G.; Porta, H.; Recht, L., The geometry of the space of selfadjoint invertible elements in a C^*-algebra, *Integral Equations Operator Theory*, **16**(1993), 333–359.

[CPR4] Corach, G.; Porta, H.; Recht, L., The geometry of spaces of projections in C^*-algebras, *Advances in Math.*, to appear.

[CD1] Cowen, M. J.; Douglas, R.G., Complex geometry and operator theory, *Acta Math.*, **141**(1978), 187–261.

[CD2] Cowen, M. J.; Douglas, R.G., Operators possessing an open set of eigenvalues, in *Colloquia Math.*, **35**, North Holland, 1980, pp. 323–341.

[FD] FELL, J. M. G.; DORAN, R. S., *Representations of $*$-Algebras, Locally Compact Groups, and Banach $*$-Algebraic Bundles*, vol. 2: *Banach $*$-Algebraic Bundles, Induced Representations, and the Generalized Mackey Analysis*, Academic Press, Inc., 1988.

[GHV] GREUB, W.; HALPERIN, S.; VANSTONE, R., *Connections, Curvature and Cohomology*, vol. 2, Academic Press, New York, 1973.

[H] HELGASON, S., *Differential Geometry and Symmetric Spaces*, Academic Press, 1962.

[KN] KOBAYASHI, S.; NOMIZU, K., *Foundations of Differential Geometry*, vol. I, Interscience, New York, 1963.

[LM] LUBOTZKY, A.; MAGID, A. R., Varietes of representations and finitely generated groups, *Mem. Amer. Math. Soc.*, **336**(1985).

[Ma] MAGID, A. R., Deformations of representations of discrete groups, in *Classical Groups and Related Topics*, Contemporary Mathematics, Vol. **82**, 1989, pp. 79–87.

[M1] MARTIN, M., Almost product structures and derivations, *Bull. Math. Soc. Sci. Math. Roumanie*, **23**(1979), 171–176.

[M2] MARTIN, M., Hermitian geometry and involutive algebras, *Math. Z.*, **188**(1985), 359–382.

[M3] MARTIN, M., An operator theoretic approach to analytic functions into the Grassmann manifold, *Math. Balkanica*, **1**(1987), 45–58.

[M4] MARTIN, M., Projective representations of compact groups in C^*-algebras, in *Operator Theory: Advances and Applications*, vol. **43**, Birkhäuser Verlag, Basel, 1990, pp. 237–253.

[MS1] MARTIN, M.; SALINAS, N., The canonical complex structure of flag manifolds in a C^*-algebra, to appear.

[MS2] MARTIN, M.; SALINAS, N., Flag manifolds and the Cowen-Douglas theory, in preparation.

[MR] MATA, L.; RECHT, L., Infinite dimensional homogeneous reductive spaces, *Acta Cientifica Venezolana*, **43**(1992). 76–90.

[Mu] MUMFORD, D., *Algebraic Geometry. I: Complexe Projective Varieties*, Springer Verlag, Berlin — Heidelberg — New York, 1976.

[N] NOMIZU, K., Invariant affine connections on homogeneous spaces, *Amer. J. Math.*, **76**(1954), 33–65.

[PR1] PORTA, H.; RECHT, L., Spaces of projections in Banach algebras, *Acta Cientifica Venezolana*, **39**(1987), 408–426.

[PR2] PORTA, H.; RECHT, L., Minimality of geodesics in Grassmann manifolds, *Proc. Amer. Math. Soc.*, **100**(1987), 464–466.

[P] POTAPOV, V. P., The multiplicative structure of J-contractive matrix functions, *Amer. Math. Soc. Transl., Ser. II*, **15**(1960), 131–244.

[S] SALINAS, N., The Grassmann manifold of a C^*-algebra and hermitian holomorphic bundles, in *Operator Theory: Advances and Applications*, vol. **28**, Birckhäuser Verlag, Basel, 1988, pp. 267–289.

[U] UPMEIER, H., *Symmetric Banach manifolds and Jordan C^*-algebras*, North-Holland Math. Studies, **104**, North-Holland, Amsterdam, 1985.

[W1] WILKINS, D. R., The Grassmann manifold of a C^*-algebra, *Proc. of the Royal Irish Acad.*, **90A**(1990), 99–116.

[W2] WILKINS, D. R., On infinite-dimensional symmetric spaces, to appear.

[W3] WILKINS, D. R., On the classification of Grassmann manifolds in C^*-algebras, to appear.

Department of Mathematics
University of Kansas
Lawrence, KS 66045
USA

1991 Mathematics Subject Classification. Primary: 46L05, 53C30.

Operator Theory:
Advances and Applications, Vol. 80

Nonlinear Equations and Inverse Spectral Problems on the Axis

L.A. Sakhnovich

For a number of nonlinear equations the method of the inverse scattering problem [1]-[3] has led to a great progress in the investigation of the Cauchy problem on the axis ($-\infty < x < \infty$). In the papers [4]-[6] the transition from the method of the scattering problem to the method of the inverse spectral problem was performed. It permitted to investigate some nonlinear equations on the half axis ($0 \leq x < \infty$).

In this paper the method of the inverse spectral problem is used for analysing the situation $-\infty < x < \infty$.

Here the evolution law of the spectral data is deduced and an isospectral property is proved. General results are applied to the analysis of the nonlinear Schrödinger equation and the modified Korteveg-de Vries equation.

1. On the Weyl-Titchmarsh matrix-function.

1. We consider a canonical system of differential equations

$$\frac{dW}{dx} = izJH(x)W(x,z), \quad W(0,z) = E_{2m} \tag{1.1}$$

where

$$J = \begin{pmatrix} 0 & E_m \\ E_m & 0 \end{pmatrix}, \quad H(x) \geq 0, \; -\infty < x < \infty.$$

To the problem on the axis (1.1) we put into correspondence the problem on the half-axis $(0,\infty)$, defined by the relations

$$\frac{dW_0}{dx} = izJ_0H_0(x)W_0(x,z), \quad W_0(0,z) = E_{4m}, \; 0 \leq x < \infty \tag{1.2}$$

where

$$J_0 = \begin{pmatrix} 0 & E_{2m} \\ E_{2m} & 0 \end{pmatrix}, \quad W_0(x,z) = T^* \begin{pmatrix} W(x,z) & 0 \\ 0 & W(-x,z) \end{pmatrix} T, \tag{1.3}$$

$$T = \frac{1}{\sqrt{2}} \begin{pmatrix} J & E_{2m} \\ -J & E_{2m} \end{pmatrix}, \quad H_0(x) = T^* \begin{pmatrix} H(x) & 0 \\ 0 & H(-x) \end{pmatrix} T. \tag{1.4}$$

2. The matrix-functions $v(z)$, for which the inequality

$$\int_0^\infty (E_{2m}, \, iv^*(z))\, W_0^*(x,z) H_0(x) W_0(x,z) \begin{pmatrix} E_{2m} \\ -iv(z) \end{pmatrix} dx < \infty, \; Im z > 0 \tag{1.5}$$

is true, we call the Weyl-Titchmarsh matrices of the problem (1.1).

Together with problem (1.1) on the axis $-\infty < x < \infty$ we study the following problems on the half axis:

$$\frac{dW_\pm(x,z)}{dx} = \pm izJH_\pm W_\pm(x,z), \quad W_\pm(0,z) = E_{2m}, \tag{1.6}$$

where

$$W_\pm(x,z) = W(\pm x, z), \quad H_+(x) = H(x), \quad H_-(x) = H(-x), \quad x \geq 0. \tag{1.7}$$

According to (1.5) the Weyl-Titchmarsh functions $v_\pm(z)$ of systems (1.6) are defined by the inequality

$$\int_0^\infty (E_{2m}, \, \pm iv_\pm^*(z))\, W_\pm^*(x,z) H_\pm(x) W_\pm(x,z) \begin{pmatrix} E_{2m} \\ \mp iv_\pm(z) \end{pmatrix} dx < \infty. \tag{1.8}$$

In order to find the connection between $v_\pm(z)$ and $v(z)$ we put down $v(z)$ in a block form

$$v(z) = \{v_{ij}(z)\}_{i,j=1}^2$$

where $v_{ij}(z)$ are matrices of order $m \times m$.

From (1.5) and (1.8) by means of algebraic transformations the following theorem is deduced.

Theorem 1.2 Let us suppose that the systems (1.6) have unique Weyl-Titchmarsh matrix-functions $v_+(z)$ and $v_-(z)$.

If the relations

$$\det v_{11}(z) \neq 0, \quad \det(E_m \pm iv_{12}(z)) \neq 0, \quad Im z > 0 \tag{1.9}$$

are fulfilled then the equations

$$v_+(z) = -(E_m - iv_{21}(z))v_{11}^{-1}(z) = v_{22}(z)(E_m - iv_{12}(z))^{-1}, \tag{1.10}$$

$$v_-(z) = -(E_m + iv_{21}(z))v_{11}^{-1}(z) = v_{22}(z)(E_m + iv_{12}(z))^{-1}, \tag{1.11}$$

$$(Jv(z))^2 = -E_m \tag{1.12}$$

are true.

Proof. From the relations (1.3), (1.4) it follows, that the inequality (1.5) is equivalent to the inequalities

$$\int_0^\infty (J \pm iv^*(z))\, W_\pm^*(x,z)H_\pm(x)W_\pm(x,z)\,(J \mp iv(z))\,dx < \infty \tag{1.13}$$

The relations (1.13) can be rewritten in the following form:

$$\int_0^\infty (\pm iv_{11}^*(z), E_m \pm iv_{21}^*(z))\, W_\pm^*(x,z)H_\pm(x)W_\pm(x,z) \begin{pmatrix} \mp iv_{11}(z) \\ E_m \mp iv_{21}(z) \end{pmatrix} < \infty, \tag{1.14}$$

$$\int_0^\infty (E_m \pm iv_{12}^*(z), E_m \pm iv_{21}^*(z))\, W_\pm^*(x,z)H_\pm(x)W_\pm(x,z) \begin{pmatrix} E_m \mp iv_{12}(z) \\ \mp iv_{22}(z) \end{pmatrix} < \infty. \tag{1.15}$$

The equalities (1.10), (1.11) follow from (1.14), (1.15). We deduce from the equalities (1.10), (1.11) that

$$v_{21}v_{11}^{-1} + v_{11}^{-1}v_{12} = 0, \qquad v_{22} = v_{21}v_{11}^{-1}v_{12} - v_{11}^{-1} \tag{1.16}$$

The relations (1.16) are equivalent to the equality (1.12). This proves the theorem.

2. SPECTRAL DATA EVOLUTION.

1. We introduce an additional parameter t and rewrite the systems (1.6) in the form

$$\frac{\partial W_\pm(x,t,z)}{\partial x} = G_\pm(x,t,z)W_\pm(x,t,z), \tag{2.1}$$

where

$$G_\pm(x,t,z) = \pm izJH_\pm(x,t). \tag{2.2}$$

Let the matrix $F_\mp(x,t,z)$ of order $2m \times 2m$ be such that

$$\frac{\partial G_\pm}{\partial t} - \frac{\partial F_\pm}{\partial x} + [G_\pm, F_\pm] = 0, \tag{2.3}$$

where the symbol $[G_1, G_2]$ denotes the commutant:

$$[G_1, G_2] = G_1G_2 - G_2G_1.$$

Setting

$$F(x,t,z) = T^{-1}\begin{pmatrix} F_+(x,t,z) & 0 \\ 0 & F_-(x,t,z) \end{pmatrix} T, \tag{2.4}$$

$$G(x,t,z) = T^{-1}\begin{pmatrix} G_+(x,t,z) & 0 \\ 0 & G_-(x,t,z) \end{pmatrix} T, \tag{2.5}$$

by (2.3) we have

$$\frac{\partial G}{\partial t} - \frac{\partial F}{\partial x} + [G, F] = 0, \tag{2.6}$$

According to (1.4), (2.2), (2.5) the system (1.2) can be written in the form

$$\frac{\partial W_0}{\partial x} = G(x,t,z)W_0(x,t,z). \tag{2.7}$$

Hence the equality (see [6])

$$W_0(x,t,z) = V(x,t,z)W_0(x,t,z)V^{-1}(0,t,z) \tag{2.8}$$

is true. Here $V(x,t,z)$ is the solution of the equation

$$\frac{\partial V}{\partial t} = F(x,t,z)V, \qquad V(x,0,z) = E_{4m}. \tag{2.9}$$

The matrix-function

$$r(t,z) = V^{*-1}(0,t,\bar{z}) = \{r_{ij}(t,z)\}_{i,j=1}^{2}, \tag{2.10}$$

where $r_{ij}(t,z)$ are matrices of order $2m \times 2m$, will play an important role in our further investigation. By (2.4), (2.10) the matrix $r(t,z)$ admits the representation

$$r(t,z) = T^{-1}\begin{pmatrix} r_+(t,z) & 0 \\ 0 & r_-(t,z) \end{pmatrix} T. \tag{2.11}$$

In accordance with (1.3) the matrix $W_0^*(l,t,\bar{z})$ admits the following block representation:

$$W_0^*(l,t,\bar{z}) = \frac{1}{2}\begin{pmatrix} Ja_0(l,t,z)J & Jb_0(l,t,z) \\ b_0(l,t,z)J & a_0(l,t,z) \end{pmatrix}, \tag{2.12}$$

where we use the notations

$$a_0(l,t,z) = W_+^*(l,t,\bar{z}) + W_-^*(l,t,\bar{z}), \tag{2.13}$$

$$b_0(l,t,z) = W_+^*(l,t,\bar{z}) - W_-^*(l,t,\bar{z}). \tag{2.14}$$

Then in the block form we write down the matrix

$$W_0^{*-1}(l,t,\bar{z})J_0W_0^{-1}(l,t,\bar{z}) = \begin{pmatrix} -\gamma(l,t,z) & \delta(l,t,z) \\ \delta^*(l,t,z) & -\nu(l,t,z) \end{pmatrix}. \tag{2.15}$$

It follows from the inequality $H_0(x,t) \geq 0$ that $\gamma(l,t,z) \geq 0$ when $Imz > 0$. If the strong inequality

$$\gamma(l,t,z) > 0, \qquad Imz > 0 \tag{2.16}$$

is valid then the fractional linear transformation [6]

$$v(l,t,z) = iJ(a_0(l,t,z)JP(z) + b_0(l,t,z)Q(z))(b_0(l,t,z)JP(z) + a_0(l,t,z)Q(z))^{-1} \quad (2.17)$$

makes sense. Here $P(z)$, $Q(z)$ are meromorphic matrix-functions such that

$$\det(P^*(z)P(z) + Q^*(z)Q(z)) \not\equiv 0, \quad (2.18)$$

(non-degeneracy)

$$P^*(z)Q(z) + Q^*(z)P(z) \geq 0, \qquad Imz > 0, \quad (2.19)$$

(J-property). The functions of form (2.17) belong to the Nevanlinna class, i.e.

$$v(t,z) = v^*(t,\bar{z}), \qquad [v(t,z) - v^*(t,z)]/(z-\bar{z}) > 0. \quad (2.20)$$

Thus $v(t,z)$ admits the representation

$$v(t,z) = \beta(t)z + \alpha(t) + \int_{-\infty}^{+\infty} (\frac{1}{u-z} - \frac{u}{1+u^2})d\tau(u,t), \quad (2.21)$$

where $\beta(t) \geq 0, \alpha(t) = \alpha^*(t)$ and with respect to u, $\tau(u,t)$ is an increasing matrix-function. If $\beta(t) = 0$ then (2.21) has the form

$$v(t,z) = \alpha(t) + \int_{-\infty}^{+\infty} (\frac{1}{u-z} - \frac{u}{1+u^2})d\tau(u,t), \quad (2.22)$$

The matrices $v(l,t,z)$ form a system of included Weyl circles. If for $l \to \infty$ the Weyl circles converge to a point, then the limit

$$v(t,z) = \lim_{l\to\infty} v(l,t,z), \qquad Imz > 0 \quad (2.23)$$

exists.

2. In the examples which we shall consider further the condition

$$F_+(0,t,z) = -JF_+^*(0,t,\bar{z})J \quad (2.24)$$

is fulfilled. From the condition (2.24) and the relations

$$\frac{dr_+}{dt} = -F_+^*(0,t,\bar{z})r_+, \qquad r_+(0,z) = E_{2m} \quad (2.25)$$

it follows that

$$r_+^*(t,z)Jr_+(t,z) = J. \quad (2.26)$$

Further we shall consider the important special case when

$$F_+(0,t,z) = F_-(0,t,z). \tag{2.27}$$

Remark 2.1. Condition (2.27) means that the solution of the nonlinear equation generated by the relation (2.6) is continuous in the point $x = 0$ (sometimes together with its derivatives, see Sections 3, 4)

Under condition (2.27) from (2.4) we obtain the following relations:

$$F(0,t,z) = \begin{pmatrix} JF_+(0,t,z)J & 0 \\ 0 & F_+(0,t,z) \end{pmatrix}, \tag{2.28}$$

$$r(t,z) = \begin{pmatrix} Jr_+(t,z)J & 0 \\ 0 & r_+(t,z) \end{pmatrix}. \tag{2.29}$$

3. The evolution law (the law of the dependence of $v(t,z)$ on t was deduced in [6]. Applying this result to the case under consideration we obtain the following assertion:

Theorem 2.1. Let $F(x,t,z)$ and $G(x,t,z)$ be defined by the relations (2.4), (2.5), (2.27) and suppose that the following requirements be fulfilled:

a) The equality (2.6) holds.

b) Starting with a certain l_0, the inequality

$$\gamma(l,0,z) > 0, \qquad l > l_0, \qquad Imz > 0 \tag{2.30}$$

is valid.

c) The limit

$$v(z) = v(0,z) = \lim_{l\to\infty} v(l,0,z) \tag{2.31}$$

exists.

d) For any $t > 0$ and for any pair $P(z), Q(z)$ satisfying the inequality

$$P^*(z)Q(z) + Q^*(z)P(z) > 0, \qquad Imz > 0, \tag{2.32}$$

there exist numbers z_0 ($Imz_0 > 0$) and l_0 such that in a certain neighbourhood of z_0 the pair

$$col[P_1(z), Q_1(z)] = V^*(l,t,\bar{z})col[P(z), Q(z)]$$

has the J-property if $l > l_0$.

Then in (2.23) the limit $v(t,z)$ exists and the equality

$$v(t,z) = Jr_+(t,z)Jv(0,z)r_+^{-1}(t,z) \tag{2.33}$$

is true.

Remark 2.2. By the equalities (2.25) the matrix $r_+(t,z)$ is invertible.

Differentiating (2.33) and taking into account (2.25) we deduce the relation

$$\frac{dv}{dt} = -JF_+^*(0,t,\bar{z})Jv + vF_+^*(0,t,\bar{z}). \tag{2.34}$$

From the Theorem 2.1 an isospectral property follows.

Corollary 2.1. Let the conditions of the Theorem 2.1 be fulfilled and suppose that $F^*(0,t,\bar{z})$ is a polynomial of the variable z. Then the singularities of the matrix-functions $v(t,z)$ and $v(0,z)$ coincide.

Corollary 2.2. Let the relations (2.33) and $[Jv(0,z)]^2 = -E_{2m}$ be valid. Then

$$[Jv(t,z)]^2 = -E_{2m}. \tag{2.35}$$

3. INVESTIGATION OF THE NSE AND THE MKdVE

1. In this paragraph we shall apply the general results from Sections 1, 2 to the investigation of the nonlinear Schrödinger equation (NSE)

$$R_t = \frac{i}{2}(R_{xx} - 2RR^*R), \qquad -\infty < x < \infty, \qquad t \geq 0 \tag{3.1}$$

and to the modified Korteveg-de Vries equation (MKdVE)

$$R_t = -\frac{1}{4}R_{xxx} + \frac{3}{4}(RR^*R_x + R_xR^*R), \qquad -\infty < x < \infty, \qquad t \geq 0, \tag{3.2}$$

where $R(x,t)$ is a matrix-function of the order $m \times m$. Following [3] we introduce the matrices

$$G_1(x,t,z) = j(izE_{2m} - \xi), \tag{3.3}$$

$$j = \begin{pmatrix} E_m & 0 \\ 0 & -E_m \end{pmatrix}, \qquad \xi = \begin{pmatrix} 0 & R \\ -R^* & 0 \end{pmatrix}, \tag{3.4}$$

$$F_1(x,t,z) = -i[jz^2 + izj\xi - \frac{1}{2}(\Omega - \frac{\partial \xi}{\partial x})], \tag{3.5}$$

where

$$\Omega = \begin{pmatrix} -RR^* & 0 \\ 0 & R^*R \end{pmatrix}. \tag{3.6}$$

The matrix $F_1(x,t,z)$ for the MKdVE has the form

$$F_1(x,t,z) = \begin{pmatrix} z^2E_m + \frac{1}{2}RR^* & 0 \\ 0 & z^2E_m + \frac{1}{2}R^*R \end{pmatrix} \begin{pmatrix} izE_m & -R \\ -R^* & -izE_m \end{pmatrix} +$$

$$+\frac{1}{4}\begin{pmatrix} R_xR^* - RR_x^* & R_{xx} + 2izR_x \\ R_{xx}^* - 2izR_x^* & R_x^*R - R^*R_x \end{pmatrix}. \tag{3.7}$$

As it is known [2] the equations (3.1), (3.2) are equivalent to the equation

$$\frac{\partial G_1}{\partial t} - \frac{\partial F_1}{\partial x} + [G_1, F_1] = 0, \tag{3.8}$$

where F_1 is defined by the formula (3.6) in case of the NSE and by (3.7) in case of the MKdVE.

2. Let us introduce the matrix

$$T_1 = \frac{1}{\sqrt{2}}\begin{pmatrix} E_m & -E_m \\ E_m & E_m \end{pmatrix}. \tag{3.9}$$

The equalities

$$T_1 j T_1^* = J, \qquad T_1^* = T_1^{-1}. \tag{3.10}$$

are easily checked. We define the matrix $A(x,t)$ by the relations

$$j\frac{\partial A}{\partial x} = -\xi A, \qquad A(0,t) = T_1^*. \tag{3.11}$$

Proposition 3.1. If $W_1(x,t,z)$ is the solution of the system

$$\frac{\partial W_1}{\partial x} = G_1(x,t,z)W_1, \qquad W_1(0,t,z) = E_{2m},$$

then

$$W(x,t,z) = A^{-1}(x,t)W_1(x,t,z)T_1^* \tag{3.12}$$

is a solution of the system

$$\frac{\partial W}{\partial x} = izJH(x,t)W, \tag{3.13}$$

where

$$H(x,t) = A^*(x,t)A(x,t). \tag{3.14}$$

The transition to system (3.13) permits us to use the results of Sections 1, 2.

Here it is necessary to take into consideration that for $x > 0$ the relations

$$G_+(x,t,z) = izJH(x,t), \tag{3.15}$$

$$F_+(x,t,z) = -A^{-1}(x,t)\frac{\partial A}{\partial t} + A^{-1}(x,t)F_1(x,t,z)A(x,t) \tag{3.16}$$

are valid. Then for the NSE the formula

$$F_+^*(0,t,\bar{z}) = iz^2J + zQ_1(t) + Q_0(t), \tag{3.17}$$

is true, where

$$Q_1(t) = T_1\xi^*(0,t)jT_1^*, \qquad Q_0(t) = \frac{1}{2}iT_1\left[\frac{\partial\xi^*(x,t)}{\partial x}\bigg|_{x=0} - \Omega(0,t)\right]T_1^*. \tag{3.18}$$

In the case of the MKdVE we have (see (3.7) and (3.16))

$$F_+^*(0,t,\bar{z}) = -iz^3J + \sum_{k=0}^{2} z^k q_k(t), \tag{3.19}$$

where, by (3.7), the equalities

$$q_2(t) = -Q_1(t), \qquad q_1(t) = -Q_0(t), \tag{3.20}$$

$$q_0(t) = \frac{1}{4}T_1\begin{pmatrix} RR_x^* - R_xR^* & R_{xx} - 2RR^*R \\ R_{xx}^* - 2R^*RR^* & R^*R_x - R_x^*R \end{pmatrix}T_1^*\bigg|_{x=0} \tag{3.21}$$

are valid.

3. Further we suppose that $v(t,z)$ in the neighbourhood of $z=\infty$ admits the expansion

$$v(t,z) = J\sum_{k=0}^{\infty}\alpha_k(t)/z^k \tag{3.22}$$

Proposition 3.2. (see [7]). If the function $R(x,t)$ is continuous with respect to x then

$$\lim_{z\to\infty} v(t,z) = iE_{2m}, \qquad 0 < \varepsilon \le arg\ z \le \pi - \varepsilon, \qquad \varepsilon > 0. \tag{3.23}$$

From (3.22) and (3.23) it follows that

$$\alpha_0(t) = iJ. \tag{3.24}$$

We say that $R(x,t)$ belongs to the regularity class P_τ if the Weyl-Titchmarsh function $v(t,z)$ of the corresponding system (3.13) admits the representation (3.22), (3.24) for $0 \le t \le \tau$.

Using formulas (3.22) and (2.34) we obtain the following relations for the NSE:

$$\alpha_k'(t) = i[\alpha_{k+2}, J] + [\alpha_{k+1}, Q_1] + [\alpha_k, Q_0]. \tag{3.25}$$

Here we have used the formulas (3.17), (3.19). The relations

$$\alpha_k'(t) = -i[\alpha_{k+3}, J] + [\alpha_{k+2}, q_2] + [\alpha_{k+1}, q_1] + [\alpha_k, q_0] \tag{3.26}$$

for the MKdVE are proved analogously. In (3.25) we have $k \ge -2$ and

$$\alpha_{-2} = \alpha_{-1} = 0, \tag{3.27}$$

and in (3.26) $k \geq -3$ and

$$\alpha_{-3} = \alpha_{-2} = \alpha_{-1} = 0. \tag{3.28}$$

From (3.28) it follows that

$$JQ_1J = -Q_1. \tag{3.29}$$

Using (3.25) for $k = -1$ and (3.29) we deduce

$$Q_1(t) = \frac{1}{2}[\alpha_1(t) - J\alpha_1(t)J] \qquad (NSE). \tag{3.30}$$

If $k = -2$ from (3.26) we have

$$q_2(t) = \frac{1}{2}[J\alpha_1(t)J - \alpha_1(t)] \qquad (MKdVE). \tag{3.31}$$

Now let us write (3.25) for $k = 0$:

$$0 = i[\alpha_2, J] + [\alpha_1, Q_1] + [\alpha_0, Q_0], \tag{3.32}$$

and (3.26) for $k = -1$ and $k = 0$:

$$0 = -i[\alpha_2, J] + [\alpha_1, q_2] + [\alpha_0, q_1] \qquad (MKdVE), \tag{3.33}$$

$$0 = -i[\alpha_3, J] + [\alpha_2, q_2] + [\alpha_1, q_1] + [\alpha_0, q_0] \qquad (MKdVE). \tag{3.34}$$

Remark 3.1. The formulas (3.30)-(3.34) connect the coefficients $\alpha_1, \alpha_2, \alpha_3$ of the asymptotic expansion of $v(t, z)$ with the Hamiltonian $H(x, t)$ of system (3.13).

This result is also interesting when $t = 0$. Indeed, with the help of the evolution law we obtain a result related to the stationary linear problem

$$\frac{dW}{dx} = izJH(x, 0)W.$$

4. The formulas (3.30) and (3.32) permit to replace the matrix-functions $Q_0(t), Q_1(t)$ in the system (3.25) through $\alpha_1(t), \alpha_2(t)$. In this case we get a nonlinear infinite system of differential equations with respect to $\alpha_1(t), \alpha_2(t), \ldots$. In the same way the formulas (3.31), (3.33), (3.34) permit to replace the matrix-functions $q_0(t), q_1(t), q_2(t)$ in the system (3.26) through $\alpha_1(t), \alpha_2(t), \alpha_3(t)$. Here again we get a nonlinear system of differential equations. Solving the obtained systems [8] we deduce the following result.

Theorem 3.1. Let the matrix-function $v_0(z)$ in a neighbourhood of $z = \infty$ admit the representation

$$v_0(z) = J\sum_{k=0}^{\infty} \alpha_k(0)/z^k, \qquad \alpha_0(0) = iJ \tag{3.35}$$

and satisfy the condition

$$[v_0(z) - v_0^*(z)]/i > 0, \qquad Imz \geq 0. \tag{3.36}$$

Then for some $\tau > 0$ each of the systems (3.25) and (3.26) has one and only one solution $v(t,z)$ $(0 \leq t \leq \tau)$, admitting the representation (3.22) and satisfying the conditions

$$V(0,z) = v_0(z), \qquad [v(t,z) - v^*(t,z)]/i > 0, \qquad Imz \geq 0. \tag{3.37}$$

As in the case of the half-axis [8] the following corollary is deduced:

Corollary 3.1. Let the inequalities

$$|\alpha_k(0)| \leq M^k \qquad k = 1,2,\dots \tag{3.38}$$

be valid. Then there exist numbers τ and M such that the functions $\alpha_k(t)$ in the expansion are analytic and the inequalities

$$|\alpha_k(t)| \leq M_\tau^k, \qquad 0 \leq t \leq \tau \tag{3.39}$$

are true.

The following important corollary is easily deduced from Theorem 3.1 and Corollary 2.1.

Corollary 3.2. Let the matrix-function $v_0(z)$ satisfy the conditions of Theorem 3.1 and let its elements be rational functions. Then the elements of the corresponding matrix $v(t,z)$ are rational with respect to the variable z.

Indeed, the singular points of $v_0(z)$ and $v(t,z)$ coincide (the point $z = \infty$ is regular).

5. As in the case of the half-axis [8] from Theorem 3.1 we deduce Theorem 3.2.

Theorem 3.2. Let the function

$$R(x,0) = f(x), \qquad -\infty < x < \infty \tag{3.40}$$

be such that the corresponding function $v_0(z)$ admits the representation (3.35) and satisfies the condition (3.36). Then for some $\tau > 0$ each of the equations (3.1) and (3.2) has one and only one solution $R(x,t)$ $(-\infty < x < \infty, 0 \leq t \leq \tau)$, belonging to the class P_τ and satisfying the condition (3.40).

References

1. Novikov S.P.(ed.) Theory of solitons, the inverse problem method. Moscow, Nauka, 1980.

2. Bullough R.K., Caudrey P.I. (eds) Solitons, New-York, 1980.

3. Zakharov V.E., Shabat A.B. Applications of the inverse scattering method 1, Funct. Anal. Appl. 8, 3, 43-53, (1974).

4. Beresanskii Yu.M. Investigation of nonlinear difference equations by the inverse spectral problem method. Dokl. AN SSSR 281, 1, 16-19, (1985).

5. Sakhnovich L.A. Factorization problems and operator identities. Russian Math. Surv. 41, 1, 1-64, (1986).

6. Sakhnovich L.A. The method of operator identities and analysis problems. Algebra and Analysis 5, 1, 3-80, (1993).

7. Sakhnovich L.A. The hyperbolic sinus-Gordon equation. Izv. Vyssh. Uchebn. Zaved. Mat. 1, 54-63, (1991).

8. Sakhnovich L.A. Integrable nonlinear equations on the half-axis. Ukr. Math. Journ. 43, 11, 1578-1584, (1991).

Pr. Dobrovolskogo 154 ap. 199,
Odessa 270111,
Ukraine.

AMS Subject Classification: 34 A 55

Operator Theory:
Advances and Applications, Vol. 80

RAYLEIGH PROBLEM AND FRIEDRICHS MODEL

Stanislaw A. Stepin

Study of linear stability for ideal fluid plane-parallel flow leads to Rayleigh equation and associated boundary eigenvalue problem being both singular and non-selfadjoint. Operatortheoretic formulation of this problem is treated within the framework of Friedrichs model. Stationary scattering theory technique is applied to obtain corresponding eigenfunction expansion theorem in the case when velocity profile of main stationary flow is close to linear. Besides, inverse scattering problem related to Rayleigh equation is considered. Local solvability of this problem and local uniqueness property are established.

1. Introduction

This paper is the expanded version of the author's talk delivered at the workshop "Operator theory and boundary eigenvalue problems". It deals with the spectral analysis of a boundary value problem appearing in hydrodynamics, the analysis being carried out within the framework of stationary scattering theory approach.

Let us consider the stability problem for a plane-parallel flow of ideal incompressible fluid between two solid walls. This motion is described by the Euler equation complemented by the incompressibility condition:

$$\begin{aligned} \frac{\partial \vec{v}}{\partial t} + (\vec{v}\nabla)\vec{v} &= -\frac{1}{\varrho}\nabla p, \\ div\vec{v} &= 0, \end{aligned} \tag{1}$$

where $\vec{v} = \vec{u} + \vec{w}$ is the velocity vector field, $\vec{u}$ being the velocity of main stationary flow, while $\vec{w}$ stands for perturbation, p and ϱ are the pressure and density of the fluid. As it is usually done for plane-parallel problems of incompressible fluid hydrodynamics we use in the sequel the representation of such a flow in terms of the so-called flow function. For this purpose we choose the Descartes coordinate system (x, y, z) so that the velocity vector field $\vec{v}$ has the form $(v_x, 0, v_z)$, then the incompressibility condition

$$\frac{\partial v_x}{\partial x} + \frac{\partial v_z}{\partial z} = 0$$

enables one to introduce flow function $\psi = \psi(x, z, t)$ such that

$$v_x = -\frac{\partial \psi}{\partial z}, \; v_z = -\frac{\partial \psi}{\partial x}.$$

Further we assume (without loss of generality) that the unperturbed velocity vector field $\vec{u}$ is directed along x-axis and depends on z-variable only, i.e. vector field $\vec{u}$ has the form $(u(z), 0, 0)$ and hence corresponding flow is completely characterized by its profile $u = u(z)$.

This permits us to separate the variables in the system that is linearized one for (1) with respect to the main stationary flow and therefore search for the flow function of the perturbed motion in the form of a so-called normal wave

$$\psi(x, z; t) = \psi(z)e^{i\alpha(x-ct)},$$

where α and c are corresponding separation constants (with the physical meaning of wave number in x-direction and phase velocity of a plane wave correspondingly). Thus the study of small oscillations of ideal fluid near stationary flow with velocity profile u leads (cf. [1]) to the following boundary value problem for the Rayleigh equation on the bounded interval $[a, b]$:

$$(u - c)(\psi'' - \alpha^2\psi) - u''\psi = 0 \tag{2}$$

with zero boundary conditions

$$\psi(a) = \psi(b) = 0; \tag{3}$$

here c stands for the spectral parameter, α is a real parameter of the problem.

Equation (2) can be obtained from the Orr-Sommerfeld equation which describes a similar problem for viscous fluid in the limiting case of vanishing viscosity. The problem (2) - (3) under consideration is singular and non-selfadjoint. The first feature is related to that the coefficient at the second derivative of ψ in equation (2) may vanish; the second one is connected with the existence of profiles u such that the problem (2) - (3) has non real eigenvalues.

In order to simplify the subsequent exposition we impose the following condition on the derivative of the velocity profile: $u'(z) > 0$ for $z \in [a, b]$.

2. Operator-theoretic formulation of the problem (2) – (3); generalized eigenfunctions

Let us transform the Rayleigh equation (2) by the substitution $\phi = \Delta\psi$, where Δ is the ordinary differential operator in $L_2(a, b)$ generated by differential expression $\frac{d^2}{dz^2} - \alpha^2$ and zero boundary conditions at the endpoints of the interval $[a, b]$. As a result of this substitution equation (2) turns into integral equation

$$u(z)\phi(z) - u''(z)\int_a^b G(z, z')\phi(z')dz' = c\phi(z),$$

where G is the Green function of the differential operator Δ. Thus boundary value spectral problem for Rayleigh equation is reduced to spectral analysis of operator $H = H_0 + V$ acting in $L_2(a,b)$, where H_0 is the operator of multiplication by u and perturbation operator V is an integral operator, in general non-selfadjoint.

According to H. Weyl theorem continuous spectrum of the operator H occupies the closed interval $[u(a), u(b)]$. As regards eigenvalues of H, all of them (if any) belong to the disc with the diameter $[u(a), u(b)]$ (see [2]).

We note here the following fundamental result due to Rayleigh (see, for example, [1]): plane-parallel flow of ideal incompressible fluid is linearly stable if the velocity profile u has no points of inflection. In other words, if u'' does not vanish then the problem (2) - (3) has no non-real eigenvalues (it is worthwhile mentioning that there are no real eigenvalues for the problem (2) - (3) in this case as well). So the problem under consideration can have no eigenfunctions at all whereas generalized (continuous spectrum) eigenfunctions do exist for every $c \in (u(a), u(b))$.

Definition. A continuous function ψ_c, $c \in (u(a), u(b))$, is called generalized eigenfunction of the problem (2) - (3), if ψ_c satisfies the Rayleigh equation on the intervals $(a, u^{-1}(c))$ and $(u^{-1}(c), b)$ as well as the boundary conditions $\psi_c(a) = \psi_c(b) = 0$.

By the conditions imposed in this definition generalized eigenfunction ψ_c is determined uniquely up to a scalar factor. Certain choice of normalization for ψ_c, natural from the viewpoint of scattering theory, is suggested below (see Proposition 1).

3. Spectral analysis of the problem (2)–(3): Friedrichs' model approach and expansion theorem

As it has been already mentioned above the problem (2)–(3) is equivalent to Friedrichs' model $H = H_0 + V$, where the perturbation V is the integral operator with the kernel

$$-u''(z)G(z, z').$$

This kernel is obviously non-symmetric and moreover can not be made symmetric by introducing weighted L_2 - scalar product, since u'' need not be sign-definite. Nevertheless under appropriate smoothness assumption concerning profile u operator formulation of boundary value problem under consideration can be treated within the framework of the non-selfadjoint Friedrichs' model.

Following [3] we introduce the class r^μ, $0 < \mu < 1$, formed by kernels $\tau(\omega, \omega')$ of finite (semi)norm

$$\|\tau\|_\mu = \sup_{\omega_1 \neq \omega_2, \omega_1' \neq \omega_2'} h^\mu(\omega_1, \omega_2) h^\mu(\omega_1', \omega_2') |\tau(\omega_1, \omega_1') - \tau(\omega_2, \omega_1') - \tau(\omega_1, \omega_2') + \tau(\omega_2, \omega_2')|,$$

where

$$h^\mu(\omega_1, \omega_2) = (1 + |\omega_1|^\mu)(1 + |\omega_2|^\mu)|\omega_1 - \omega_2|^{-\mu}.$$

The elements of this class satisfy the Hölder condition with exponent μ in both variables ω and ω'. Let us denote by $v(\omega, \omega')$ the kernel of V in the spectral representation of the operator H_0 (in other words instead of z we choose $\omega = u(z)$ to be new "independent" variable). By direct verification the following statement can be established.

Lemma 1. If $u \in C^{2+\nu}[a,b]$ for $\nu > 0$ then the kernel v belongs to the class r^{μ}, where $\mu = \min\{1/2, \mu\}$.

In the situation described above application of the scattering theory technique elaborated by Friedrichs under assumption that $\|v\|_{\mu}$ is sufficiently small enables us to construct explicitly the continuous spectrum eigenfunctions for the problem (2)–(3) and prove corresponding expansion theorem.

In the selfadjoint case mathematically rigorous relationship between generalized eigenfunction expansions and scattering theory has been first established by A. Povzner for Schrödinger equation (see [4]). The synthesis of Povzner's and Friedrich's methods was carried out by O. Ladyzhenskaya and L. Faddeev [5], [6] (see also Rejto [7]). This approach enabled them in the selfadjoint Friedrichs' model to remove the resriction that the perturbation operator is small, though at the expense of the following requirement: the perturbation V is an integral operator with kernel v of the Hölder class with exponent $\lambda > 1/2$. For our problem which is nonselfadjoint the kernel of perturbation operator belongs to r^{μ} where μ can possibly be less than $1/2$ and therefore the condition that v is small becomes essential. Under these assumptions following Friedrichs we produce in our setting wave operators, realizing similarity of H and H_0, and write out corresponding scattering operator.

The question concerning similarity of H and H_0 can be reduced (see [3]) to the solvability (with respect to τ) of the following equation

$$\tau = v - v\gamma\tau, \tag{4}$$

where composite kernel $v\gamma\tau$ is defined by the formula

$$(v\gamma\tau)(\omega,\omega') = v.p. \int_{-\infty}^{\infty} \frac{v(\omega,\tilde{\omega})\tau(\tilde{\omega},\omega')}{\tilde{\omega}-\omega'} d\tilde{\omega} + i\pi\tau(\omega',\omega')v(\omega,\omega').$$

Provided that v is sufficiently small in r^{μ} the equation (4) is solvable in the class r^{μ} and moreover its solution can be represented by the series

$$\tau = v - v\gamma v + v\gamma(v\gamma v) - \dots \tag{5}$$

converging exponentially in r^{μ}.

Further the obtained kernel $\tau(\omega,\omega')$ is used to construct operator $U = E - \Gamma R$, where singular integral operator ΓR is given by the formula

$$(\Gamma R\phi)(\omega) = v.p. \int_{-\infty}^{\infty} \frac{\tau(\omega,\omega')}{\omega-\omega'} \phi(\omega')d\omega' + i\pi\tau(\omega,\omega)\phi(\omega).$$

Defined initially on Hölder continuous functions the operator U due to the following estimate for L_2 - operator norm of ΓR

$$\|\Gamma R\phi\| \leq const\|\tau\|_\mu\|\phi\|$$

can be extended by continuity to L_2 - space. Thus constructed operator U has L_2 - bounded inverse and satisfies intertwining relation

$$HU = UH_0$$

(according to Friedrichs' terminology, U represents a transformation operator for the pair $\{H_0, H_0 + V\}$). Similarity of H to the operator of multiplication by the "independent" variable $\omega = u(z)$ leads us to explicit construction of continuous spectrum eigenfunctions of the problem (2)-(3).

Proposition 1. Function

$$\psi_c(z) = \frac{1}{u''(z)}\tau(u(z), c), \tag{6}$$

where $\tau(\omega, \omega')$ satisfies equation (4), represents a generalized eigenfunction for the problem (2)-(3) (corresponding to continuous spectrum point c).

Notice that in the case under consideration spectrum of our problem is purely continuous (in spite of the fact that values of u'' may change sign). In this situation the question naturally arises concerning continuous spectrum eigenfunction expansion associated with the problem (2)-(3). We construct such expansions using the approach from the viewpoint of Friedrichs' model. Application of the scattering theory technique to the problem considered assumes the fulfilment of the condition that perturbation operator kernel v is small with respect to the norm in r^μ. The latter is provided by the requirement that profile u is close to linear function in an appropriate norm. As regards the corresponding expansion formula it can be obtained from standard δ-function expansion (i.e. expansion in terms of continuous spectrum eigenfunctions of H_0) by the action of the wave operator U associated with the pair $\{H_0, H\}$, followed by the action of the Green integral operator with kernel G.

Theorem 1. Suppose that for some $\nu > 0$ the function $u(z)$ is sufficiently close in $C^{2+\nu}[a,b]$, to a linear function. Then there exists an operator W, bounded and boundedly invertible in $L_2(a,b)$ such that each function $f \in C^{2+\epsilon}[a,b]$, $\epsilon > 0$, satisfying the boundary conditions $f(a) = f(b) = 0$ has the integral representation

$$f = \int_a^b \left[W(f'' - \alpha^2 f)\right](z)\psi_{u(z)}du(z) \tag{7}$$

in terms of the generalized eigenfunctions associated with the problem (2)-(3).

We add here that the operator W involved in the statement of the theorem is similar to (in fact, coincides with) U^{-1}. Coefficients of expansion in formula (7) can be interpreted as scalar products of f with elements of the system which is biorthogonal to generalized eigenfunctions system.

Expansion formula (7) enables us to apply Fourier method in order to solve corresponding initial-boundary value problem for the non-stationary Rayleigh equation:

$$\left(\frac{\partial}{\partial t} + i\alpha u\right)\left(\frac{\partial^2 \psi}{\partial z^2} - \alpha^2 \psi\right) - i\alpha u'' \psi = 0$$

$$\psi(a,t) = \psi(b,t) = 0, \ \psi(z,0) = \psi_0(z)$$

and study linear stability of the ideal fluid flow with velocity profile u. For analytic profiles the stability of such flows was studied earlier by Dikii [8] and Case [9]. It should be mentioned that boundary value problems on an interval with continuous spectrum also arise in magnetohydrodynamics [10] and the theory of shell oscillations [11]. Sometimes expansions in terms of generalized eigenfunctions can be obtained by contour integration of the resolvent. For the above mentioned problems (in the selfadjoint situation) this approach has been carried out in [10,12,13].

In conclusion of this section we underline that considerations above deal with the case of close-to-linear profiles. The general case in which the profile u need not be close to linear one will be considered in our paper to be published.

4. Scattering problem for Friedrichs' model related to the Rayleigh equation

First of all we note in accordance with considerations above the kernel of the scattering operator for the pair $\{H_0, H_0 + V\}$ has the form (see [3]):

$$[1 - 2\pi i \tau(\omega,\omega')]\,\delta(\omega - \omega')$$

Explicit formula for continuous spectrum eigenfunctions ψ_c obtained in the previous section (see Proposition 1) permits us to prove that scattering is trivial (i.e. $\tau(\omega,\omega') \equiv 0$) if and only if the function u is linear.

In this situation the question concerning solvability of inverse scattering problem for close-to-linear profiles naturally arises. We suggest here the setting of the problem just for Friedrichs' model of the type considered. Recall, first of all, that the kernel of perturbation operator V has the form

$$v(\omega,\omega') = h(\omega)Q(\omega,\omega'),$$

where Q is the Green kernel of selfadjoint differential operator of second order, function h characterizes profile u (or, to be more exact, its closeness to a linear profile). Notice that in our assumptions h belongs to the class h^ν of functions satisfying Hölder condition with exponent $\nu > 0$; in other words

$$\|h\|_\nu = \sup_{\omega_1 \neq \omega_2} h^\nu(\omega_1,\omega_2)|h(\omega_1) - h(\omega_2)| < \infty,$$

when $u \in C^{2+\nu}[a,b]$ and $u'(z) > 0,\ z \in [a,b]$, as before.

Version of the scattering problem that we consider here is to describe the relationship between function h specifying perturbation and the kernel of corresponding scattering operator (or what is equivalent, restriction of the kernel $\tau(\omega,\omega')$ satisfying equation (4) to diagonal $\omega = \omega'$):

$$h(\omega) \leftrightarrow \tau(\omega,\omega).$$

Solution of the direct ($\rightarrow$) problem under assumption that h is small with respect to the Hölder norm $\|\cdot\|_\nu$ (i.e. for profiles u close to linear) is given by the formula (5).

As regards inverse ($\leftarrow$) problem of reconstruction of function h by given scattering data, it consists in 1) proving the existence and uniqueness theorem and 2) obtaining suitable reconstruction procedure. (We recall that in connection with the formula (6) the "uniqueness part" of this problem has been already touched upon in the particular case of trivial scattering.)

5. Solvability class for inverse scattering problem. Local existence and uniqueness theorem

In this section we consider nonlinear mapping

$$S:\ h(\omega) \rightarrow \tau(\omega,\omega'),$$

where $h \in h^\nu,\ 0 < \nu < 1/2$, from the viewpoint of application of implicit function theorem to solvability of inverse scattering problem in the case when $\|h\|_\nu$ is small enough and, hence, corresponding kernel $\tau(\omega,\omega')$ belongs to r^ν.

Remark. Restriction of the kernel $\tau \in r^\nu$ to the diagonal $\omega = \omega'$ belongs to Hölder class with the same exponent ν.

This simple observation enables us to consider S-matrix as the mapping of the space $h^\nu,\ 0 < \nu < 1/2$, into itself and compute its differential at zero. To this end we shall use explicit form for the solution of Friedrichs' equation:

$$(S(h))\,(\omega) = \upsilon(\omega,\omega) - (\upsilon\gamma\upsilon)(\omega,\omega) + \dots. \tag{8}$$

Next we note that due to the certain estimate (see [3])

$$\|\upsilon\gamma w\|_\nu \leq const\|\upsilon\|_\nu\|w\|_\nu \tag{9}$$

it turns out that the sum of all the terms in (8) but the first one is $o(\|h\|_\nu)$; therefore we have:

$$(S(h))\,(\omega) = h(\omega)Q(\omega,\omega) + o(\|h\|_\nu).$$

Lemma 2. Differential of the mapping $S:\ h^\nu \rightarrow h^\nu$ at zero has the form

$$dS|_0:\ h(\omega) \rightarrow h(\omega)Q(\omega,\omega) = \upsilon(\omega,\omega).$$

This formula shows that the image space of mapping S coincides with the following class

$$X = \{\phi \in h^\nu : \frac{\phi(\omega)}{Q(\omega,\omega)} \in h^\nu\} \subset h^\nu,$$

which is a Banach space with respect to the induced norm

$$\|\phi\|_X = \|\frac{\phi(\omega)}{Q(\omega,\omega)}\|_\nu.$$

The key fact that makes possible the application of inverse function theorem to the mapping S is the property specific for solutions of equation (4). This property is described by

Proposition 2. If $h \in h^\nu$ with $\nu < 1/4$ then the quotient $\tau(\omega,\omega)[Q(\omega,\omega)]^{-1}$ belongs to h^ν, i.e. $\tau(\omega,\omega) \in X$.

We shall outline the idea of the proof of latter statement since it reveals the role of our restriction for Hölder exponent: $\nu < 1/4$. First, introduce auxiliary kernel

$$\rho(\omega,\omega') = \frac{\tau(\omega,\omega')}{\sqrt{Q(\omega,\omega)}\sqrt{Q(\omega',\omega')}}$$

which can be represented according to (5) as the series with the terms of the form (up to sign $+$ or $-$):

$$p\gamma(v\gamma(v\gamma(\ldots(v\gamma q)\ldots))),$$

where

$$p(\omega,\omega') = h(\omega)\frac{Q(\omega,\omega')}{\sqrt{Q(\omega,\omega)}}$$

and

$$q(\omega,\omega') = h(\omega)\frac{Q(\omega,\omega')}{\sqrt{Q(\omega',\omega')}}$$

Since $h \in h^\nu$ it can be straightforwardly verified that kernels p and q belong to the space r^μ, $\mu = \min\{\nu, 1/4\}$. Therefore if Hölder exponent ν is less than $1/4$ then p, q and $v \in r^\nu$ and according to well-known properties of operation γ the series representing $\rho(\omega,\omega')$ converges in r^ν exponentially if $\|v\|_\nu$ is sufficiently small. So we establish that kernel ρ belongs to r^ν and hence its restriction to the diagonal $\tau Q^{-1}|_{\omega=\omega'} \in h^\nu$, $\nu < 1/4$.

Returning to the statement of Lemma 2 we note that differential $dS|_0$ is the operator of multiplication by $Q(\omega,\omega)$ acting from h^ν into X; besides it is obviously continuous, $\mathrm{Ker}(dS|_0) = 0$ (since $dS|_0$ is isometric) and $\mathrm{Im}(dS|_0) = X$. Summing up we get

Lemma 3. Mapping $S: h^\nu \to X$ is differentiable at zero, the corresponding differential being continuous and boundedly invertible as an operator from h^ν into X.

The last step on the way to (local) invertibility of S is the verification of its smoothness near zero. To prove this we compute corresponding differential $dS|_h$: consider increment $S(h,\delta) - S(h)$, where h and δ are sufficiently close to zero in h^ν and by straightforward calculations separate the linear in δ part of this increment. As a result of appropriate estimations of the remainder terms we establish differentiability of S near zero an simultaneously get the following explicit form of the differential

$$\begin{aligned} dS|_h\delta = \delta Q - [(hQ)\gamma(\delta Q) + (\delta Q)\gamma(hQ)] + \\ + [(hQ)\gamma((hQ)\gamma(\delta Q)) + (hQ)\gamma((\delta Q)\gamma(hQ)) + \\ + (\delta Q)\gamma((hQ)\gamma(hQ))] + \dots \end{aligned} \tag{10}$$

where after the resriction on the diagonal $\omega = \omega'$ dependence on ω is meant and for shortening the following notations are used:

$$hQ = h(\omega)Q(\omega,\omega'),\ \delta Q = \delta(\omega)Q(\omega,\omega').$$

Remark. In view of the estimate (9) convergence of the series (10) is provided by the condition that we confine h to a small neighbourhood of zero in h^ν. Assuming this and using the obtained representation for $dS|_h$ it is easy to verify that operator $dS|_h : h^\nu \to X$ is bounded.

Consider now the family of operators $dS|_h : h^\nu \to X$ parametrized by $h \in h^\nu$. Using the abovementioned technique of estimations one can study the h-dependence of the family $dS|_h$ in appropriate operator norm.

Lemma 4. Let h_1 and h_2 belong to sufficiently small neighbourhood of zero in h^ν. Then there exists a constant $D > 0$ such that for any $\delta \in h^\nu$ the following estimate holds:

$$\|dS|_{h_1}\delta - dS|_{h_2}\delta\|_X \le D\|h_1 - h_2\|_\nu \|\delta\|_\nu.$$

In other words the operator family $dS|_h : h^\nu \to X$ satisfies the Lipschitz condition with respect to h in the vicinity of zero and hence mapping S is continuously differentiable there.

So all the conditions of inverse function theorem are checked for the mapping $S : h^\nu \to X$ and therefore we can assert that S is locally invertible near zero and, besides, inverse mapping is continuous.

Theorem 2. For any $\nu \in (0, 1/4)$ one can find $\epsilon_1 > 0$ and $\epsilon_2 > 0$ such that for any $\tau \in X$, $\|\tau\|_X < \epsilon_1$, there exists a unique function $h \in h^\nu$, $\|h\|_\nu < \epsilon_2$, which is a solution of the equation $S(h) = \tau$; thus arising local inverse $S^{-1} : X \to h^\nu$ is continuous.

Summarising facts stated above we come to the conclusion that for the model under consideration it turns out possible by given scattering data to reconstruct the kernel of perturbation operator. As regards a concrete computational procedure of this reconstruction we indicate successive approximation method in the form used to prove the inverse function theorem. In conclusion we write out corresponding formula which is in fact a variant of perturbation theory series:

$$h(\omega) = \frac{1}{Q(\omega,\omega)} \sum_{n=1}^{\infty} v_n(\omega,\omega);$$

here terms v_n are defined recursively

$$v_n(\omega,\omega') = \sum_{k_1+\ldots+k_t=n} (-1)^t v_{k_1}\gamma(v_{k_2}\gamma(\ldots(v_{k_{t-1}}\gamma v_{k_t})\ldots)), \quad n>1,$$

(summation is taken over all possible partitions of $n \in N$ into sums $k_1+\ldots+k_t$ of positive integers) and

$$v_1(\omega,\omega') = \frac{\tau(\omega,\omega)}{Q(\omega,\omega)}Q(\omega,\omega').$$

Complicated form of this inversion formula deserves special commentary. Contrary to the case of Sturm–Liouville problem in our situation transformation integral operator for the pair $\{H_0, H_0+V\}$ can not be chosen to be of Volterra type. Therefore Gelfand - Levitan - Marchenko approach to solution of inverse scattering problem does not work in our setting.

References

[1] C.C. Lin, Theory of Hydrodynamic Stability, Cambridge Univ. Press Cambridge (1953).

[2] L.N. Howard, J. fl. mech., 10, No.4, 509-512 (1961).

[3] K.O. Friedrichs, Perturbation of Spectra in Hilbert Space, Lectures in Applied Math., Vol.3, Amer. Math. Soc., Providence, Rhode Island (1965).

[4] A.Ya. Povzner, Dokl. Akad. Nauk SSSR, 104, No.3, 360-363 (1955).

[5] O.A. Ladyzhenskaya and L.D. Faddeev, Dokl. Akad. Nauk SSSR, 120, No.6, 1187-1190 (1958).

[6] L.D. Faddeev, Tr. Math. Inst. Akad. Nauk SSSR, 73, 292-313 (1964).

[7] P.A. Rejto, Comm. Pure Appl. Math., 16, 279-303 (1963); 17, 257-292 (1964).

[8] L.A. Dikii, Dokl. Akad. Nauk SSSR, 135, No.5, 1068-1071 (1960).

[9] K.M. Case, Phys. Fluids, 3, No.2, 143-148 (1960).

[10] A.L. Krylov and E.N. Fedorov, Dokl. Akad. Nauk SSSR, 231, No.1, 68-70 (1976).

[11] A.G. Aslanyan and V.B. Lidskii, Distribution of Fundamental Frequences of Thin Elastic Shells (Russian), Nauka, Moscow (1974).

[12] G.G. Tarposhyan, Funct. Anal. Prilozhen., 11, No.1, 83-84 (1977).

[13] A.E. Lifschitz, Funct. Anal. Prilozhen., 17, No.1, 77-78 (1983).

Department of Mathematics
Moscow State University
119899 Moscow, Russia

AMS Subject Classification: 34 B 30, 47 A 56

Operator Theory:
Advances and Applications, Vol. 80

YET ANOTHER FACE OF THE CREATION OPERATOR

F.H.SZAFRANIEC

Włodzimierz Mlak in memoriam

If one defines *the creation operator* as an abstract weighted shift of weights $\{\sqrt{n+1}\}_{n=0}^{\infty}$, then the operator

$$\frac{1}{\sqrt{2}}(x - \frac{d}{dx})$$

appears as a face (read: unitary image) of it while the operator of multiplication by the independent variable in the Segal-Bargmann space does as another. In [11] a finite difference operator

$$\sqrt{x}f(x-1) - \sqrt{a}f(x), \quad a > 0$$

was recorded. It may be considered as yet another face of the creation one. Our intension here is to invite attention to this operator by considering the case in detail.

1. The classical paper of Bargmann [1] is devoted to the operator

$$\frac{1}{\sqrt{2}}(x - \frac{d}{dx}) \tag{1}$$

acting in $\mathcal{L}^2(\mathbb{R})$; its domain can be taken as the linear span of Hermite functions. Besides its meaning in Quantum Mechanics this operator serves as an important example of an *unbounded subnormal operator* (look at [8] and [9] for an account of this matter). It was shown there that $\mathcal{L}^2(\mathbb{R})$ is unitarily equivalent to the (closed) subspace[1], say

The author's research was supported by a grand of the *Komitet Badań Naukowych, Warsaw*

[1] This space bears names of Bargmann, Segal and Fisher or even Fock, in various combinations, and, despite its importance, has no canonical notation so far.

$\mathcal{A}^2(\mathbb{C}, \pi^{-1}\exp(-|z|^2)dxdy)$, of all analytic functions in $\mathcal{L}^2(\mathbb{C}, \pi^{-1}\exp(-|z|^2)dxdy)$ (here $z = x + iy$): the monomials

$$\frac{z^n}{\sqrt{n!}}, \quad n = 0, 1, \ldots$$

form an orthonormal basis in $\mathcal{A}^2(\mathbb{C}, \pi^{-1}\exp(-|z|^2)dxdy)$ and they are precisely images of the Hermite functions under the unitary operator established therein. Moreover, because the operator (1) shifts, with the weights $\{\sqrt{n+1}\}_{n=0}^{\infty}$, the Hermite functions, its unitary image is just the operator of multiplication by the independent variable. The unitary operator from $\mathcal{L}^2(\mathbb{R})$ to $\mathcal{A}^2(\mathbb{C}, \pi^{-1}\exp(-|z|^2)dxdy)$ is an integral operator whose kernel is the generating function for the Hermite functions. In this note we show how ℓ^2 can be made unitarily isomorphic to $\mathcal{A}^2(\mathbb{C}, \pi^{-1}\exp(-|z|^2)dxdy)$ in such a way that the finite difference operator in question becomes the operator of multiplication by the independent variable in the latter space.

2. Recall (cf. [2] and also [5]) that the Charlier (or Poisson-Charlier) polynomials $\{C_n^{(a)}\}_{n=0}^{\infty}$, $a > 0$ are orthogonal with respect to a nonnegative integer supported measure as follows

$$\sum_{x=0}^{\infty} C_m^{(a)}(x) C_n^{(a)}(x) \frac{e^{-a} a^x}{x!} = \delta_{mn} a^n n!, \quad m, n = 0, 1, \ldots. \tag{2}$$

They are related to their generating function by

$$e^{-az}(1+z)^x = \sum_{n=0}^{\infty} C_n^{(a)}(x) \frac{z^n}{n!}. \tag{3}$$

Because $|C_n^{(a)}(x)| \leq (x+a)^n$ for $x = 0, 1, \ldots$, cf. [5], the right hand side of the above converges absolutely and uniformly in z on compat sets.

The three term recurrence relation

$$C_{n+1}^{(a)}(x) = (x - n - a) C_n^{(a)}(x) - an C_{n-1}^{(a)}(x)$$

and the difference equation

$$\Delta C_{n+1}^{(a)}(x) = (n+1) C_n^{(a)}(x)$$

lead directly to[2]

$$C_{n+1}^{(a)}(x+1) = (x+1) C_n^{(a)}(x) - a C_n^{(a)}(x+1). \tag{4}$$

[2] We ought to point out that a relation *like* this appears in [3, Th.2].

More precisely, one has to put the difference relation written for n and $n-1$ in the three term recurrence relation so as to get (4).

Define the Charlier functions $\tilde{c}_n^{(a)}$, $n = 0, 1, \dots$ as

$$\tilde{c}_n^{(a)}(x) = a^{-\frac{n}{2}}(n!)^{-\frac{1}{2}} C_n^{(a)}(x) \mathrm{e}^{-\frac{a}{2}} a^{\frac{x}{2}} \begin{cases} (x!)^{-\frac{1}{2}}, & \text{for } x \geq 0 \\ 1 & \text{for } x < 0. \end{cases}$$

It is a matter of direct calculation to check that for the Charlier functions we get from (4)

$$\begin{aligned} \sqrt{n+1}\,\tilde{c}_{n+1}^{(a)}(x+1) &= \sqrt{x+1}\,\tilde{c}_n^{(a)}(x) - \sqrt{a}\,\tilde{c}_n^{(a)}(x+1), \quad \text{for } x \geq 0, \\ \sqrt{n+1}\,\tilde{c}_{n+1}^{(a)}(0) &= -\sqrt{a}\,\tilde{c}_n^{(a)}(0). \end{aligned} \tag{5}$$

Now consider the Charlier functions as functions in discrete variable $x = 0, 1, \dots$, that is set

$$c_n^{(a)} = \tilde{c}_n^{(a)}|_{\mathbb{N}}, \quad n = 0, 1, \dots .$$

Due to (2), the sequence $\{c_n^{(a)}\}_{n=0}^{\infty}$ is orthonormal in ℓ^2 and, because it is complete (cf. [7] and also [4]), it forms a basis[3] in ℓ^2. Set $\mathcal{D}_a = \operatorname{lin}\{c_n^{(a)}; n = 0, 1, \dots\}$. Define an operator S_a as

$$\mathcal{D}(S_a) = \mathcal{D}_a, \quad S_a f = g$$

where, for $f \in \mathcal{D}_a$,

$$g(x) = \sqrt{x} f(x-1) - \sqrt{a} f(x), \quad x = 1, 2, \dots \quad \text{and} \quad g(0) = -\sqrt{a} f(0).$$

Now formulae (5) enable us to state

Theorem 1. $S_a c_n^{(a)} = \sqrt{n+1}\, c_{n+1}^{(a)}$ *for* $n = 0, 1, \dots$.

3. The companion of the creation operator S_a, the *anihilation* one S_a^+, that is the operator which satisfies

$$< S_a f, g > = < f, S_a^+ g >, \quad f, g \in \mathcal{D}(S_a),$$

(or, in other words, which is the restriction of the adjoint of S_a to $\mathcal{D}(S_a)$, cf. the subsequent section), is given explicitely as

$$(S_a^+ f)(x) = \sqrt{x+1} f(x+1) - \sqrt{a} f(x), \quad x = 0, 1, \dots .$$

Notice that, because S_a is a (weighted) shift and S_a^+ is a (weighted) backward shift with respect to the same basis $\{c_n^{(a)}\}_{n=0}^{\infty}$, the linear subspace $\mathcal{D}_a$ is invariant for S_a as well as for S_a^+. Thus, the others are:

$$\begin{aligned} (N_a f)(x) &= (S_a S_a^+ f)(x) \\ &= \begin{cases} (x+a) f(x) - \sqrt{ax} f(x-1) - \sqrt{a(x+1)} f(x+1) & x = 1, 2, \dots \\ a f(0) - \sqrt{a} f(1) & x = 0 \end{cases}, \\ & \qquad f \in \mathcal{D}_a, \end{aligned}$$

[3] This is a rather rare situation when in ℓ^2 a basis different from the usual zero-one one is considered.

the *number operator* and

$$(H_a f)(x) = ((S_a S_a^+ + \frac{1}{2}I)f)(x)$$
$$= \begin{cases} (x+a+\frac{1}{2})f(x) - \sqrt{ax}f(x-1) - \sqrt{a(x+1)}f(x+1) & x = 1,2,\dots \\ (a+\frac{1}{2})f(0) - \sqrt{a}f(1) & x = 0 \end{cases},$$
$$f \in \mathcal{D}_a,$$

the *Hamiltonian*. They both are difference operators of second order.

This is the way in which the *quantum harmonic oscilator*[4] acts on the Charlier polynomials.

4. Notice first that because, for $a > 0$ and $b > 0$,

$$(\sum_{x=1}^{\infty} |\sqrt{x}f(x-1) - \sqrt{b}f(x)|^2)^{\frac{1}{2}}$$
$$\leq (\sum_{x=1}^{\infty} |\sqrt{x}f(x-1) - \sqrt{a}f(x)|^2)^{\frac{1}{2}} + |\sqrt{a} - \sqrt{b}|(\sum_{x=1}^{\infty} |f(x)|^2)^{\frac{1}{2}},$$

the linear space

$$\mathcal{D} = \{f \in \ell^2;\ \{\sqrt{a}f(0), \sqrt{x}f(x-1) - \sqrt{a}f(x),\ x = 1,2,\dots\} \in \ell^2\}$$

is independent of a. Define the operator $S_{a,\max}$ by

$$\mathcal{D}(S_{a,\max}) = \mathcal{D}, \quad S_{a,\max}f = g$$

where, for $f \in \mathcal{D}$,

$$g(x) = \sqrt{x}f(x-1) - \sqrt{a}f(x), \quad x = 1,2,\dots \quad g(0) = -\sqrt{a}f(0).$$

The operator $S_{a,\max}$ is closed (the argument: $f_\nu \to f$ and $S_{a,\max}f_\nu \to g$ imply $f_\nu(x) \to f(x)$ and $S_{a,\max}f_\nu(x) \to g(x)$ for every $x = 0,1,\dots$).

For the same reason as before the linear space

$$\mathcal{D}^+ = \{f \in \ell^2;\ \{\sqrt{x+1}f(x+1) - \sqrt{a}f(x)\}_{x=0}^{\infty} \in \ell^2\}$$

does not depend on a either. Define the operator $S_a^{+,\max}$ as

$$\mathcal{D}(S_a^{+,\max}) = \mathcal{D}^+$$
$$S_a^{+,\max}f = \{\sqrt{x+1}f(x+1) - \sqrt{a}f(x)\}_{x=0}^{\infty}.$$

[4] We would like to recomend here, by the way, a beautiful overview [6] of the story of the quantum harmonic oscilator written by mathematicians.

Theorem 2. *$S_a^- = S_{a,\max}$ and $S_a^* = S_a^{+,\max}$.*

Here the dash $^-$ stands for closure of an operator.

Proof. Suppose $g \in \mathcal{D}(S_a^*)$. This means

$$|< S_a f, g >| \le C_g \|f\|, \quad f \in \mathcal{D}_a.$$

Because

$$(6) \quad \sum_{x=1}^{\infty}(\sqrt{x}f(x+1) - \sqrt{a}f(x))\overline{g(x)} - \sqrt{a}f(0)\overline{g(0)}$$

$$= \sum_{x=0}^{\infty} f(x)(\overline{\sqrt{x+1}\,g(x+1)} - \overline{\sqrt{a}\,g(x)})$$

and $\mathcal{D}_a$ is dense in ℓ^2, $\{\sqrt{x+1}f(x+1) - \sqrt{a}f(x)\}_{x=0}^{\infty} \in \ell^2$ and, consequently, $g \in \mathcal{D}(S_a^{+,\max})$. Using again (6) for $g \in \mathcal{D}(S_a^{+,\max})$ we get

$$|< S_{a,\max} f, g >| = |< f, S_a^{+,\max} g >| \le \|S_a^{+,\max} g\| \|f\|, \quad f \in \mathcal{D}(S_{a,\max}).$$

Thus $g \in \mathcal{D}(S_{a,\max}^*)$. Finally we get

$$S_a^* \subset S_a^{+,\max} \subset S_{a,\max}^*.$$

Since $S_{a,\max}^* \subset S_a^*$ automaticaly, we have

$$S_a^* = S_a^{+,\max} = S_{a,\max}^*,$$

which proves the second half of the conclusion. Now, because $S_{a,\max}$ is closed,

$$S_a^- = S_a^{**} = S_{a,\max}^{**} = S_{a,\max}$$

and we get the other. □

The canonical zero-one basis $\{e_n\}_{n=0}^{\infty}$ defined as

$$e_n = \{\delta_{n,x}\}_{x=0}^{\infty}, \quad n = 0, 1, \ldots$$

is apparently *not* in $\mathcal{D}_a$ however it is in $\mathcal{D}$. Then, with a little use of Theorem 2,

$$(S_a^- e_n)(x) = \begin{cases} -\sqrt{a} & x = n \\ \sqrt{n+1} & x = n+1 \\ 0 & \text{otherwise.} \end{cases}$$

So

$$S_a^- e_n = \sqrt{n+1}e_{n+1} - \sqrt{a}e_n$$

and, after introducing the unitary operator $V_a : \ell^2 \mapsto \ell^2$ defined as

$$V_a c_n^{(a)} = e_n, \quad n = 0, 1, \dots,$$

this means precisely that

$$V_a^{-1} S_a^- V_a c_n^{(a)} = \sqrt{n+1}c_{n+1}^{(a)} - \sqrt{a}c_n^{(a)}$$

or, in other words,

$$V_a^{-1} S_a^- V_a|_{\mathcal{D}_a} = S_a - \sqrt{a}I|_{\mathcal{D}_a}.$$

After taking closure in this and setting $\mathcal{D}_0 = \text{lin}\{e_n;\, n = 0, 1, \dots\}$ we come to

Corollary. *$\mathcal{D}_0$ is a core*[5] *for S_a^- and*

$$V_a^{-1} S_a^- V_a = S_a^- - \sqrt{a}I|_{\mathcal{D}}.$$

It is a right time to make a definition of an abstract creation operator more precise. First we recall that a densely defined operator T in $\mathcal{H}$ is a *weighted shift* with respect to a basis $\{e_n\}_{n=0}^{\infty}$ with weights $\{\lambda_n\}_{n=0}^{\infty}$ if $\mathcal{D}(T) = \text{lin}\{e_n;\, n = 0, 1, \dots\}$ and $Te_n = \lambda_n e_{n+1}$. Then we say that a densely defined operator S is the *creation operator* if there is a basis $\{e_n\}_{n=0}^{\infty}$ in $\mathcal{H}$ such that $S|_{\text{lin}\{e_n;\, n=0,1,\dots\}}$ is a weighted shift with respect to this basis and with the weights $\{\sqrt{n+1}\}_{n=0}^{\infty}$.

Thus the statement of Corollary can be rephrased as: *the creation operator is unitarily equivalent to itself "plus" a multiply of the identity operator.* Something like this would never happen for bounded operators (the spectral mapping theorem!).

5. Set

$$A^{(a)}(x, z) = \mathrm{e}^{-\frac{a}{2} - \sqrt{a}z}(\sqrt{a} + z)^x (x!)^{-\frac{1}{2}}, \quad x = 0, 1, \dots, \quad z \in \mathbb{C}.$$

It is a matter of direct calculation to check that

$$\sum_{x=0}^{\infty} A^{(a)}(x, z)\overline{A^{(a)}(x, w)} = \mathrm{e}^{z\overline{w}}. \tag{7}$$

Now take $f \in \ell^2$, $z_1, \dots, z_k \in \mathbb{C}$, $\xi_1, \dots, \xi_k \in \mathbb{C}$ and use the ℓ^2-Schwarz inequality together with (7) to write

$$\begin{aligned}|\sum_{i=1}^{k}(\sum_{x=0}^{\infty} A^{(a)}(x, z_i)f(x))\xi_i|^2 &\le \sum_{x=1}^{\infty} |f(x)|^2 \sum_{x=0}^{\infty} |\sum_{i=1}^{k} A^{(a)}(x, z_i)\xi_i|^2 \\ &= \|f\|^2 \sum_{x=0}^{\infty} \sum_{i,j=1}^{k} A^{(a)}(x, z_i)\overline{A^{(a)}(x, z_j)} = \|f\|^2 \sum_{i,j=0}^{k} \mathrm{e}^{z_i\overline{z_j}}\xi_i\overline{\xi_j}.\end{aligned}$$

[5] $\mathcal{D}$ is a core for an operator A if $(A|_{\mathcal{D}})^- = A^-$.

This implies, by the RKHS test (cf. Sec. 7(b) below), that the function

$$F(z)=\sum_{x=0}^{\infty}A^{(a)}(x,z)f(x),\quad z\in\mathbb{C}$$

belongs to $\mathcal{A}^2(\mathbb{C},\pi^{-1}\exp(-|z|^2)dxdy)$ and $\|F\|\le\|f\|$. Consequently, the operator U_a given by

$$(U_af)(z)=F(z)=\sum_{x=0}^{\infty}A^{(a)}(x,z)f(x),\quad f\in\ell^2$$

maps ℓ^2 into $\mathcal{A}^2(\mathbb{C},\pi^{-1}\exp(-|z|^2)dxdy)$. We want to show that

$$U_a:c_n^{(a)}\mapsto\frac{z^n}{\sqrt{n!}},\quad n=0,1,\dots.$$

Because ℓ^2 is a reproducing kernel Hilbert space it follows from Sec. 7(a) (cf. also [4]) that

$$\sum_{n=0}^{\infty}c_n^{(a)}(x)c_n^{(a)}(y)=\delta_{x,y}. \tag{8}$$

We show that the series

$$\sum_{n=0}^{\infty}c_n^{(a)}(x)\frac{z^n}{\sqrt{n!}}$$

converges absolutely and uniformly in z on compact sets of $\mathbb{C}$ for $x=0,1,\dots$. Indeed, the ℓ^2-Schwarz inequality, due to (8), implies

$$|\sum_{n=0}^{\infty}|c_n^{(a)}(x)\frac{z^n}{\sqrt{n!}}||^2\le\sum_{n=0}^{\infty}|c_n^{(a)}(x)|^2\sum_{n=0}^{\infty}|\frac{z^n}{\sqrt{n!}}|^2=e^{|z|^2}. \tag{9}$$

Now, under the notations we have introduced in the meantime, the relation (3) for the sequence $\{c_n^{(a)}\}_{n=0}^{\infty}$ takes the form

$$A^{(a)}(x,z)=\sum_{n=0}^{\infty}c_n^{(a)}(x)\frac{z^n}{\sqrt{n!}},\quad x=0,1,\dots\ z\in\mathbb{C}.$$

However, from (9) and (8) we get (the Schwarz inequality again)

$$\left|\sum_{x=0}^{\infty}|A^{(a)}(x,z)c_n^{(a)}(x)|\right|^2=\left|\sum_{x=0}^{\infty}|c_n^{(a)}(x)||\sum_{m=0}^{\infty}c_m^{(a)}(x)\frac{z^m}{\sqrt{m!}}|\right|^2$$

$$\le\left|\sum_{x=0}^{\infty}|c_n^{(a)}(x)|(\sum_{m=0}^{\infty}|c_m^{(a)}(x)|^2)^{\frac12}(\sum_{m=0}^{\infty}\frac{|z|^{2m}}{m!})^{\frac12}\right|^2\le\sum_{x=0}^{\infty}|c_n^{(a)}(x)e^{\frac{|z|^2}{2}}|^2=e^{|z|^2}.$$

Due to orthogonality of $\{c_n^{(a)}\}_{n=0}^{\infty}$, changing sumations, we have

$$\sum_{x=0}^{\infty} A^{(a)}(x,z)c_n^{(a)}(x) = \sum_{x=0}^{\infty} c_n^{(a)}(x) \sum_{m=0}^{\infty} c_m^{(a)}(x)\frac{z^m}{\sqrt{m!}}$$
$$= \sum_{m=0}^{\infty} \frac{z^m}{\sqrt{m!}} \sum_{x=0}^{\infty} c_n^{(a)}(x)c_m^{(a)}(x) = \frac{z^n}{\sqrt{n!}}, \quad x = 0,1,\ldots\ z \in \mathbb{C}.$$

Because

$$\frac{z^n}{\sqrt{n!}}, \quad n = 0,1,\ldots$$

is a basis in $\mathcal{A}^2(\mathbb{C}, \pi^{-1}\exp(-|z|^2)dxdy)$, we arrive at

Theorem 3. *The operator*

$$U_a : \ell^2 \mapsto \mathcal{A}^2(\mathbb{C}, \pi^{-1}\exp(-|z|^2)dxdy)$$

is unitary.

Because $A^{(a)}(x,\cdot)$ is in $\mathcal{A}^2(\mathbb{C}, \pi^{-1}\exp(-|z|^2)dxdy)$ provided F is therein, we can write

$$(W_aF)(x) = \pi^{-1}\int_{\mathbb{C}} \overline{A^{(a)}(x,z)}F(z)\exp(-|z|^2)dxdy. \tag{10}$$

Since $A^{(a)}(x,\cdot) = U_a e_x$ and U_a is unitary, for $f \in \ell^2$ we have

$$(W_aU_af)(x) = \int_{\mathbb{C}} \overline{A^{(a)}(x,z)}(U_af)(z)\mu(dz) = \int_{\mathbb{C}} \overline{(U_ae_x)(z)}(U_af)(z)\mu(dz)$$
$$=< f, e_x >= f(x), \quad x = 0,1,\ldots,$$

where $\mu(dz) = \pi^{-1}\exp(-|z|^2)dxdy$. Thus we are ready to state

Theorem 4. *The operator $W_a : F \mapsto W_aF$ defined by formula (10) maps $\mathcal{A}^2(\mathbb{C}, \pi^{-1}\exp(-|z|^2)dxdy)$ onto ℓ^2 and*

$$W_a = U_a^{-1}.$$

6. There is one more occasion we would like to point out when the creation operator appears. In $\mathcal{L}^2(\mathbb{R})$ we have

$$\sqrt{2}x = \frac{1}{\sqrt{2}}(x - \frac{d}{dx}) + \frac{1}{\sqrt{2}}(x + \frac{d}{dx}),$$

which says that the operator of multiplication in $\mathcal{L}^2(\mathbb{R})$ by the independent variable (multiplied by $\sqrt{2}$) is the sum of the creation and the anihilation operators. Under our circumstances, the corresponding relation is a bit more involved. The orthonormal Charlier polynomials $\{\tilde{C}_n^{(a)}\}_{n=0}^{\infty}$ satisfy the three term recurrence relation

$$x\tilde{C}_n^{(a)}(x) = \sqrt{a}\sqrt{n+1}\tilde{C}_{n+1}^{(a)}(x) + (n+a)\tilde{C}_n^{(a)}(x) + \sqrt{a}\sqrt{n}\tilde{C}_{n-1}^{(a)}(x) \tag{10}$$

and the same holds for the Charlier functions $\{c_n^{(a)}\}_{n=0}^{\infty}$. The latter relation, after denoting by M the multiplication operator defined as

$$\mathcal{D}(M) = \mathcal{D}_a, \qquad (Mf)(x) = xf(x), \quad x = 0, 1, \ldots, \quad f \in \mathcal{D}_a,$$

means that

$$M = \sqrt{a}S + N + aI + \sqrt{a}S^{+}, \tag{11}$$

where N is the number operator considered in Sec. 3. Thus, in contrast to the $\mathcal{L}^2(\mathbb{R})$ case, the operator of multiplication in ℓ^2 does not seem to be a right candidate for the position operator.

Another way of looking at the relation (10) is to attach to it the (infinite) Jacobi matrix

$$\begin{pmatrix} a & \sqrt{a} & 0 & 0 & 0 & 0 & \ldots \\ \sqrt{a} & 1+a & \sqrt{2a} & 0 & 0 & 0 & \ldots \\ 0 & \sqrt{2a} & 2+a & \sqrt{3a} & 0 & 0 & \ldots \\ 0 & 0 & \sqrt{3a} & 3+a & \sqrt{4a} & 0 & \ldots \\ \vdots & \vdots & \vdots & \vdots & \vdots & \vdots & \ddots \end{pmatrix}.$$

Denote by J the (unbounded) operator, with $\mathcal{D}(J) = \mathcal{D}_0$, it defines. Then

$$J = V_a M V_a^{-1}$$

and the decomposition (11) can be written down explicitely for this distinguished object of operator theory as well.

7. Let $K : X \times X \mapsto \mathbb{C}$ be a positive definite kernel. Let $\mathcal{H}_K$ denote its reproducing kernel Hilber space (RKHS). In this paper we utilize the following two facts:

(a) if $\{e_n\}_{n=0}^{\infty}$ is a basis in $\mathcal{H}_K$, then

$$K(x, y) = \sum_{n=0}^{\infty} e_n(x)\overline{e_n(y)}, \quad x, y \in X,$$

(b) RKHS test, cf. [10]: a function f belongs to $\mathcal{H}_K$ if and only if for any $x_1, \ldots, x_k$ in X and any $\xi_1, \ldots, \xi_k$ in $\mathbb{C}$ the following inequality holds true

$$|\sum_{i=1}^{k} f(x_i)\xi_i|^2 \le C \sum_{i,j=1}^{k} K(x_i, x_j)\xi_i\overline{\xi_j}$$

with some C. If this happens, $\|f\|^2 \le C$.

It has to be noticed that both ℓ^2 and $\mathcal{A}^2(\mathbb{C}, \pi^{-1}\exp(-|z|^2)dxdy)$ are RKHS's. Their kernels are:

$$\mathbb{N} \times \mathbb{N} \ni (x, y) \mapsto \delta_{x,y} \in \mathbb{C},$$
$$\mathbb{C} \times \mathbb{C} \ni (z, w) \mapsto e^{z\overline{w}} \in \mathbb{C},$$

respectively.

REFERENCES

1. Bargmann, V., *On a Hilbert space of analytic functions and an associated integral transform*, Comm. Pure App. Math. **14** (1961), 187-214.
2. Chichara, T.S., *An introduction to orthogonal polynomials*, Gordon and Breach, New York, N.Y., 1978.
3. de Branges, L., Trutt, D., *Charlier spaces of entire functions*, Proc. Amer. Math. Soc. **20** (1969), 134-140.
4. Eagleson, G.K., *A duality relation for discrete orthogonal systems*, Studia Sc. Math. Hungarica **3** (1968), 127-136.
5. Meixner, J., *Erzuegende Funktionen der Charlierschen Polynome*, Math. Z. **44** (1939), 331-335.
6. Mlak, W., Słociński, M., *Quantum phase and circular operators*, Univ. Iagell. Acta Math. **24** (1992), 133-144.
7. Schmidt, E., *Über die Charlier-Jordansche Entwiklung einer willkürchen Funktion nach der Poissonsche Funktion und ihren Ableitungen*, Zeitschrift angew. Math. Mech. **13** (1933), 139-142.
8. Stochel, J., Szafraniec, F.H., *On normal extensions of unbounded operators. II*, Acta Sci. Math. (Szeged) **53** (1989), 153-177.
9. ———, *On normal extensions of unbounded operators. III. Spectral properties*, Publ. RIMS, Kyoto Univ. **25** (1989), 105-139.
10. Szafraniec, F.H., *Interpolation and domination by positive definite kernels*, Complex Analysis - Fifth Romanian-Finish Seminar, Part 2, Proc., Bucarest (Romania), 1981, eds. C.Andrean Cazacu, N.Boboc, M.Jurchescu and I.Suciu, Lecture Notes in Math., vol. 1014, pp. 291-295, Springer, Berlin-Heidelberg, 1983..
11. ———, *Orthogonal polynomials and subnormality of related shift operators*, Orthogonal polynomials and their applications, Proc., Segovia (Spain), 1986, ed. R.-C. Palacios, Monogr. Acad. Cienc. Zaragoza, vol. 1, pp. 15-155, 1988.

INSTYTUT MATEMATYKI UJ, UL. REYMONTA 4, PL-30059 KRAKÓW

MSC: 47B37

Operator Theory:
Advances and Applications, Vol. 80

ON TRANSFORMATIONS OF CANONICAL SYSTEMS

by H. Winkler

0. Introduction

In this note we consider transformations of a *canonical system* of the form

$$\mathbf{J}y'(x) = -z\mathbf{H}(x)y(x), \;\; x \in [0,\infty), \tag{0.1}$$

where $\mathbf{H}$ is a real symmetric nonnegative measurable 2×2 matrix function on the interval $[0,\infty)$ with trace $\mathbf{H}(x) = 1$ (a.e.), $\mathbf{J} = \begin{pmatrix} 0 & -1 \\ 1 & 0 \end{pmatrix}$ and z is a complex parameter. For the system (0.1) we study the initial value problem $y(0) \in l.s. \begin{pmatrix} 0 \\ 1 \end{pmatrix}$. Its *fundamental matrix* function $\mathbf{W}$ is the solution of the problem:

$$\frac{d\mathbf{W}(x,z)}{dx}\mathbf{J} = z\mathbf{W}(x,z)\mathbf{H}(x), \;\; \mathbf{W}(0,z) = \mathbf{I}. \tag{0.2}$$

Let N be the set of *Nevanlinna* functions, i.e. the set of all functions which are analytic on the upper half plane C^+ and map C^+ into $C^+ \cup R$. Let $\tilde{N} := N \cup \{\infty\}$. It is well known (see [dB2]) that the assumption trace $\mathbf{H} \equiv 1$ on $[0,\infty)$ implies that for an arbitrary $t \in \tilde{N}$ and $z \in C^+$ the limit

$$Q(z) := \lim_{x\to\infty} \frac{w_{11}(x,z)t(z) + w_{12}(x,z)}{w_{21}(x,z)t(z) + w_{22}(x,z)} \tag{0.3}$$

exists, is independent of t and belongs to the class $\tilde{N}$ as a function of z. The function Q is called the *Titchmarsh-Weyl coefficient* of the canonical system (0.1). If $Q \neq \infty$, it has a unique *spectral representation*

$$Q(z) = bz + a + \int_{-\infty}^{+\infty} \left(\frac{1}{\lambda - z} - \frac{\lambda}{1+\lambda^2}\right) d\sigma(\lambda), \;\; b \geq 0, \; a \in R, \; \int_{-\infty}^{+\infty} \frac{d\sigma(\lambda)}{1+\lambda^2} < \infty, \tag{0.4}$$

where σ is a nonnegative measure which is called the *spectral measure* of the canonical system given by (0.1). We call the canonical system *discrete* if its spectral measure σ is a finite sum of point measures.

It follows from results of L. de Branges that to any function $Q \in \tilde{N}$ there exists a unique *Hamiltonian* $\mathbf{H}$ on $[0, \infty)$, such that Q is the Titchmarsh-Weyl coefficient of the canonical system corresponding to $\mathbf{H}$ (see [dB1-5] and [W]). The proof that there is a bijective correspondence between canonical systems (0.1) and their Titchmarsh-Weyl coefficients is not constructive and the corresponding Hamiltonian is known explicitely only for a few Nevanlinna functions Q. Therefore it seems to be of interest, to give some general rules, how the Hamiltonian changes if the Titchmarsh-Weyl coefficient (or its spectral measure) undergoes certain transformations. In this note we prove some results of this type. They can be considered as generalizations of corresponding results of M. G. Krein for strings and their spectral measures (see [K1-2], [DM], [DK1-2], [dB5]).

If only the constants b and a in the representation (0.4) of Q are changed, the corresponding transformations for the Hamiltonian were given in [W]. In this note we consider transformations concerning the spectral measure σ.

The following intervals (see [Ka], [dB1-4]) play a special role in the further considerations: Let

$$\xi_\phi := \begin{pmatrix} \cos\phi \\ \sin\phi \end{pmatrix}.$$

The open interval $I \subset [0, \infty)$ is called an $\mathbf{H}$ - *indivisible interval* if the relation

$$\mathbf{H}(x) = \xi_\phi \xi_\phi^T \quad \text{for all } x \in I$$

holds. Here ϕ is called the *type* of I.

1. Some transformation rules

Here and in the following we always suppose that a canonical system with Hamiltonian $\mathbf{H}$, fundamental matrix $\mathbf{W}$, Titchmarsh-Weyl coefficient Q and spectral measure σ is given. Starting from this canonical system we construct another system by corresponding transformations of the Hamiltonian $\mathbf{H}$ and the spectral measure σ. In this connection the variables of the transformed system are marked with a $\bullet$ as in [DM].

Rule 1. If K is a positive constant, for the canonical system corresponding to

$$\mathbf{H}^\bullet(x^\bullet) = \mathbf{H}(x) \quad \text{and} \quad x^\bullet = Kx \tag{1.1}$$

the following relations hold:

$$\begin{aligned} \mathbf{W}^\bullet(x^\bullet, z) &= \mathbf{W}(x, Kz), \\ Q^\bullet(z) &= Q(Kz), \\ d\sigma^\bullet(\lambda) &= d\sigma(K\lambda). \end{aligned} \tag{1.2}$$

Proof: It must be shown that for the transformed system the equations (0.2) and (0.3) hold with $\mathbf{W}^\bullet$ instead of $\mathbf{W}$, $x^\bullet$ instead of x etc.. E.g. the relation (0.2) becomes

$$\frac{d\mathbf{W}^\bullet(x^\bullet, z)}{dx}\mathbf{J} = z\mathbf{W}^\bullet(x^\bullet, z)\mathbf{H}^\bullet(x^\bullet)\frac{dx^\bullet}{dx}, \tag{1.3}$$

which can be checked by direct computation.

Rule 2: If K is a positive constant, for the canonical system corresponding to

$$\mathbf{H}^\bullet(x^\bullet)dx^\bullet = \mathbf{P}\mathbf{H}(x)\mathbf{P}^T dx, \quad \mathbf{P} = \begin{pmatrix} K & 0 \\ 0 & K^{-1} \end{pmatrix}, \quad K > 0 \tag{1.4}$$

it holds:

$$\begin{aligned} \mathbf{W}^\bullet(x^\bullet, z) =& \mathbf{P}\mathbf{W}(x,z)\mathbf{P}^{-1}, \\ Q^\bullet(z) =& K^2 Q(z), \\ d\sigma^\bullet(\lambda) =& K^2 d\sigma. \end{aligned} \tag{1.5}$$

The proof is as easy as that of Rule 1.

For the "shift" of the spectral measure σ the following holds:

Rule 3: If a is real, for the canonical system corresponding to

$$\begin{aligned} \mathbf{H}^\bullet(x^\bullet)dx^\bullet =& \mathbf{W}(x,-a)\mathbf{H}(x)\mathbf{W}(x,-a)^T dx, \\ x^\bullet(x) =& \operatorname{trace}\left(\int_0^x \mathbf{W}(t,-a)\mathbf{H}(t)\mathbf{W}(t,-a)^T dt\right) \end{aligned} \tag{1.6}$$

the following relations hold:

$$\begin{aligned} \mathbf{W}^\bullet(x^\bullet, z) =& \mathbf{W}(x, z-a)\mathbf{W}(x,-a)^{-1}, \\ Q^\bullet(z) =& Q(z-a), \\ d\sigma^\bullet(\lambda) =& d\sigma(\lambda - a). \end{aligned} \tag{1.7}$$

Proof: An elementary computation shows, that the transformed system satisfies the equation (1.3). If $t \in R$ and

$$R(x,t) = \frac{w_{22}(x,-a)t - w_{12}(x,-a)}{-w_{21}(x,-a)t + w_{11}(x,-a)}$$

it follows that

$$\frac{w_{11}^\bullet(x^\bullet,z)t + w_{12}^\bullet(x^\bullet,z)}{w_{21}^\bullet(x^\bullet,z)t + w_{22}^\bullet(x^\bullet,z)} = \frac{w_{11}(x,z-a)R(x,t) + w_{12}(x,z-a)}{w_{21}(x,z-a)R(x,t) + w_{22}(x,z-a)}. \tag{1.8}$$

If x tends to ∞, as $R(x,t) \in R \cup \{\infty\}$ the right hand side of (1.8) tends for all $t \in R$ to the limit $Q(z-a)$. It remains to show that $\lim_{x\to\infty} x^\bullet(x) = \infty$. Otherwise, if $\lim_{x\to\infty} x^\bullet(x) < \infty$, it would follow that for each t the transformed system continued by the $\mathbf{H}$ - indivisible interval of type ϕ with $\cot\phi = t$ would have the same Titchmarsh-Weyl coefficient $Q(z-a)$. This is a contradiction to the bijective correspondence between Hamiltonians and their corresponding Titchmarsh-Weyl coefficients. Therefore the transformed system is defined on $[0,\infty)$.

■

Now we show how the canonical system changes if a (positive or negative) point mass is added to the spectral measure at zero. In the sequel, let δ_x be the unit measure at the point x.

Rule 4: If $m \in R$ is such that $m + \sigma([0]) \geq 0$ we define:

$$\begin{aligned} S(x) =& 1 + m \int_0^x h_{22}(t)dt, \\ A(x) =& 2 \int_0^x S(t)h_{12}(t)dt, \\ \mathbf{P}(x) =& \begin{pmatrix} S(x) & -mS(x)A(x), \\ 0 & S(x)^{-1} \end{pmatrix}. \end{aligned} \tag{1.9}$$

Further, let

$$\begin{aligned} \mathbf{H}^\bullet(x^\bullet)dx^\bullet =& \mathbf{P}(x)\mathbf{H}(x)\mathbf{P}(x)^T dx, \\ x^\bullet(x) =& \operatorname{trace}\left(\int_0^x \mathbf{P}(t)\mathbf{H}(t)\mathbf{P}(t)^T dt \right), \\ l^\bullet =& \lim_{x \to \infty} x^\bullet(x). \end{aligned} \tag{1.10}$$

Then the following relations hold:

$$\mathbf{W}^\bullet(x^\bullet, z) = \begin{pmatrix} 1 & -mz^{-1} \\ 0 & 1 \end{pmatrix} \mathbf{W}(x, z) \begin{pmatrix} S(x)^{-1} & mS(x)^{-1}A(x) + mz^{-1} \\ 0 & S(x) \end{pmatrix} \tag{1.11}$$

and

$$\begin{aligned} Q^\bullet(z) =& Q(z) - mz^{-1}, \\ \sigma^\bullet =& \sigma + m\delta_0. \end{aligned} \tag{1.12}$$

By (1.10) the Hamiltonian $\mathbf{H}^\bullet$ is only defined on $[0, l^\bullet]$. If $l^\bullet < \infty$, on $(l^\bullet, \infty)$ it holds

$$\mathbf{H}^\bullet = \begin{pmatrix} 1 & 0 \\ 0 & 0 \end{pmatrix}.$$

Proof: A direct computation shows, that $\mathbf{H}^\bullet$ and $\mathbf{W}^\bullet$ given by (1.10) and (1.11) satisfy the equation (1.3) on $[0, l^\bullet)$.

Now we suppose $\sigma([0]) + m > 0$, then it holds $S(x) > 0$ for all $x > 0$. From (1.11) it follows

$$\frac{w_{11}^\bullet(x^\bullet, z)}{w_{21}^\bullet(x^\bullet, z)} = \frac{w_{11}(x, z)}{w_{12}(x, z)} - mz^{-1},$$

and we get:

$$\lim_{x\to\infty}\frac{w_{11}^{\bullet}(x^{\bullet},z)}{w_{21}^{\bullet}(x^{\bullet},z)}=Q(z)-mz^{-1}.$$

This shows, that $\mathbf{H}^{\bullet}$ given by (1.10) must be continued in the case $l^{\bullet}<\infty$ with an indivisible interval of type 0. Then it holds

$$Q^{\bullet}(z)=\frac{w_{11}^{\bullet}(l^{\bullet},z)}{w_{21}^{\bullet}(l^{\bullet},z)}=Q(z)-mz^{-1}.$$

If $\sigma([0])+m=0$ and if (l_0,∞) is $\mathbf{H}$ - indivisible of type 0 the relation

$$\int_0^{l_0} h_{22}(t)dt=\sigma([0])^{-1}$$

holds. It follows $S(l_0)=0$ and as

$$x^{\bullet}\geq\int_0^{x^{\bullet}} h_{22}^{\bullet}(t)dt=S(x)^{-1}\int_0^{x} h_{22}(t)dt$$

we get $x^{\bullet}\to\infty$ for $x\to l_0$. But in this case it holds $Q(z)=\dfrac{w_{11}(l_0,z)}{w_{12}(l_0,z)}$ which implies the relation (1.12).

■

A combination of the Rules 3 and 4 implies a transformation rule for the case that a point measure $m\delta_x$ is added to σ at any point $x\in R$.

2. Changes of the spectral density by rational factors

In this section we prove transformation rules for a change of the spectral measure of the form $d\sigma^{\bullet}(\lambda)=p(\lambda)d\sigma$, where $p(\lambda)$ or $p(\lambda)^{-1}$ is a nonnegative polynomial of second degree. Let G be the following polynomial:

$$G(z):=|z_0|^2-2z\Re z_0+z^2. \tag{2.1}$$

If the spectral measure σ is finite, let $s_0=\int_{-\infty}^{+\infty}d\sigma$. A simple computation gives the following relation:

$$\begin{aligned}\int_{-\infty}^{+\infty}\left(\frac{1}{\lambda-z}-\frac{\lambda}{1+\lambda^2}\right)G(\lambda)d\sigma=s_0z+\int_{-\infty}^{+\infty}\frac{\lambda(|z_0|^2-1)-2\Re z_0}{1+\lambda^2}d\sigma+\\+G(z)\int_{-\infty}^{+\infty}\frac{d\sigma}{\lambda-z}.\end{aligned} \tag{2.2}$$

This shows, that for a Nevanlinna function Q with the representation

$$Q(z) = \int_{-\infty}^{+\infty} \frac{d\sigma}{\lambda - z} \tag{2.3}$$

the function $Q^\bullet$ with $Q^\bullet(z) = zs_0 + G(z)Q(z)$ is again a Nevanlinna function with the spectral measure $d\sigma^\bullet(\lambda) = G(\lambda)d\sigma$. If Q has a representation of the form (2.3), the corresponding Hamiltonian has the indivisible interval $(0, s_0^{-1})$ of type $\frac{\pi}{2}$ (see [KL]). With a simple transformation of the Hamiltonian of a canonical system with a finite spectral measure we can always get that the corresponding Titchmarsh-Weyl coefficient Q is of the form (2.3).

Rule 5: Assume $s_0 = \int_{-\infty}^{+\infty} d\sigma < \infty$ and $\mathbf{H}(x) = \begin{pmatrix} 0 & 0 \\ 0 & 1 \end{pmatrix}$ for $0 \leq x \leq s_0^{-1}$.

If $z_0 \in C$ with $\Im z_0 < 0$ and $x \geq s_0^{-1}$ define:

$$C(x) = \sqrt{-\Im\left(\frac{w_{22}(x, z_0)}{w_{21}(x, z_0)}\right)}, \quad S(x) = -\Re\left(\frac{w_{22}(x, z_0)}{w_{21}(x, z_0)}\right) C(x)^{-1}, \tag{2.4}$$

$$\mathbf{P}(x) = \begin{pmatrix} \Re z_0 - \Im z_0 \dfrac{S(x)}{C(x)} & -\dfrac{\Im z_0}{|z_0|^2 C(x)^2} \\ \Im z_0 \left(S(x)^2 + C(x)^2\right) & |z_0|^{-2}\left(\Re z_0 + \Im z_0 \dfrac{S(x)}{C(x)}\right) \end{pmatrix}, \tag{2.5}$$

$$x^\bullet(x) = \operatorname{trace}\left(\int_{s_0^{-1}}^{x} \mathbf{P}(t)^T \mathbf{H}(t) \mathbf{P}(t) dt\right), \quad l^\bullet = \lim_{x \to \infty} x^\bullet(x), \tag{2.6}$$

$$\mathbf{H}^\bullet(x^\bullet) dx^\bullet = \mathbf{P}(x)^T \mathbf{H}(x) \mathbf{P}(x) dx \tag{2.7}$$

and

$$\mathbf{R}(x, z) = \begin{pmatrix} 1 & 0 \\ 0 & |z_0|^2 \end{pmatrix} + z\mathbf{J}\mathbf{P}(x)\mathbf{J} =$$

$$= \begin{pmatrix} 1 - \dfrac{z}{|z_0|^2}\left(\Im z_0 \dfrac{S(x)}{C(x)} + \Re z_0\right) & z\Im z_0 \left(S(x)^2 + C(x)^2\right) \\ -\dfrac{z\Im z_0}{|z_0|^2 C(x)^2} & |z_0|^2 + z\left(\Im z_0 \dfrac{S(x)}{C(x)} - \Re z_0\right) \end{pmatrix}$$

Let t be defined by

$$t := \lim_{x \to \infty} -\Re\left(\frac{w_{22}(x, z_0)}{w_{21}(x, z_0)}\right).$$

If $l^\bullet < \infty$, we put with $\cot\phi := t|z_0|^2$:

$$\mathbf{H}^\bullet(x^\bullet) = \begin{pmatrix} \cos^2\phi & \sin\phi\cos\phi \\ \sin\phi\cos\phi & \sin^2\phi \end{pmatrix}, \quad x^\bullet \in (l^\bullet, \infty),$$

that is, $(l^\bullet, \infty)$ is $\mathbf{H}^\bullet$ - indivisible of type ϕ.

The the following relations hold:

$$\begin{aligned} \mathbf{W}^\bullet(x^\bullet, z) &= \begin{pmatrix} 1 & s_0 z G(z) \\ 0 & G(z) \end{pmatrix} \mathbf{W}(x,z)\mathbf{R}(x,z), \; x^\bullet \in [0, l^\bullet), \\ \mathbf{W}^\bullet(x^\bullet, z) &= \mathbf{W}^\bullet(l^\bullet, z)(\mathbf{I} - z(x^\bullet - l^\bullet)\mathbf{H}^\bullet(x^\bullet)\mathbf{J}), \; x^\bullet \in (l^\bullet, \infty), \end{aligned} \tag{2.8}$$

and

$$\begin{aligned} Q^\bullet(z) &= s_0 z + G(z)Q(z), \\ d\sigma^\bullet(\lambda) &= (|z_0|^2 - 2\Re z_0 \lambda + \lambda^2) d\sigma. \end{aligned} \tag{2.9}$$

Proof: A lengthy but elementary computation shows that the transformed system satisfies the equation (1.3). With $\mathbf{R}(x,z) = \begin{pmatrix} r_{11}(x,z) & r_{12}(x,z) \\ r_{21}(x,z) & r_{22}(x,z) \end{pmatrix}$ and

$$M(\omega) := \frac{r_{11}(x,z)\omega + r_{12}(x,z)}{r_{21}(x,z)\omega + r_{22}(x,z)}$$

for $\omega \in C$ it follows from (2.6):

$$\frac{w^\bullet_{11}(x^\bullet, z)\omega + w^\bullet_{12}(x^\bullet, z)}{w^\bullet_{21}(x^\bullet, z)\omega + w^\bullet_{22}(x^\bullet, z)} = G(z)\frac{w_{11}(x,z)M(\omega) + w_{12}(x,z)}{w_{21}(x,z)M(\omega) + w_{22}(x,z)} + s_0 z \tag{2.10}$$

Next we determine those values $\omega \in C^+$, for which $M(\omega) \in C^+$. To this end we consider the inverse transformation

$$M^{-1}(\omega) = \frac{r_{22}(x,z)\omega - r_{12}(x,z)}{-r_{21}(x,z)\omega + r_{11}(x,z)}.$$

It maps the upper half plane C^+ onto a disk $D_{a,r}(x)$ with center

$$a(x) = -\frac{i}{2}\frac{|z_0|^2 C(x)^2}{\Im z \Im z_0}\left[(\Re z_0 - \Re z)^2 + (\Im z)^2 + (\Im z_0)^2\right] + |z_0|^2 S(x) C(x)$$

and radius

$$r(x) = -\frac{|G(z)||z_0|^2 C(x)^2}{2\Im z \Im z_0}.$$

The relation $(\Re z_0 - \Re z)^2 + (\Im z)^2 + (\Im z_0)^2 > |G(z)|$ implies that $D_{a,r}(x)$ lies in the upper half plane.

Now we have to distinguish two cases. First let $\lim_{x\to\infty} C(x)^2 > 0$. Then $\lim_{x\to\infty} r(x) > 0$ for $z \neq \bar{z}_0$. If $\omega \in D_{a,r}(\infty)$ and $l^\bullet < \infty$ the limit for $x \to \infty$ at the right hand side of (2.10) is $Q^\bullet(z)$ and the Hamiltonian $\mathbf{H}^\bullet$ should be continued on $(l^\bullet, \infty)$ in such a way that the Titchmarsh-Weyl coefficient corresponding to the restriction of $\mathbf{H}^\bullet$ on $(l^\bullet, \infty$ is equal to ω. As for ω there can be chosen any point in $D_{a,r}(\infty)$, this is a contradiction with the uniqueness between Hamiltonian and corresponding Titchmarsh-Weyl coefficient. Consequently, $l^\bullet = \infty$ and the relation (2.9) is shown.

If $\lim_{x\to\infty} C(x)^2 = 0$ then $D_{a,r}(x)$ converges for $x \to \infty$ to the point $|z_0|^2 t$. If $l^\bullet < \infty$ the Hamiltonian $\mathbf{H}^\bullet$ must be continued such that $(l^\bullet, \infty)$ is the corresponding indivisible interval. This happens if the system is discrete, i.e. if σ is a finite sum of point measures.

■

If in Rule 5 we choose $z_0 = iy$ and let $y \to 0$ the following Rule 6 can be derived. Here we observe the relations

$$\lim_{y\to 0} -\Re\left(\frac{w_{22}(x,iy)}{w_{21}(x,iy)}\right) = \frac{2\int_0^x\int_0^t h_{22}(s)ds h_{12}(t)dt}{\left(\int_0^x h_{22}(t)dt\right)^2}$$

and

$$\lim_{y\to 0} -y\Im\left(\frac{w_{22}(x,iy)}{w_{21}(x,iy)}\right) = -\int_0^x h_{22}(t)dt.$$

Rule 6: Assume $s_0 = \int_{-\infty}^{+\infty} d\sigma < \infty$ and $\mathbf{H}(x) = \begin{pmatrix} 0 & 0 \\ 0 & 1 \end{pmatrix}$ for $0 \le x \le s_0^{-1}$. If $x \ge s_0^{-1}$ we define:

$$A(x) := \int_0^x h_{22}(t)dt, \quad B(x) := 2\int_0^x M(t)h_{12}(t)dt \tag{2.11}$$

$$\mathbf{P}(x) := \begin{pmatrix} 0 & A(x) \\ -A(x)^{-1} & -B(x)A(x)^{-1} \end{pmatrix}, \tag{2.12}$$

$$x^\bullet(x) := \text{trace}\left(\int_{s_0^{-1}}^x \mathbf{P}(t)^T\mathbf{H}(t)\mathbf{P}(t)dt\right), \quad l^\bullet := \lim_{x\to\infty} x^\bullet(x), \tag{2.13}$$

$$\mathbf{H}^\bullet(x^\bullet)dx^\bullet := \mathbf{P}(x)^T\mathbf{H}(x)\mathbf{P}(x)dx \tag{2.14}$$

If $l^\bullet < \infty$ let

$$\mathbf{H}^\bullet(x^\bullet) = \begin{pmatrix} 0 & 0 \\ 0 & 1 \end{pmatrix}, \quad x^\bullet \in (l^\bullet, \infty),$$

that is $(l^\bullet, \infty)$ is $\mathbf{H}^\bullet$ - indivisible of type $\frac{\pi}{2}$. The the following relations hold:

$$\begin{aligned} \mathbf{W}^\bullet(x^\bullet, z) &= \begin{pmatrix} 1 & s_0 z^{-1} \\ 0 & z^{-2} \end{pmatrix} \mathbf{W}(x,z) \begin{pmatrix} 1 + zB(x)A(x)^{-1} & -zA(x)^{-1} \\ zA(x) & 0 \end{pmatrix}, \\ \mathbf{W}^\bullet(x^\bullet, z) &= \mathbf{W}^\bullet(l^\bullet, z) \begin{pmatrix} 1 & 0 \\ -z(x^\bullet - l^\bullet) & 1 \end{pmatrix}, \quad x^\bullet \in (l^\bullet, \infty), \end{aligned} \tag{2.15}$$

$$Q^\bullet(z) = s_0 z + z^2 Q(z), \quad d\sigma^\bullet(\lambda) = \lambda^2 d\sigma. \tag{2.16}$$

Proof: A computation shows that the transformed system satisfies the equation (1.3). From (2.15) it follows

$$\frac{w^\bullet_{12}(x^\bullet, z)}{w^\bullet_{22}(x^\bullet, z)} = z^2\frac{w_{11}(x,z)}{w_{21}(x,z)} + s_0 z.$$

If $l^\bullet < \infty$, we get

$$\frac{w_{12}^\bullet(l^\bullet, z)}{w_{22}^\bullet(l^\bullet, z)} = z^2 Q(z) + s_0 z,$$

such that the interval $(l^\bullet, \infty)$ must be of type $\frac{\pi}{2}$ if (2.16) holds.

■

Another transformation rule follows from the inverse transformation of the Rule 5. The relation (2.8) implies

$$w_{12}^\bullet(x^\bullet, z) - s_0 z w_{22}^\bullet(x^\bullet, z) = w_{11}(x, z) r_{12}(x, z) + w_{12}(x, z) r_{22}(x, z),$$
$$w_{11}^\bullet(x^\bullet, z) - s_0 z w_{21}^\bullet(x^\bullet, z) = w_{11}(x, z) r_{11}(x, z) + w_{12}(x, z) r_{21}(x, z).$$

As $\det \mathbf{R}(x, z_0) = G(z_0) = 0$ it holds

$$\frac{r_{12}(x, z_0)}{r_{11}(x, z_0)} = \frac{r_{22}(x, z_0)}{r_{21}(x, z_0)} = |z_0|^2 \left(iC(x)^2 - S(x)C(x) \right),$$

and we get expressions for $C(x)^2$ and $S(x)C(x)$ in the terms of the transformed system:

$$C(x)^2 = \frac{1}{|z_0|^2} \Im \left(\frac{w_{12}^\bullet(x^\bullet, z_0) - s_0 z_0 w_{22}^\bullet(x^\bullet, z_0)}{w_{11}^\bullet(x^\bullet, z_0) - s_0 z_0 w_{21}^\bullet(x^\bullet, z_0)} \right),$$
$$S(x)C(x) = -\frac{1}{|z_0|^2} \Re \left(\frac{w_{12}^\bullet(x^\bullet, z_0) - s_0 z_0 w_{22}^\bullet(x^\bullet, z_0)}{w_{11}^\bullet(x^\bullet, z_0) - s_0 z_0 w_{21}^\bullet(x^\bullet, z_0)} \right).$$

Doing the same with the relations (2.5)-(2.9), we get the following rule:

Rule 7: Assume that

$$s_0 := \int_{-\infty}^{+\infty} (|z_0|^2 - 2\Re z_0 \lambda + \lambda^2)^{-1} d\sigma < \infty.$$

and that the linear term in the representation (0.4) of the Titchmarsh-Weyl coefficient Q is chosen so, that the function $z \to G(z)(Q(z) - s_0 z)$ is a Nevanlinna function. Because of the relation (2.2) by a simple transformation such a choice is always possible. Define

$$C(x)^2 := \frac{1}{|z_0|^2} \Im \left(\frac{w_{12}(x, z_0) - s_0 z_0 w_{22}(x, z_0)}{w_{11}(x, z_0) - s_0 z_0 w_{21}(x, z_0)} \right),$$
$$S(x)C(x) := -\frac{1}{|z_0|^2} \Re \left(\frac{w_{12}(x, z_0) - s_0 z_0 w_{22}(x, z_0)}{w_{11}(x, z_0) - s_0 z_0 w_{21}(x, z_0)} \right), \tag{2.17}$$

$$\mathbf{P}(x) := \begin{pmatrix} \frac{1}{|z_0|^2} \left(\Re z_0 + \Im z_0 \frac{S(x)}{C(x)} \right) & \frac{\Im z_0}{|z_0|^2 C(x)^2} \\ -\Im z_0 \left(S(x)^2 + C(x)^2 \right) & \Re z_0 - \Im z_0 \frac{S(x)}{C(x)} \end{pmatrix}, \tag{2.18}$$

$$\mathbf{R}(x, z) := \begin{pmatrix} |z_0|^2 + z \left(-\Re z_0 + \Im z_0 \frac{S(x)}{C(x)} \right) & -z \Im z_0 \left(S(x)^2 + C(x)^2 \right) \\ \frac{z \Im z_0}{|z_0|^2 C(x)^2} & 1 - \frac{z}{|z_0|^2} \left(\Re z_0 + \Im z_0 \frac{S(x)}{C(x)} \right) \end{pmatrix}.$$

For the canonical system corresponding to

$$Q^{\bullet}(z) := G(z)\left(Q(z) - s_0 z\right) \tag{2.19}$$

the following relations hold:

$$d\sigma^{\bullet}(\lambda) = \left(|z_0|^2 - 2\Re z_0 \lambda + \lambda^2\right)^{-1} d\sigma, \tag{2.20}$$

$$\mathbf{W}^{\bullet}(x^{\bullet}, z) = \begin{pmatrix} 1 & 0 \\ -zx^{\bullet} & 1 \end{pmatrix}, \quad \mathbf{H}^{\bullet}(x^{\bullet}) = \begin{pmatrix} 0 & 0 \\ 0 & 1 \end{pmatrix} \text{ if } 0 \le x^{\bullet} \le s_0^{-1} \tag{2.21}$$

$$\mathbf{H}^{\bullet}(x^{\bullet})dx^{\bullet} = \mathbf{P}(x)^T \mathbf{H}(x) \mathbf{P}(x) dx, \tag{2.22}$$

$$x^{\bullet}(x) = s_0^{-1} + \operatorname{trace}\left(\int_0^x \mathbf{P}(t)^T \mathbf{H}(t) \mathbf{P}(t) dt\right), \tag{2.23}$$

$$\mathbf{W}^{\bullet}(x^{\bullet}, z) = \begin{pmatrix} G(z) & -s_0 z G(z) \\ 0 & 1 \end{pmatrix} \mathbf{W}(x, z) \mathbf{R}(x, z). \tag{2.24}$$

If we choose $z_0 = iy$ and let y tend to 0, the Rule 8 below follows. First we prove

Lemma 2.1. Assume that the Titchmarsh-Weyl coefficient Q has the representation

$$Q(z) = a + \int_{-\infty}^{+\infty} \left(\frac{1}{\lambda - z} - \frac{\lambda}{1+\lambda^2}\right) d\sigma(\lambda). \tag{2.25}$$

If

$$s_{-2} := \int_{-\infty}^{+\infty} \frac{d\sigma(\lambda)}{\lambda^2} < +\infty \tag{2.26}$$

and

$$a = -\int_{-\infty}^{+\infty} \frac{d\sigma(\lambda)}{\lambda(1+\lambda^2)}$$

it holds

$$\int_0^{\infty} h_{11}(t)dt = s_{-2}. \tag{2.27}$$

If $a \neq -\int_{-\infty}^{+\infty} \frac{d\sigma(\lambda)}{\lambda(1+\lambda^2)}$ or $s_{-2} = \infty$, it holds

$$\int_0^{\infty} h_{11}(t)dt = +\infty. \tag{2.28}$$

Proof. If (2.26) is satisfied we have

$$-\int_{-\infty}^{+\infty} \frac{d\sigma(\lambda)}{\lambda(1+\lambda^2)} + \int_{-\infty}^{+\infty} \left(\frac{1}{\lambda - z} - \frac{\lambda}{1+\lambda^2}\right) d\sigma(\lambda) = z \int_{-\infty}^{+\infty} \frac{d\sigma(\lambda)}{\lambda^2} + z^2 \int_{-\infty}^{+\infty} \frac{d\sigma(\lambda)}{(\lambda - z)\lambda^2}.$$

It follows that Q has an angular derivative at $z = 0$ (see [Ca]) with $Q(0) = 0$. Now we consider a canonical system with the property that (l, ∞) is $\mathbf{H}$ - indivisible of type $\frac{\pi}{2}$ for some $l > 0$. In this case

$$Q(z) = \frac{w_{12}(l,z)}{w_{22}(l,z)},$$

and differentiation with respect to z gives at $z = 0$

$$Q'(0) = \int_{-\infty}^{+\infty} \frac{d\sigma(\lambda)}{\lambda^2} = \int_0^l h_{11}(t)dt.$$

This observation proves (2.27) in the particular case of discrete systems with $\sigma(\{0\}) = 0$ and $Q(0) = 0$, as these systems end with an indivisible interval of type $\frac{\pi}{2}$. In the general case we use a continuity principle of L. de Branges (see [dB2], Proof of Theorem 12). Any Titchmarsh-Weyl coefficient Q with the representation (2.25) can be approximated by a sequence Q_n of Titchmarsh-Weyl coefficients of discrete systems of the form above wich converge to Q locally uniformly in the upper half plane and also converge at $z = 0$. Then for all bounded sets $B \subset [0, \infty)$ it holds

$$\int_0^x \mathbf{H}_n(t)dt \to \int_0^x \mathbf{H}(t)dt \quad \text{uniformly for} \quad x \in B, \tag{2.29}$$

which proves (2.27) in the general case. Any other discrete system, for which $Q(0) \neq 0$ or Q has a pole at $z = 0$, satisfies the relation (2.28), and any Titchmarsh-Weyl coefficient which is not of the form (2.25) can be approximated by Titchmarsh-Weyl coefficients of discrete systems of this form. From (2.29) follows in this case the relation (2.28). ■

Rule 8: Assume that

$$s_{-2} = \int_{-\infty}^{+\infty} \frac{d\sigma}{\lambda^2} < \infty \tag{2.30}$$

and that the linear term in the representation (0.4) of the Titchmarsh-Weyl coefficient Q is chosen such that the function $z \to z^{-2}(Q(z) - s_{-2}z)$ is a Nevanlinna function.

Define

$$L(x) := s_{-2} - \int_0^x h_{11}(t)dt, \quad B(x) := 2\int_0^x L(t)h_{12}(t)dt, \tag{2.31}$$

$$\mathbf{P}(x) := \begin{pmatrix} B(x)L(x)^{-1} & -L(x)^{-1} \\ L(x) & 0 \end{pmatrix}. \tag{2.32}$$

For the canonical system corresponding to the Titchmarsh-Weyl coefficient $Q^\bullet(z)$ with

$$Q^\bullet(z) = z^{-2}\left(Q(z) - s_{-2}z\right) \tag{2.33}$$

the following relations hold:

$$d\sigma^\bullet(\lambda) = \lambda^{-2}d\sigma, \tag{2.34}$$

$$\mathbf{W}^\bullet(x^\bullet, z) = \begin{pmatrix} 1 & 0 \\ -zx^\bullet & 1 \end{pmatrix}, \quad \mathbf{H}^\bullet(x^\bullet) = \begin{pmatrix} 0 & 0 \\ 0 & 1 \end{pmatrix}, \text{ if } 0 \le x^\bullet \le s_{-2}^{-1},$$

$$\mathbf{H}^\bullet(x^\bullet)dx^\bullet = \mathbf{P}(x)^T\mathbf{H}(x)\mathbf{P}(x)dx, \tag{2.35}$$

$$x^\bullet(x) = s_{-2}^{-1} + \operatorname{trace}\left(\int_0^x \mathbf{P}(t)^T\mathbf{H}(t)\mathbf{P}(t)dt\right), \tag{2.36}$$

$$\mathbf{W}^\bullet(x^\bullet, z) = \begin{pmatrix} z^{-2} & -s_{-2}z^{-1} \\ 0 & 1 \end{pmatrix}\mathbf{W}(x,z)\begin{pmatrix} 0 & zL(x) \\ -zL(x)^{-1} & 1 - zB(x)L(x)^{-1} \end{pmatrix}. \tag{2.37}$$

Proof: A computation shows that the transformed system satisfies the relation (1.3). From (2.37) the relation

$$\frac{w_{11}^\bullet(x^\bullet, z)(-B(x)) + w_{12}^\bullet(x^\bullet, z)}{w_{21}^\bullet(x^\bullet, z)(-B(x)) + w_{22}^\bullet(x^\bullet, z)} = z^{-2}\left(\frac{w_{11}(x,z)zL(x) + w_{12}(x,z)}{w_{21}(x,z)zL(x) + w_{22}(x,z)} + s_{-2}z\right) \tag{2.38}$$

follows and from (2.34) and (2.35) we conclude that

$$x^\bullet(x) \ge \int_0^x L(t)^{-2}h_{11}dt + s_{-2}^{-1} = L(x)^{-1}.$$

By Lemma 2.1, $L \ge 0$ and if $L > 0$ it follows that $x^\bullet(x) \to \infty$ if $x \to \infty$ and we get the relation (2.31) from (2.38). If there exists an $l > 0$ with $L(l) = 0$ we have $x^\bullet(x) \to \infty$ if $x \to l$. But then the interval (l, ∞) is $\mathbf{H}$ - indivisible of type $\frac{\pi}{2}$, which means that

$$Q(z) = \frac{w_{12}(l,z)}{w_{22}(l,z)}$$

holds, and from (2.38) the relation (2.31) follows.

References

[Ca] Caratheodory, C. Über die Winkelderivierten von beschränkten analytischen Funktionen. *Sitzungsber. der Preuß. Akad. d. Wiss., Phys.-mathem. Klasse.* (1929), 39–54.

[dB1-4] de Branges, L. Some Hilbert spaces of entire functions. *Trans. Amer. Math. Soc.* **96** (1960), 259–295; **99** (1961), 118–152; **100** (1960), 73–115; **105** (1962), 43–83.

[dB5] de Branges, L. *Hilbert Spaces of Entire Functions.* Prentice Hall, Englewood Cliffs, N.J., 1968.

[DK1] Dym, H., Kravitzky, N. On the inverse spectral problem for the string equation. *Integral Equations Operator Theory* **1/2** (1978).

[DK2] Dym, H., Kravitzky, N. On recovering the mass distribution of a string from its spectral function, in: *Topics in Functional Analysis* (I. Gohberg and M. Kac, eds.), Academic Press, New York, 1978.

[DM] Dym, H., McKean, H.P. *Gaussian Processes, Function Theory, and the Inverse Spectral Problem.* Academic Press, New York, 1976.

[Ka] Kac, I.S. Linear relations, generated by a canonical differential equation on an interval with a regular endpoint, and expansibility in eigenfunctions(Russian). UDK 517.9, Odessa, 1984.

[K1] Krein, M.G. On a generalization of investigations of Stieltjes (Russian). *Dokl. Akad. Nauk. SSSR* **87** (1952), 881-884.

[K2] Krein, M.G. On some cases of the effective determination of the density of a non-homogeneous string from its spectral funktion (Russian). *Dokl. Akad. Nauk. SSSR* **93** (1953), 617-620.

[K3] Krein, M.G. On a fundamental approximation problem in the theory of extrapolation and filtration of stationary random processes (Russian). *Dokl. Akad. Nauk. SSSR* **94** (1954), 13-16.

[KL] Krein, M.G., Langer H. Continuation of Hermitian positive definite functions and related questions, unpublished manuscript.

[W] Winkler, H. The inverse spectral problem for canonical systems, to appear in IEOT.

Institut für Analysis,
Technische Mathematik und
Versicherungsmathematik
TU Wien,
Wiedner Hauptstr. 8 - 10,
A - 1040 Wien.

AMS Subject Classification: 34 A 55

Operator Theory:
Advances and Applications, Vol. 80

COMPLEMENTARY TRIANGULAR FORMS FOR INFINITE MATRICES

R.A. Zuidwijk

If A and Z are complex, square finite matrices, and if one of them is diagonable, then there exists an invertible matrix S, such that $S^{-1}AS$ is upper triangular, and $S^{-1}ZS$ is lower triangular. This paper presents analogues of this result for pairs of bounded operators, acting on the separable Hilbert space $l_2(\mathbf{Z}^+)$. The main result states that there exist two diagonable operators A and Z acting on $l_2(\mathbf{Z}^+)$, that are not simultaneously similar respectivily to an upper triangular operator and a lower triangular operator. The example is based on the existence of a unitary operator on $l_2(\mathbf{Z}^+)$, that does not admit lower-upper factorization, even after independently permuting rows and columns. On the other hand, for pairs of bounded operators, where one of the operators is of finite rank, positive results are obtained.

1 Introduction

In the study of complete factorization of proper rational matrix functions, one encounters the following problem: Given two $m \times m$ matrices A and Z, under what circumstances does there exist an invertible $m \times m$ matrix S, such that $S^{-1}AS$ is upper triangular and $S^{-1}ZS$ is lower triangular? In other words, when do the matrices A and Z *admit simultaneous reduction to complementary triangular forms*? The following theorem provides a sufficient condition for a pair of matrices to admit simultaneous reduction to complementary triangular forms.

Theorem 1.1 *Let A and Z be $m \times m$ matrices. If A or Z is diagonable, then A and Z admit simultaneous reduction to complementary triangular forms.*

In terms of complete factorization, this result was already stated in [2]. In explicit form as above, it appeared in [1]. The proof presented in Section 3, is taken from [3].

In this paper, we will investigate versions of Theorem 1.1 for a pair of bounded operators, acting on the separable Hilbert space $l_2(\mathbf{Z}^+)$, with standard orthonormal basis $\{e_k \mid k \in \mathbf{Z}^+\}$. The analogues to Theorem 1.1 turn out to be the following: Let A and Z be bounded operators acting on $l_2(\mathbf{Z}^+)$, and assume that either A or Z is diagonable. If in addition, one of the operators A or Z is of finite rank, then A and Z admit simultaneous reduction to complementary triangular forms, if an obvious necessary condition is met. This is shown in Propositions 3.5 and 3.6, and the remarks after their proofs. On the other hand, there exist two diagonable operators, that do not admit simultaneous reduction to complementary triangular forms (Theorem 3.3). These operators can even be taken self-adjoint and trace-class: See Corollary 3.4.

Section 2 describes the role of lower-upper factorization in simultaneous reduction to complementary triangular forms. This notion is used in the construction of a pair of bounded diagonable operators, that do not admit simultaneous reduction to complementary triangular forms.

Section 3 deals with the Hilbert space analogues of Theorem 1.1.

We will end the introduction with some preliminaries. For $m \in \mathbf{Z}^+$, the orthoprojector (of rank m) onto the subspace span$\{e_1, \ldots, e_m\}$ will be denoted by E_m. A bounded operator A is called *upper triangular*, if $AE_m = E_m A E_m$ for all $m \in \mathbf{Z}^+$. It is called *lower triangular*, if $E_m A = E_m A E_m$ for all $m \in \mathbf{Z}^+$, and *diagonal*, if it is both upper and lower triangular. A bounded operator A is called upper triangularizable, if there exists an invertible operator S, such that $S^{-1}AS$ is upper triangular. In the same fashion, we define lower triangularizable and diagonable operators. Two bounded operators A and Z admit *simultaneous reduction to complementary triangular forms*, if there exists an invertible operator S,

such that $S^{-1}AS$ is upper triangular, and $S^{-1}ZS$ is lower triangular. We will denote the collection of such pairs of bounded operators on $l_2(\mathbf{Z}^+)$ by $\mathcal{C}(l_2(\mathbf{Z}^+))$.

Unlike the situation for finite matrices (see [3], Remark 2), the property of complementary triangular forms is not symmetric in its arguments: $(A, Z) \in \mathcal{C}(l_2(\mathbf{Z}^+))$ does not imply that $(Z, A) \in \mathcal{C}(l_2(\mathbf{Z}^+))$. For example, let A be the upper triangular shift; $Ae_1 = 0$, and $Ae_{k+1} = e_k$ for $k \in \mathbf{Z}^+$. Then A is not lower triangularizable. It follows, that $(A, A^*) \in \mathcal{C}(l_2(\mathbf{Z}^+))$, but $(A^*, A) \notin \mathcal{C}(l_2(\mathbf{Z}^+))$.

We will use the short hand notation $\mathcal{C} = \mathcal{C}(l_2(\mathbf{Z}^+))$.

Lemma 1.2 *Let A and Z be bounded operators acting on $l_2(\mathbf{Z}^+)$, and let T be an invertible operator on $l_2(\mathbf{Z}^+)$. Then the following are equivalent:*

1. $(A, Z) \in \mathcal{C}$,
2. $(Z^*, A^*) \in \mathcal{C}$,
3. $(T^{-1}AT, T^{-1}ZT) \in \mathcal{C}$.

Now, we will introduce some notation, that is well-known for finite matrices. If T is a bounded operator acting on $l_2(\mathbf{Z}^+)$, its infinite matrix with respect to $\{e_k \mid k \in \mathbf{Z}^+\}$ is given by $(T_{kl})_{k,l=1}^{\infty}$, where $T_{kl} = e_k^* T e_l$. The *diagonal* of the infinite matrix of T is given by

$$\operatorname{diag}(T) = (\ T_{11},\ T_{22},\ T_{33},\ \cdots\ )^T .$$

If τ is a permutation (i.e., bijection) on $\mathbf{Z}^+$, then the *permutation operator* U_τ is defined as $U_\tau e_k = e_{\tau(k)}$

for $k \in \mathbf{Z}^+$. An invertible operator S *admits lower-upper factorization*, if there exist invertible operators L and R, such that $S = LR$, where L, L^{-1} are lower triangular and R, R^{-1} are upper triangular operators.

2 Lower-Upper Factorization

In this section, we will discuss how lower-upper factorization of invertible operators is used in the context of simultaneous reduction to complementary triangular forms. The following matrix result is well-known.

Proposition 2.1 *Given an invertible complex $m \times m$ matrix A, there exists an $m \times m$ permutation matrix U_τ such that AU_τ admits lower-upper factorization.*

In [3], Proposition 2.1 was used to prove Theorem 1.1 as follows: We may assume without loss of generality, that A is a diagonable $m \times m$ matrix, and Z is any $m \times m$ matrix. Let T, V be invertible $m \times m$ matrices, such that $T^{-1}AT$ is diagonal, and $V^{-1}ZV$ is lower triangular. By Proposition 2.1, there exists an $m \times m$ permutation matrix U_τ, such that $V^{-1}TU_\tau = LR^{-1}$, where L is an invertible lower triangular matrix, and R is an invertible upper triangular matrix. Define $S = VL = TU_\tau R$. Then

$$S^{-1}AS = R^{-1}(U_\tau^{-1}T^{-1}ATU_\tau)R$$

is upper triangular, since $U_\tau^{-1}T^{-1}ATU_\tau$ is a diagonal matrix, and R, R^{-1} are upper triangular matrices. Further,

$$S^{-1}ZS = L^{-1}(V^{-1}ZV)L$$

is the product of lower triangular matrices, and hence lower triangular. This proves Theorem 1.1. We may rewrite part of this method of proof in the form of a proposition as follows:

Proposition 2.2 *Let A and Z be bounded operators acting on the Hilbert space $l_2(\mathbf{Z}^+)$, then the following are equivalent:*

1. *A and Z admit simultaneous reduction to complementary triangular forms.*
2. *There exist invertible operators T and V, such that $T^{-1}AT$ is upper triangular, $V^{-1}ZV$ is lower triangular, and $V^{-1}T$ admits lower-upper factorization.*

Although Proposition 2.2 is formulated for operators on an infinite-dimensional Hilbert space, a straightforward analogue of Theorem 1.1 does not follow. The reason is, that Proposition 2.1 fails to extend to the infinite-dimensional setting, as the following theorem shows.

Theorem 2.3 *There exists a unitary operator U on $l_2(\mathbf{Z}^+)$, such that for all permutation operators U_ρ and U_σ the operator $U_\rho^* U U_\sigma$ does not admit lower-upper factorization.*

Before we give the proof of theorem 2.3, we state the following auxilary lemma. The lemma is probably well-known and based on techniques used in [4].

Lemma 2.4 *If an invertible operator S admits lower-upper factorization, then for each $m \in \mathbf{Z}^+$, the operator $E_m S E_m + I - E_m$ is invertible. Moreover,*

$$\sup_{m \in \mathbf{Z}^+} \|(E_m S E_m + I - E_m)^{-1}\| < \infty.$$

Proof Write $S = LR$, with L, L^{-1} lower triangular and R, R^{-1} upper triangular. Fix $m \in \mathbf{Z}^+$ and write

$$E_m S E_m + (I - E_m) = E_m L R E_m + (I - E_m) = E_m L E_m E_m R E_m + (I - E_m) =$$

$$(E_m L E_m + I - E_m)(E_m R E_m + I - E_m).$$

Note that

$$(E_m L E_m + I - E_m)(E_m L^{-1} E_m + I - E_m) = E_m L L^{-1} E_m + I - E_m = I$$

and

$$(E_m R E_m + I - E_m)(E_m R^{-1} E_m + I - E_m) = E_m R R^{-1} E_m + I - E_m = I.$$

Consequently,

$$(E_m S E_m + I - E_m)^{-1} = (E_m R^{-1} E_m + I - E_m)(E_m L^{-1} E_m + I - E_m).$$

Since

$$\|E_m L^{-1} E_m + (I - E_m)\| \leq \|L^{-1}\| + 1,$$

and

$$\|E_m R^{-1} E_m + (I - E_m)\| \leq \|R^{-1}\| + 1,$$

we get

$$\|(E_m S E_m + I - E_m)^{-1}\| \leq \|E_m R^{-1} E_m + (I - E_m)\| \|E_m L^{-1} E_m + (I - E_m)\| \leq$$

$$(\|R^{-1}\| + 1)(\|L^{-1}\| + 1) < \infty.$$

The lemma is proved. □

We now prove Theorem 2.3.

Proof Define the unitary $n \times n$ matrix $U(n)$ as

$$U(n) = \left(\frac{\sqrt{n}}{n} e^{\frac{2\pi i(u-1)(v-1)}{n}} \right)_{u,v=1}^{n},$$

and define the unitary operator U acting on $l_2(\mathbf{Z}^+)$ with respect to the basis $\{e_k \mid k \in \mathbf{Z}^+\}$ as the infinite diagonal block-matrix

$$U = U(1) \oplus U(2) \oplus U(3) \oplus \cdots.$$

Let ρ, σ be permutations on $\mathbf{Z}^+$ and U_ρ, U_σ be the corresponding permutation operators. Define the unitary operator

$$V = U_\rho^* U U_\sigma.$$

It follows that $(k, l \in \mathbf{Z}^+)$

$$V_{kl} = U_{\rho(k),\sigma(l)}, \quad U_{kl} = V_{\rho^{-1}(k),\sigma^{-1}(l)}.$$

We need to investigate whether V admits lower-upper factorization. If $E_m V E_m + I - E_m$ is not invertible for certain $m \in \mathbf{Z}^+$, then, by Lemma 2.4, V does not admit lower-upper factorization. For that reason, we may assume that all the operators $E_m V E_m + I - E_m$ are invertible ($m \in \mathbf{Z}^+$). We claim that this fact leads to the following restriction on the permutations ρ and σ:

$$W = U_\rho U_\sigma^* = W(1) \oplus W(2) \oplus W(3) \oplus \cdots,$$

where $W(n)$ denotes an $n \times n$ permutation matrix ($n \in \mathbf{Z}^+$). This is more than we need for the proof of the theorem.

Fix $n \in \mathbf{Z}^+$ and write $\kappa_n = \frac{n(n-1)}{2}$. Let $\mathcal{I}_n = \{\kappa_n + 1, \ldots, \kappa_n + n\}$. For the theorem, it suffices to prove that $\min \rho^{-1}(\mathcal{I}_n) = \min \sigma^{-1}(\mathcal{I}_n)$, but we will even prove that

$$\rho^{-1}(\mathcal{I}_n) = \sigma^{-1}(\mathcal{I}_n). \tag{1}$$

Write

$$\rho^{-1}(\mathcal{I}_n) = \{k_1, \ldots, k_n\}, \quad \sigma^{-1}(\mathcal{I}_n) = \{l_1, \ldots, l_n\},$$

with $k_{u-1} < k_u$ and $l_{u-1} < l_u$ for $u = 2, \ldots, n$. Put $k = \kappa_n + u$ and $l = \kappa_n + v$ with $1 \le u, v \le n$. Then

$$V_{\rho^{-1}(k),\sigma^{-1}(l)} = U_{kl} = \frac{\sqrt{n}}{n} e^{\frac{2\pi i(u-1)(v-1)}{n}}.$$

Further, $V_{\rho^{-1}(k),s} = 0$ if $s \notin \sigma^{-1}(\mathcal{I}_n)$, and $V_{r,\sigma^{-1}(l)} = 0$ if $r \notin \rho^{-1}(\mathcal{I}_n)$.

To prove (1), we will show that $k_u = l_u$ for $u = 1, \ldots, n$.

As a first step, we prove that $k_1 = l_1$. Indeed, write $k = k_1$ and $l = l_1$. Then $|V_{kl}| = \frac{\sqrt{n}}{n}$. If $s < l$, we obtain that $s \notin \rho^{-1}(\mathcal{I}_n)$ and hence that $V_{ks} = 0$. In the same fashion it follows that $V_{rl} = 0$ if $r < k$. Assume that $k < l$. Then the k-th row of $E_kVE_k + I - E_k$ consists of zero elements only, since $V_{ks} = 0$ for $1 \leq s \leq k < l$. The operator $E_kVE_k + I - E_k$ is not invertible for that reason, a contradiction. Therefore, $k \geq l$. Next, assume that $k > l$. The l-th column of $E_lVE_l + I - E_l$ consists of zero elements only, since $V_{rl} = 0$ for $1 \leq r \leq l < k$. The operator $E_lVE_l + I - E_l$ is not invertible and again a contradiction has been obtained. We conclude that $k = l$.

Second, fix $2 \leq p \leq n$ and assume that $k_u = l_u$ for $u = 1, \ldots, p-1$. We will prove that $k_p = l_p$. First assume that $k = k_p < l_p$. The p rows of the operator $E_kVE_k + I - E_k$ labelled $k_1, \ldots, k_p$ have nonzero entries exactly at the $p-1$ column positions $l_1, \ldots, l_{p-1}$. These rows are linear dependent for that reason and the operator $E_kVE_k + I - E_k$ is not invertible. On the other hand, if $k_p > l_p = l$, the operator $E_lVE_l + I - E_l$ is not invertible. Indeed, the p columns of this operator labelled $l_1, \ldots, l_p$ have nonzero entries exactly at the $p-1$ row positions $k_1, \ldots, k_{p-1}$. Again the columns are linear dependent. It follows that $k_p = l_p$ and by induction on p, we get (1).

To prove the theorem, fix $n \in \mathbf{Z}^+$ and let

$$k = \min \rho^{-1}(\mathcal{I}_n) = \min \sigma^{-1}(\mathcal{I}_n).$$

Note that $|V_{kk}| = \frac{\sqrt{n}}{n}$ and that $V_{kj} = V_{jk} = 0$ for $j = 1, \ldots, k-1$. Identify E_kVE_k with the $k \times k$ matrix of the operator with respect to the first k standard basis vectors, to obtain

$$E_kVE_k = \begin{pmatrix} E_{k-1}VE_{k-1} & O \\ O & V_{kk} \end{pmatrix}.$$

This $k \times k$ matrix is invertible with inverse

$$(E_kVE_k)^{-1} = \begin{pmatrix} (E_{k-1}VE_{k-1})^{-1} & O \\ O & V_{kk}^{-1} \end{pmatrix}.$$

It follows that

$$\|(E_kVE_k)^{-1}\| \geq |V_{kk}^{-1}| = \sqrt{n}.$$

But $n \in \mathbf{Z}^+$ was chosen arbitrary, so

$$\sup_{k \in \mathbf{Z}^+} \|(E_kVE_k + I - E_k)^{-1}\| \geq \sup_{k \in \mathbf{Z}^+} \|(E_kVE_k)^{-1}\| = \infty.$$

The theorem now follows from Lemma 2.4. □

3 Hilbert Space Analogues

In this section, we study analogues of Theorem 1.1 for a pair of bounded operators acting on the separable Hilbert space $l_2(\mathbf{Z}^+)$. First, we construct pairs of diagonable operators, that do not admit simultaneous reduction to complementary triangular forms. Then, we consider pairs of bounded operators, that contain an operator of finite rank, in Propositions 3.5 and 3.6 below.

Lemma 3.1 *Let D be a bounded diagonal operator, acting on $l_2(\mathbf{Z}^+)$, given by*

$$De_k = \delta_k e_k, \quad k \in \mathbf{Z}^+,$$

with mutually distinct diagonal elements: $\delta_k \neq \delta_l$ if $k \neq l$. If M is a finite dimensional invariant subspace (of dimension m) of D, then there exist distinct positive integers $\tau(1), \ldots, \tau(m)$, such that

$$M = \text{span}\{e_{\tau(1)}, \ldots, e_{\tau(m)}\}.$$

Proof First of all, note that $\text{Ker}(D - \delta) \neq (0)$, if and only if $\delta = \delta_k$ for some $k \in \mathbf{Z}^+$. In that case, $\text{Ker}(D - \delta) = \text{span}\{e_k\}$. Let M be an m dimensional invariant subspace for D, and let D_M denote the restriction of D to M. If $\delta \in \sigma(D_M)$, then $\text{Ker}(D_M - \delta) \neq (0)$. Since $\text{Ker}(D_M - \delta) = \text{Ker}(D - \delta) \cap M$, it follows that $\text{Ker}(D - \delta) \neq (0)$, so $\delta = \delta_k$ for some $k \in \mathbf{Z}^+$. This proves that $\sigma(D_M) \subseteq \{\delta_k \mid k \in \mathbf{Z}^+\}$.

To prove that D_M has m distinct eigenvalues, assume that this is not the case, i.e., there exists $k \in \mathbf{Z}^+$, such that $\dim\text{Ker}(D_M - \delta_k)^2 \geq 2$. It then follows that $\dim\text{Ker}(D - \delta_k)^2 \geq 2$. On the other hand, $\text{Ker}(D - \delta_k)^2 = \text{span}\{e_k\}$, a contradiction. Let $\tau(1), \ldots, \tau(m)$ denote the distinct positive integers, such that

$$\sigma(D_M) = \{\delta_{\tau(1)}, \ldots, \delta_{\tau(m)}\}.$$

Note that $\text{Ker}(D_M - \delta_{\tau(j)}) = \text{Ker}(D - \delta_{\tau(j)}) = \text{span}\{e_{\tau(j)}\}$. We may conclude that

$$M = \text{span}\{e_{\tau(1)}, \ldots, e_{\tau(m)}\},$$

and the lemma is proved. □

Lemma 3.2 *Let D be a bounded diagonal operator with mutually distinct diagonal elements. Let S be an invertible operator, such that $C = S^{-1}DS$ is upper triangular, with $\text{diag}(C) = (\gamma_1, \gamma_2, \gamma_3, \ldots)^T$. Then there exists a permutation τ on $\mathbf{Z}^+$, such that $\gamma_k = \delta_{\tau(k)}$, and such that the invertible operator $U_\tau^* S$ is upper triangular.*

Proof Fix $m \in \mathbf{Z}^+$; the subspace $M_m = \text{span}\{e_1, \cdots, e_m\}$ satisfies $DSM_m \subseteq SM_m$. By Lemma 3.1, there exist distinct integers $\tau_m(1), \ldots, \tau_m(m)$, such that

$$SM_m = \text{span}\{e_{\tau_m(1)}, \ldots, e_{\tau_m(m)}\},$$

and $\sigma(D\mid_{SM_m}) = \{\delta_{\tau_m(1)}, \ldots, \delta_{\tau_m(m)}\}$. Since the proper inclusion $SM_m \subset SM_{m+1}$ holds for $m \in \mathbf{Z}^+$, there exists an injective mapping τ on $\mathbf{Z}^+$, such that

$$SM_m = \text{span}\{e_{\tau(1)}, \ldots, e_{\tau(m)}\},$$

for $m \in \mathbf{Z}^+$. Since $\bigcup_{m=1}^{\infty} SM_m \subseteq l_2(\mathbf{Z}^+)$ is dense, the mapping τ is also surjective, and hence a permutation of the
positive integers. Since

$$\sigma(D\mid_{SM_m}) = \sigma(C\mid_{M_m}) = \{\gamma_1, \ldots, \gamma_m\},$$

it follows that $\delta_{\tau(m)} = \gamma_m$ for $m \in \mathbf{Z}^+$. Further, $U_\tau^* SM_m = M_m$ for all $m \in \mathbf{Z}^+$, so $U_\tau^* S$ is upper triangular. □

Theorem 3.3 *Let D_1 and D_2 be bounded diagonal operators, each with mutually distinct diagonal elements, and let U be the unitary operator as defined in Theorem 2.3. Let $A = D_1$ and $Z = U^* D_2 U$. Then A and Z do not admit simultaneous reduction to complementary triangular forms.*

Proof Assume there exists an invertible operator S acting on $l_2(\mathbf{Z}^+)$, such that $S^{-1}AS$ is upper triangular and $S^{-1}ZS$ is lower triangular. Apply Lemma 3.2 on $A = D_1$ to obtain a permutation operator U_σ such that $R_1 = U_\sigma^* S$ is upper triangular. Further use that $(S^{-1}ZS)^* = (S^{-1}U^*D_2US)^* = S^*U^*D_2^*US^{-*}$ is upper triangular, and apply Lemma 3.2 on D_2^* to obtain a permutation operator U_ρ such that $R_2 = U_\rho^* US^{-*}$ is upper triangular. Note that $R = R_1^{-1} = S^{-1}U_\sigma$ is upper triangular and $L = R_2^{-*} = U_\rho^* US$ is lower triangular. Then

$$LR = U_\rho^* USS^{-1}U_\sigma = U_\rho^* UU_\sigma,$$

i.e., the unitary operator U admits lower-upper factorization after permutations of rows and columns. A contradiction has been obtained and the theorem is proved. □

Corollary 3.4 *There exist two self-adjoint trace-class operators A and Z on $l_2(\mathbf{Z}^+)$, such that A and Z do not admit simulataneous reduction to complementary triangular forms.*

Proof Let A and Z be as in Theorem 3.3, but assume in addition that the diagonals of both D_1 and D_2 are l_1-sequences consisting of positive numbers. □

We now turn to pairs of bounded operators A and Z, that contain an operator of finite rank. An obvious necessary condition on A and Z to admit simultaneous reduction to complementary triangular forms, is that A is upper triangularizable, and that Z is lower triangularizable. In Proposition 3.6 and the remark thereafter, we will have to impose one of these conditions on the operator, that is neither of finite rank nor diagonable.

Proposition 3.5 *Let Z be a bounded diagonable operator, and A an operator of finite rank, both acting on $l_2(\mathbf{Z}^+)$. Then A and Z admit simultaneous reduction to complementary triangular forms.*

Proof By Lemma 1.2, we may assume without loss of generality that Z is diagonal; say $Ze_k = \zeta_k e_k$ for $k \in \mathbf{Z}^+$. Write rank $A = m$, and let $M = \operatorname{Ran} A$. There exist vectors $b_1, \ldots, b_m \in M$, and integers $n_1 < \cdots < n_m$, such that

$$b_k = \sum_{j=n_k}^{\infty} \beta_{kj} e_j, \quad \beta_{kn_k} \neq 0,$$

for $k = 1, \ldots, m$. The vectors are linearly independent and hence form a basis in M; $M = \operatorname{span}\{b_1, \ldots, b_m\}$. We claim that if $N = \operatorname{span}\{e_{n_1}, \ldots, e_{n_m}\}^\perp$, then

$$M \oplus N = l_2(\mathbf{Z}^+). \tag{2}$$

It suffices to prove that $M \cap N = (0)$. Let $x = \sum_{k=1}^{m} \xi_k b_k \in M \cap N$, and assume that $x \neq 0$. Then there exists an integer $1 \leq p \leq m$, such that $\xi_1 = \ldots = \xi_{p-1} = 0$, and $\xi_p \neq 0$. Then $e^*_{n_p} x = \beta_{n_p} \xi_p \neq 0$. On the other hand, $x \in N$, so $e^*_{n_p} x = 0$. A contradiction has been obtained. Therefore, the decomposition (2) indeed holds. With respect to this decomposition,

$$A = \begin{pmatrix} A_1 & A_{12} \\ O & O \end{pmatrix}, \quad Z = \begin{pmatrix} Z_1 & O \\ Z_{21} & Z_2 \end{pmatrix}.$$

We claim that Z_1 is diagonable: Let the m dimensional subspace $\hat{M}$ be given by $\text{span}\{e_{n_1}, \ldots, e_{n_m}\}$. Define the invertible operator T on $l_2(\mathbf{Z}^+)$ as

$$Te_{n_j} = b_j, \quad j = 1, \ldots, m, \quad Te_i = e_i, \quad i \notin \{n_1, \ldots, n_m\}.$$

Then

$$T = \begin{pmatrix} T_1 & O \\ O & I_N \end{pmatrix} : \hat{M} \oplus N \longrightarrow M \oplus N,$$

with T_1 an invertible $m \times m$ matrix. Since Z is diagonal with respect to $\{e_k \mid k \in \mathbf{Z}^+\}$, we get $ZT = T\hat{Z}$, where $\hat{Z}$ is of the form

$$\hat{Z} = \begin{pmatrix} D_1 & O \\ O & D_2 \end{pmatrix} : \hat{M} \oplus N \longrightarrow \hat{M} \oplus N,$$

with D_1 a diagonal $m \times m$ matrix (and D_2 a diagonal operator). Rewrite $ZT = T\hat{Z}$ as

$$\begin{pmatrix} Z_1 T_1 & O \\ Z_{21} T_1 & Z_2 \end{pmatrix} = \begin{pmatrix} T_1 D_1 & O \\ O & D_2 \end{pmatrix}.$$

In particular, $Z_1 = T_1 D_1 T_1^{-1}$, so Z_1 is diagonable.

It follows, by Theorem 1.1, that A_1 and Z_1 admit simultaneous reduction to complementary triangular forms: There exists a basis $s_1, \ldots, s_m$ for M, such that $(k = 1, \ldots, m)$

$$A_1 s_k \in \text{span}\{s_1, \ldots, s_k\}, \quad Z_1 s_k \in \text{span}\{s_k, \ldots, s_m\}.$$

Define the invertible operator S on $l_2(\mathbf{Z}^+)$ as

$$Se_j = \begin{cases} s_j, & j = 1, \ldots, m \\ e_{\pi(j)}, & j = m+1, \ldots, n_m \\ e_j, & j > n_m \end{cases},$$

where $\pi : \{m+1, \ldots, n_m\} \longrightarrow \{1, \ldots, n_m\} \backslash \{n_1, \ldots, n_m\}$ is any bijection. Write $L = \text{span}\{e_1, \ldots, e_m\}$. Then

$$S = \begin{pmatrix} S_1 & O \\ O & S_2 \end{pmatrix} : L \oplus L^{\perp} \longrightarrow M \oplus N.$$

Further, with respect to $L \oplus L^{\perp}$, we get

$$S^{-1}AS = \begin{pmatrix} S_1^{-1}A_1S_1 & S_1^{-1}A_{12}S_2 \\ O & O \end{pmatrix}, \quad S^{-1}ZS = \begin{pmatrix} S_1^{-1}Z_1S_1 & O \\ S_2^{-1}Z_{21}S_1 & S_2^{-1}Z_2S_2 \end{pmatrix},$$

where $S_1^{-1}A_1S_1$ is an upper triangular $n_m \times n_m$ matrix, $S_1^{-1}Z_1S_1$ is a lower triangular $n_m \times n_m$ matrix, and $S_2^{-1}Z_2S_2$ is a diagonal infinite matrix. Therefore, $(A, Z) \in \mathcal{C}$. The proposition is proved. □

The case when A is a bounded diagonable operator and Z is of finite rank is dealt with as follows: Apply Proposition 3.5 to the operators Z^* and A^* and use Lemma 1.2.

Proposition 3.6 *Let A be a diagonable operator of finite rank, and let Z be a bounded operator, which is lower triangularizable. Then A and Z admit simultaneous reduction to complementary triangular forms.*

Proof By Lemma 1.2, we may assume without loss of generality, that Z is lower triangular. By assumption, there exists an invertible operator V, such that $V^{-1}AV$ is diagonal. Let the diagonal of $V^{-1}AV$ be given by

$$\text{diag}(V^{-1}AV) = (\ \alpha_1, \alpha_2, \alpha_3, \dots\ )^T.$$

Since A is of finite rank, it holds that $m = \max\{k \mid k \in \mathbf{Z}^+, \alpha_k \neq 0\} < \infty$. Write $V(\text{Ran } E_m) = M$, and $V(\text{Ker } E_m) = N$. Define $d(t) = \dim(M \cap \text{Ker } E_t)$ for $t \in \mathbf{Z}^+$, then $d : \mathbf{Z}^+ \longrightarrow \{0, \dots, m\}$ is decreasing, and $\lim_{t\to\infty} d(t) = 0$. Indeed, if $\lim_{t\to\infty} d(t) > 0$, there exists $0 \neq x \in M$, such that $x \in \text{Ker } E_t$ for all $t \in \mathbf{Z}^+$, a contradiction. Let $\tau \in \mathbf{Z}^+$, such that $d(\tau) = 0$. Since $M \cap \text{Ker } E_\tau = (0)$, and $M + \text{Ker } A = l_2(\mathbf{Z}^+)$, there exists a finite dimensional subspace $R \subseteq \text{Ker } A$, with $M \oplus R \oplus \text{Ker } E_\tau = l_2(\mathbf{Z}^+)$. The vectors $y_k = Ve_k$ for $k = 1, \dots, m$ form a basis in M. In addition, let $y_{m+1}, \dots, y_\tau$ be a basis in R. Note that $Ay_k \in \text{span}\{y_k\}$ for $k = 1, \dots, \tau$. Therefore, the restriction of A to $M_\tau = M \oplus R$ is diagonable. With respect to the decomposition $M_\tau \oplus \text{Ker } E_\tau = l_2(\mathbf{Z}^+)$, we get

$$A = \begin{pmatrix} A_1 & A_{12} \\ O & O \end{pmatrix}, \quad Z = \begin{pmatrix} Z_1 & O \\ Z_{21} & Z_2 \end{pmatrix},$$

where A_1 is a diagonable $\tau \times \tau$ matrix and Z_2 is a

lower triangular operator. By Theorem 1.1, there exists a basis $s_1, \dots s_\tau$ for M_τ, such that $A_1 s_k \in \text{span}\{s_1, \dots, s_k\}$ and $Z_1 s_k \in \text{span}\{s_k, \dots, s_\tau\}$ for $k = 1, \dots, \tau$. The invertible operator S, defined by

$$Se_j = \begin{cases} s_j, & 1 \leq j \leq \tau \\ e_j, & j > \tau \end{cases},$$

puts A and Z into complementary triangular forms. The proposition is proved. □

If A is a bounded operator, which is upper triangularizable, and Z is diagonable and of finite rank, then apply Proposition 3.6, to obtain that $(Z^*, A^*) \in \mathcal{C}$. Next apply Lemma 1.2 to obtain $(A, Z) \in \mathcal{C}$.

References

[1] H. Bart, Transfer Functions and Operator Theory, *Lin. Alg. Appl.* 84: 33-61 (1986).

[2] H. Bart, I. Gohberg, M.A. Kaashoek, *Minimal Factorization of Matrix and Operator Functions*, Operator Theory: Adv. Appl. 1, Birkhäuser Verlag, Basel (1979).

[3] H. Bart, H. Hoogland, Complementary Triangular Forms of Pairs of Matrices, Realizations with Prescribed Main Matrices, and Complete Factorization of Rational Matrix Functions, *Lin. Alg. Appl.* 103: 193-228 (1988).

[4] I. Gohberg, M.G. Krein, *Theory and Applications of Volterra Operators in Hilbert Space*, Transl. Math. Monographs 24, A.M.S, Providence, RI (1969).

[5] P. Lancaster, M. Tismenetsky, *The Theory of Matrices, Second Edition with Applications*, Academic Press, Orlando, Fl. (1985).

Tinbergen Institute
Oostmaaslaan 950-952
NL-3063 DM Rotterdam
The Netherlands

MSC: 47B99

LIST OF PARTICIPANTS

V.M. ADAMJAN
Francuzskij Bul'var 12, Korp. 1, kw. 4, Odessa 270044, Ukraine;
Dept. of Theoretical Physics, Odessa University, Petra Velikogo 2,
70100 Odessa, Ukraine
email: vma@dtp.odessa.ua; odessa.va@ussr.eu.net

D. ALPAY
Dept. of Mathematics, P.O. Box 653, Ben Gurion University,
Beer Sheva 84105 Israel
email: dany@black.bgu.ac.il

T. ANDO
Research Inst. of Electronic Science, Hokkaido University,
Sapporo 060, Japan
email: ando@splab.elsip.hokudai.ac.jp

D.Z. AROV
Prospekt Dimitrova 15/8, 270104 Odessa, Ukraine
email: odessaeco@glas.apc.org

T.J. AZIZOV
Khol'zunova, 40-b, kv. 37, 294068 Voronezh - 68, Russia
email: azizov@imath.vucnit.voronezh.su

J.A. BALL
Dept. of Mathematics, Virginia Polytechnic Institute and State
University Blacksburg, Virginia 24061-0123 U.S.A.
email: ball@math.vt.edu

L. BARATCHART
Projet MIAUO, INRIA-Sophia-antipolis, B.P. 93,
F-06902 Sophia-antipolis, France
email: France.Limouzis@sophia.inria.fr

H. BART
Econometrics Institute, Erasmus University Rotterdam, Postbus 1738,
NL-3000 DR Rotterdam, The Netherlands
email: bart@wis.few.eur.nl

A. BEN-ARTZI
School of Mathematical Sciences, Tel-Aviv University,
69978 Tel Aviv, Israel
email: benartzi@math.tau.ac.il

C. BINDER
Inst. f. Analysis, Techn. Math. u. Versicherungsmath., TU Vienna,
Wiedner Hauptstr. 8-10/1141, A-1040 Vienna, Austria
email: chbinder@email.tuwien.ac.at

P. BINDING
Dept. of Mathematics, University of Calgary,
Calgary, AB, T2N1N4, Canada
email: binding@acs.ucalgary.ca

M. BLÜMLINGER
Inst. f. Analysis, Techn. Math. u. Versicherungsmath., TU Vienna,
Wiedner Hauptstr. 8-10/1141, A-1040 Vienna, Austria
email: mbluemli@email.tuwien.ac.at

B. BODENSTORFER
Inst. f. Analysis, Techn. Math. u. Versicherungsmath., TU Vienna,
Wiedner Hauptstr. 8-10/1144, A-1040 Vienna, Austria

J. BOGNÁR
Mathematical Institute, Hungarian Academy of Sciences,
P.O.B. 127, H-1364 Budapest, Hungary

L. DE BRANGES
Le Hameau de lÝvette, Batiment D, Chemin des Graviers,
F-91190 Gif-sur-Yvette, France

C. DAVIS
Dept. of Mathematics, University of Toronto,
Toronto, Ontario, M5S 1A1, Canada
email: davis@math.toronto.edu

G. DERFEL
Dept. of Mathematics, Ben-Gurion University of the Negev,
P.O.B. 653, Beer Sheva 84105, Israel
email: derfel@bengus.bitnet

M. DRITSCHEL
Purdue University, Math. Sciences Bldg.
West Lafayette, IN 47906 U.S.A.
email: mad@cs.wm.edu

H. DYM
Dept. of Theoretical Mathematics, The Weizmann Institute of Science,
P.O. Box 26, Rehovot 76100, Israel
email: mtdym@weizmann.weizmann.ac.il

M. FAIERMAN
Dept. of Mathematics, University of Witwatersrand, Private Bag 3,
Johannesburg, WITS 2050, South Africa

I.A. FELDMAN
Dept. of Mathematics, Bar-Ilan University, Ramat-gan 52900, Israel
email: feldmni@bimacs.cs.biu.ac.il

A. FLEIGE
FB Mathematik, Univ. Essen, Universitätsstr. 2,
D-45117 Essen, Germany
email: MATD01@VM.HRZ.UNI-ESSEN.DE

C. FOIAS
Dept. of Mathematics, Indiana University, Swain Hall,
East Bloomington, Indiana 47408, U.S.A.
email: foias@ucs.indiana.edu

B. FRITZSCHE
FB Mathematik/Informatik, Universität Leipzig,
Augustuspl. 10, D-04109 Leipzig, Germany

D. GASPAR
Dept. of Mathematics, University of Timisoara,
Bul.V.Parvan nr. 4, RO-1900 Timisoara, Romania

M. GEBEL
FB Mathematik, Universität Halle-Wittenberg, Gimritzer Damm,
Postfach, D-06099 Halle (Saale), Germany

A. GHEONDEA
Institute of Mathematics of the Romanian Academy,
P.O.Box 1-764, Romania

I. GOHBERG
School of Mathematical Sciences, Tel Aviv University,
Ramat Aviv 69978, Israel
email: gohberg@math.tau.ac.il

J. GOTTLIEB
Inst. für Bodenmechanik und Felsmechanik,
Universität (TH) Karlsruhe, Richard Willstädter-Allee,
D-76128 Karlsruhe 1, Postfach 6980, Germany
email: GN27@IBM3090.RZ.uni-karlsruhe.dbp.de

E. GRINSHPUN
Dept. of Mathematics and Comp. Sci., Ben Gurion University of the
Negev, P.O.B. 653, Beer Sheva 84105, Israel
email: edward@indigo.BGU.AC.IL

S. HASSI
Dept. of Statistics, University of Helsinki, Aleksanterinkatu 7,
SF-00100 Helsinki 7, Finland
email: hassi@cc.Helsinki.FIw

G. HEINIG
Bruno-Granz-Str. 46, D-09122 Chemnitz, Germany
email: ghein@hadrian.hrz.tu-chemnitz.de

J.W. HELTON
Dept. of Mathematics, University of California at San Diego,
La Jolla, California 92037, U.S.A.
email: helton@osiris.ucsd.edu

D. HINTON
Dept. of Mathematics, University of Tennessee,
Knoxville, Tennessee, 37 996-1300, U.S.A.

J. HU
Dept. of Mathematics, Hong Kong Univ. of Science and Techn.,
Clear Water Bay, Kowloon, Hong Kong
email: majhu@ustsu3.ust.hk

C.R. JOHNSON
Dept. of Mathematics, The College of William and Mary,
Williamsburg, Virginia 23185, U.S.A.
email: crjohn@wmvm1.bitnet; crjohnson@cs.wm.edu (after Dec 31, 1993)

P. JONAS
Neltestr. 12, D-12489 Berlin, Germany

M.A. KAASHOEK
Faculteit Wiskunde en Informatica, Vrije Universiteit,
De Boelelaan 1081 a, NL-1081 HV Amsterdam, The Netherlands
email: kaash@cs.vu.nl

W. KABALLO
FB Mathematik, Univ. Dortmund, Postfach 500 500,
D-44309 Dortmund 50, Germany

V.E. KATSNELSON
Dept. of Theoretical Mathematics, The Weitzmann Inst, Rehovot 76100,
Israel; Fachbereich Mathematik, Universität Leipzig,
D-04103 Leipzig, Germany
email: katze@wisdom.weizmann.ac.il

B. KIRSTEIN
FB Mathematik, Universität Leipzig,
Augustuspl. 10, D-04109 Leipzig, Germany

J. KOS
Faculteit Wiskunde en Informatica, De Boelelaan 1081 a,
Vrije Universiteit. NL-1081 HV Amsterdam, The Netherlands
email: jkos@cs.vu.nl

A. KOZHEVNIKOV
Dept. of Math. and Comp.Sc., University of Haifa, Haifa 31905, Israel
email: RSMA201@HAIFAUVM.HAIFA.AC.IL

N. Krupnik
Dept. of Mathematics, Bar-Ilan University, Ramat-Gan 52900, Israel
email: krupnik@bimacs.cs.biu.ac.il

P. Lancaster
Dept. of Mathematics, University of Calgary,
Calgary, AB, T2N1N4, Canada
email: lancaste@acs.ucalgary.ca

H. Langer
Inst. f. Analysis, Techn. Math. u. Versicherungsmath., TU Vienna,
Wiedner Hauptstr. 8-10/1141, A-1040 Vienna, Austria
email: hlanger@email.tuwien.ac.at

H. Lehninger
Inst. f. Analysis, Techn. Math. u. Versicherungsmath., TU Vienna,
Wiedner Hauptstr. 8-10/1141, A-1040 Vienna, Austria

H.-J. Linden
FB Mathematik, Fernuniv. Hagen, Postfach 940,
D-58084 Hagen 1, Germany

M. Lundquist
Dept. of Mathematics, Brigham Young Univeristy,
Provo, Utah 84057, U.S.A.
email: mike@math.byu.edu

S.A.M. Marcantognini
Universidad Simon Bolivar, Depto. de matematicas puras y apl.,
Apartado Postal 89000, Caracas 1096-A, Venezuela
email: stefania@usb.ve

A.S. Markus
Dept. of Mathematics, Ben Gurion University, P.O. Box 653,
Beer Sheva 84105, Israel
email: markus@black.bgu.ac.il

V.I. Matsaev
Dept. of Mathematics, Tel-Aviv University, Tel-Aviv,
Ramat-Aviv 69978, Israel
email: matsaev@math.tau.ac.il

R. Mennicken
FB Mathematik, Univ. Regensburg, Universitätstr. 31,
D-93053 Regensburg, Germany
email: mennicken@vax1.rz.uni-regensburg.dbp.de

A. Mingarelli
Dept. of Math. and Statistics, Carleton University,
710 Dunton Tower, Ottawa, K1S 5B6, Canada

A. MLADENKA
Inst. f. Analysis, Techn. Math. u. Versicherungsmath., TU Vienna,
Wiedner Hauptstr. 8-10/1141, A-1040 Vienna, Austria

M. MÖLLER
Dept. of Mathematics, University of the Witwatersrand,
Private Bag 3, Johannesburg, WITS 2050, South Africa
email: 036MHM@witsvma.wits.ac.za

B. NAJMAN
Dept. of Mathematics, University of Zagreb, Bijenička 30,
41000 Zagreb, Croatia
email: najman@math.hr

T. NAKAZI
Dept. of Mathematics, Hokkaido University, Sapporo 060, Japan

A.A. NUDELMAN
Ul. Frunze 141/187, 270005, Odessa-5, Ukraine
email: odessaeco@glas.apc.org

A. OCTAVIO
IVIC M-S81, P.O.Box 020010, Miami, FL 33102-0010, U.S.A.
email: aoctavio@cauchy.jvnc.net; aoctavio@ivic.ivic.ve

V.V. PELLER
Branch of the V.A.Steklov Mathematical Institute, Fontanka 27,
191011 St.Petersburg, Russia; Dept. of Mathematics,
University of Hawaii, Honolulu, HI 96822, U.S.A.
email: peller@math.hawaii.edu

POMET
Projet MIAUO, INRIA-Sophia-antipolis, B.P. 93,
F-06902 Sophia-antipolis, France

M. RAKOWSKI
Dept. Mathematics, The Ohio State University, 231 West, 18th Avenue,
Columbus, OH 43210, U.S.A.
email: rakowski@math.mps.ohio-state.edu

A.C.M. RAN
Faculteit Wiskunde en Informatica, Vrije Universiteit,
De Boelelaan 1081, NL-1081 HV Amsterdam, The Netherlands
email: ran@cs.vu.nl

S. ROCH
Ulmenstr. 29, D-09112 Chemnitz (TU Chemnitz)

L. RODMAN
Dept. of Mathematics, The College of William an Mary,
Williamsburg, Virginia 23185-8795, U.S.A.
email: lxrodm@wmvm1.bitnet

C.S. SADOSKY
Dept. of Mathematics, Howard University,
Washington DC 20059-0001, U.S.A.
email: cora@msri.org; cs@scsla.howard.edu

L.A. SAKHNOVICH
pr. Dobrovolskogo 154, ap. 199, Odessa, 270111, Ukraine

N. SALINAS
Dept. of Mathematics, University of Kansas, ,405 Snow Hall,
Lawrence, KS 66045-2142, U.S.A.
email: norberto@ukanvax.bitnet; norberto@kuhnb.cc.ukans.edu

K. SEDDIGHI
Dept. of Mathematics, Shiraz University, Iran;
ICTP, P.O.Box 586, I-34100 Trieste, Italy
email: seddighi@ictp.trieste.it

A.A. SHKALIKOV
Dept. of Mathematics, Moscow State University,
Moscow 119899, Russia; FB Mathematik, Univ. Regensburg,
Universitätstr. 31, D-93053 Regensburg, Germany
email: shkalikov@vax1.rz.uni-regensburg.dbp.de

E. SIMEONOV
Inst. f. Analysis, Techn. Math. u. Versicherungsmath., TU Vienna,
Wiedner Hauptstr. 8-10/1141, A-1040 Vienna, Austria

H.S.V. DE SNOO
Afdeling Wiskunde en Informatica, Rijksuniversiteit Groningen,
Postbus 800, NL-9700 AV Groningen, The Netherlands
email: desnoo@math.rug.nl

M. SONNTAG
Inst. f. Analysis, Techn. Math. u. Versicherungsmath., TU Vienna,
Wiedner Hauptstr. 8-10/1144, A-1040 Vienna, Austria

S. STEPIN
Dept. of Mathematics, Moscow State University,
Moscow 119899, Russia
email: shkal@compnet.npimsu.msk.su

I. SUCIU
Inst. of Math., Romanian Academy, Calea Grivitei 21, P.O. Box 1-764,
RO-70700 Bucarest, Romania
email: isuciu@roimar.bitnet

F. SZAFRANIEC
Mathematical Inst., University of Cracow, ul. Reymonta 4,
PL-30059 Cracow, Polen; Universidao Carlos III de Madrid,
Avda. Mediterraneo 20, E-28913 Leganes (Madrid), Spain
email: fhszafra@ing.uc3m.es

D. TEMME
Faculteit Wiskunde en Informatica, De Boelelaan 1081 a,
Vrije Universiteit, NL-1081 HV Amsterdam, The Netherlands
email: dirkt@cs.vu.nl

P. THIJSSE
Econometrics Institute, Erasmus University Rotterdam,
NL-3000 DR Rotterdam, The Netherlands

S. TREIL
Dept. of Mathematics, Michigan State University,
East Lansing, MI 48824-1027, U.S.A.
email: treil@math.msu.edu

CH. TRETTER
FB Mathematik, Univ. Regensburg, Universitätstr. 31,
D-93053 Regensburg, Germany
email: mennicken@vax1.rz.uniregensburg.dbp.de

I. VALUSESCU
Inst. of Mathematics of the Romanian Academy, P.O. Box 1-764,
RO-70700 Bucarest, Romania
email: ilieval@roimar.bitnet

K. VESELIČ
FB Mathematik, Fernuniv. Hagen, Postfach 940,
D-58084 Hagen 1, Germany
email: MA704@DHAFEU11.bitnet

V. VINNIKOV
Dept. of Theoretical Mathematics, Weizmann Institute of Science,
Rehovot 76100, Israel
email: vinnikov@wisdom.weizmann.ac.il

F. VOGL
Inst. f. Analysis, Techn. Math. u. Versicherungsmath., TU Vienna,
Wiedner Hauptstr. 8-10/1141, A-1040 Vienna, Austria

R. VONHOFF
Fachbereich Mathematik, Universität Dortmund, Postfach 500 500,
D-44309 Dortmund, Germany
email:vonhoff@hausdorff.mathematik.uni-dortmund.de

H. WINKLER
Inst. f. Analysis, Techn. Math. u. Versicherungsmath., TU Vienna, Wiedner Hauptstr. 8-10/1141, A-1040 Vienna, Austria

H. WOERDEMAN
Dept. of Mathematics, The College of William and Mary, Williamsburg, Virginia 23187, U.S.A.
email: hugo@cs.wm.edu

H. WORACEK
Inst. f. Analysis, Techn. Math. u. Versicherungsmath., TU Vienna, Wiedner Hauptstr. 8-10/1141, A-1040 Vienna, Austria

J. ZEMANEK
Inst. of Mathematics, Polish Academy of Sciences, PO Box 137, P-00950 Warschau, Polen
email: zemanek@impan.impan.gov.pl

P. ZIZLER
Dept. of Mathematics, University of Calgary, 2500 University Drive N.W., Calgary, AB, T2N1N4, Canada
email: zizler@acs.ucalgary.ca

N. ZORBOSKA
Dept. of Math. amd Astronomy, University of Manitoba, Winnipeg, Manitoba R3T 2N2, Canada
email: zorbosk@ccu.umanitoba.ca

R. ZUIDWIJK
Tinbergen Institute, Oostmaaslaan 950-952, NL-3063 DM Rotterdam, The Netherlands
email: zuidwijk@tir.few.eur.nl

LIST OF LECTURES

V.M. ADAMJAN *(Odessa, Ukraine)*
On eigenvector expansions associated with operator-valued R-functions.

D. ALPAY *(Beer Sheva, Israel)*
Reproducing kernel spaces on Riemann surfaces.

T. ANDO *(Sapporo, Japan)*
Matrix inequalities and norm inequalities.

D.Z. AROV *(Odessa, Ukraine)*
Computation the resolvent matrices for the generalized bitangential Schur- and Caratheodory- Nevanlinna-Pick problems in the strictly completely indeterminant case.

T.J. AZIZOV *(Voronezh, Russia)*
On some development of the S. Krein pencil theory.

J.A. BALL *(Blacksburg, Virginia, U.S.A.)*
Zero-pole interpolation problems for meromorphic matrix functions on Riemann surfaces.

L. BARATCHART *(Sophia-antipolis, France)*
Traces of Hardy functions of the circle and bounded extremal problems.

H. BART *(Rotterdam, The Netherlands)*
Logarithmic residues and zero sums of idempotents in Banach algebras.

A. BEN-ARTZI *(Tel Aviv, Israel)*
Orthogonal polynomials and interpolation in Hilbert modules.

P. BINDING *(Calgary, Canada)*
Eigenparameter dependent boundary conditions for Sturm-Liouville problems.

J. BOGNÁR *(Budapest, Hungary)*
Spectral radius and fundamental norms in infinite dimensions.

L. DE BRANGES *(Gif Sur Yvette, France)*
Pontryagin spaces of analytic functions.

G. DERFEL *(Beer Sheva, Israel)*
Functional and functional-differential equations with rescaling.

M. DRITSCHEL *(West Lafayette, Indiana, U.S.A.)*
The Agler boundary and linear extreme points of the numerical radius unit ball, II.

H. DYM *(Rehovot, Israel)*
On the zeros of a class of matrix functions and some related extension problems.

M. FAIERMAN *(Johannesburg, South Africa)*
On the spectral theory of an elliptic boundary value problem involving an indefinite weight.

I.A. FELDMAN *(Ramat-gan, Israel)*
Explicit factorization of some 2×2 matrix functions.

A. FLEIGE *(Essen, Germany)*
Turning point conditions of Beals for indefinite Sturm-Liouville problems.

C. FOIAS *(Bloomington, Indiana, U.S.A.)*
Extensions of intertwining contractions and causality in the commutant lifting theorem.

D. GASPAR *(Timisoara, Rumania)*
Invariant subspaces in the bitorus.

M. GEBEL *(Halle-Wittemberg, Germany)*
Elementary proof of the spectral theorem for bounded definitizable operators.

A. GHEONDEA *(Bucharest, Rumania)*
On the signatures of the selfadjoint pencil $\lambda G - F$.

I. GOHBERG *(Tel Aviv, Israel)*
On a new class of infinite matrices.

J. GOTTLIEB *(Karlsruhe, Germany)*
Measures in differential equations, eigenvalue problems and Krein spaces.

E. GRINSHPUN *(Beer Sheva, Israel)*
Asymptotic of spectra under weak nonselfadjoint perturbations.

S. HASSI *(Helsinki, Finland)*
On projections in a space with indefinite metric.

G. HEINIG *(Leipzig, Germany)*
Generalized inversion of Toeplitz-like operators.

J.W. HELTON *(La Jolla, California, U.S.A.)*
Optimization over spaces of analytic functions and resulting problems in operator theory.

J. HU *(Hongkong)*
A general study of the radiation loss: asymptotics beyond all orders.

C.R. JOHNSON *(Williamsburg, Virginia, U.S.A.)*
Matrix completion problems: recent advances.

P. JONAS *(Berlin, Germany)*
On selfadjoint extensions of nonnegative linear relations in Krein spaces.

M.A. KAASHOEK *(Amsterdam, The Netherlands)*
Two-sided Nudelman interpolation for input-output operators of time-varying systems.

W. KABALLO *(Dortmund, Germany)*
Multiplicative decompositions of holomorphic Fredholm functions and Ψ^*-algebras.

V.E. KATSNELSON *(Tel Aviv, Israel)*
Weight approximation of pseudocontinuable functions by rational functions with prescribed poles.

B. KIRSTEIN *(Leipzig, Germany)*
On Arov-singular interpolation Problems.

J. KOS *(Amsterdam, The Netherlands)*
The Nehari-Takagi problem for input-output operators of time-varying continuous time systems.

A. KOZHEVNIKOV *(Haifa, Israel)*
General elliptic systems and some questions regarding their spectral theory.

N. KRUPNIK *(Ramat-gan, Israel)*
Extension theorems for Fredholm symbols.

P. LANCASTER *(Calgary, Canada)*
Bounded operators and operator polynomials with real spectrum.

H.-J. LINDEN *(Hagen, Germany)*
On the trace of certain rational and meromorphic operator functions of a general type.

S.A.M. MARCANTOGNINI *(Caracas, Venezuela)*
Commuting unitary Hilbert space extensions of a pair of Krein space isometries.

A.S. MARKUS *(Beer Sheva, Israel)*
Factorization of a selfadjoint nonanalytic operator-valued function.

V.I. MATSAEV *(Tel-Aviv, Israel)*
A property of the Hilbert transform.

R. MENNICKEN *(Regensburg, Germany)*
Non-standard boundary eigenvalue problems.

A. MINGARELLI *(Ottawa, Ontario, Canada)*
A class of maps in an algebra with indefinite metric and applications.

M. MÖLLER *(Johannesburg, South Africa)*
Orthogonal systems of root functions for symmetric operators.

B. NAJMAN *(Zagreb, Croatia)*
Regularity of the elliptic eigenvalue problems with indefinite weights.

T. NAKAZI *(Sapporo, Japan)*
$(A2)$ condition and Carleson inequalities in Bergman spaces.

A.A. NUDELMAN*(Odessa, Ukraine)*
Canonical solutions of the matrix Hamburger and Stieltjes moment problems.

A. OCTAVIO*(Miami, Florida, U.S.A.)*
Some problems in the theory of dual algebras generated by commuting contractions.

V.V. PELLER*(St. Petersburg, Russia)*
Superoptimal H^∞ approximations of matrix functions.

M. RAKOWSKI*(Columbus, Ohio, U.S.A.)*
Rational matrix functions with co-isometric values on the line or circle.

A.C.M. RAN*(Amsterdam, The Netherlands)*
Pseudo-canonical factorization in certain classes of matrix functions.

L. RODMAN*(Williamsburg, Virginia, U.S.A.)*
Inertia of operator polynomials.

C.S. SADOSKY*(Washington D.C., U.S.A.)*
Nehari and Nevanlinna-Pick theorems in the polydisc.

L.A. SAKHNOVICH*(Odessa, Ukraine)*
Spectral problems and nonlinear equations.

N. SALINAS*(Lawrence, Kansas, U.S.A.)*
The canonical complex structure of flag manifolds in a C^*-algebra.

K. SEDDIGHI*(Shivaz, Iran)*
Two parameter asymptotic spectra in the uniformly ellliptic case.

A.A. SHKALIKOV*(Moscow, Russia)*
Factorization theorems for selfadjoint operator pencils and their applications.

H.S.V. DE SNOO*(Groningen, The Netherlands)*
On the sum of two Q-functions.

S. STEPIN*(Moscow, Russia)*
Rayleigh equation: scattering problem and eigenfunction expansions.

I. SUCIU*(Bucarest, Rumania)*
A class of examples of operator valued positive definite functions.

F. SZAFRANIEC*(Cracow, Polen)*
Yet another face of the creation operator.

D. TEMME*(Amsterdam, The Netherlands)*
Dissipative matrices and invariant maximal semidefinite subspaces.

P. THIJSSE*(Rotterdam, The Netherlands)*
Symmetry relations for partial mulitiplicities of products of matrix functions.

S. TREIL*(East Lansing, Michigan, U.S.A.)*
On superoptimal approximations by analytic and meromorphic functions.

CH. TRETTER *(Regensburg, Germany)*
The Kamke problem — properties of the eigenfunctions.

K. VESELIČ *(Hagen, Germany)*
On the condition of J-orthonormal eigenvectors.

V. VINNIKOV *(Rehovot, Israel)*
Functional models for commuting nonselfadjoint operators and realization of functions on a compact Riemann surface.

R. VONHOFF *(Dortmund, Germany)*
On a class of regular quasi-leftdefinite boundary eigenvalue problems.

H. WINKLER *(Vienna, Austria)*
On the inverse spectral problem for canonical systems.

H. WOERDEMAN *(Williamsburg, Virginia, U.S.A.)*
The Agler boundary and linear extreme points of the numerical radius unit ball.

H. WORACEK *(Vienna, Austria)*
Degenerated Nevanlinna-Pick problem and solutions with poles.

J. ZEMANEK *(Warsaw, Poland)*
Iterates of operators and the spectrum.

N. ZORBOSKA *(Winnipeg, Manitoba, Canada)*
Composition operators and the determining function.

R. ZUIDWIJK *(Rotterdam, The Netherlands)*
Simultaneous triangular forms for pairs of operators.

Titles previously published in the series

OPERATOR THEORY: ADVANCES AND APPLICATIONS

BIRKHÄUSER VERLAG

Edited by
I. Gohberg,
School of Mathematical Sciences, Tel-Aviv University, Ramat Aviv, Israel

This series is devoted to the publication of current research in operator theory, with particular emphasis on applications to classical analysis and the theory of integral equations, as well as to numerical analysis, mathematical physics and mathematical methods in electrical engineering.

30. **Yu.I. Lyubich:** Introduction to the Theory of Banach Representations of Groups, 1988, (3-7643-2207-1)
31. **E.M. Polishchuk:** Continual Means and Boundary Value Problems in Function Spaces, 1988, (3-7643-2217-9)
32. **I. Gohberg** (Ed.): Topics in Operator Theory. Constantin Apostol Memorial Issue, 1988, (3-7643-2232-2)
33. **I. Gohberg** (Ed.): Topics in Interplation Theory of Rational Matrix-Valued Functions, 1988, (3-7643-2233-0)
34. **I. Gohberg** (Ed.): Orthogonal Matrix-Valued Polynomials and Applications, 1988, (3-7643-2242-X)
35. **I. Gohberg, J.W. Helton, L. Rodman** (Eds.): Contributions to Operator Theory and its Applications, 1988, (3-7643-2221-7)
36. **G.R. Belitskii, Yu.I. Lyubich:** Matrix Norms and their Applications, 1988, (3-7643-2220-9)
37. **K. Schmüdgen:** Unbounded Operator Algebras and Representation Theory, 1990, (3-7643-2321-3)
38. **L. Rodman:** An Introduction to Operator Polynomials, 1989, (3-7643-2324-8)
39. **M. Martin, M. Putinar:** Lectures on Hyponormal Operators, 1989, (3-7643-2329-9)
40. **H. Dym, S. Goldberg, P. Lancaster, M.A. Kaashoek** (Eds.): The Gohberg Anniversary Collection, Volume I, 1989, (3-7643-2307-8)
41. **H. Dym, S. Goldberg, P. Lancaster, M.A. Kaashoek** (Eds.): The Gohberg Anniversary Collection, Volume II, 1989, (3-7643-2308-6)
42. **N.K. Nikolskii** (Ed.): Toeplitz Operators and Spectral Function Theory, 1989, (3-7643-2344-2)
43. **H. Helson, B. Sz.-Nagy, F.-H. Vasilescu, Gr. Arsene** (Eds.): Linear Operators in Function Spaces, 1990, (3-7643-2343-4)

44. **C. Foias, A. Frazho:** The Commutant Lifting Approach to Interpolation Problems, 1990, (3-7643-2461-9)
45. **J.A. Ball, I. Gohberg, L. Rodman:** Interpolation of Rational Matrix Functions, 1990, (3-7643-2476-7)
46. **P. Exner, H. Neidhardt** (Eds.): Order, Disorder and Chaos in Quantum Systems, 1990, (3-7643-2492-9)
47. **I. Gohberg** (Ed.): Extension and Interpolation of Linear Operators and Matrix Functions, 1990, (3-7643-2530-5)
48. **L. de Branges, I. Gohberg, J. Rovnyak** (Eds.): Topics in Operator Theory. Ernst D. Hellinger Memorial Volume, 1990, (3-7643-2532-1)
49. **I. Gohberg, S. Goldberg, M.A. Kaashoek:** Classes of Linear Operators, Volume I, 1990, (3-7643-2531-3)
50. **H. Bart, I. Gohberg, M.A. Kaashoek** (Eds.): Topics in Matrix and Operator Theory, 1991, (3-7643-2570-4)
51. **W. Greenberg, J. Polewczak** (Eds.): Modern Mathematical Methods in Transport Theory, 1991, (3-7643-2571-2)
52. **S. Prössdorf, B. Silbermann:** Numerical Analysis for Integral and Related Operator Equations, 1991, (3-7643-2620-4)
53. **I. Gohberg, N. Krupnik:** One-Dimensional Linear Singular Integral Equations, Volume I, Introduction, 1992, (3-7643-2584-4)
54. **I. Gohberg, N. Krupnik:** One-Dimensional Linear Singular Integral Equations, Volume II, General Theory and Applications, 1992, (3-7643-2796-0)
55. **R.R. Akhmerov, M.I. Kamenskii, A.S. Potapov, A.E. Rodkina, B.N. Sadovskii:** Measures of Noncompactness and Condensing Operators, 1992, (3-7643-2716-2)
56. **I. Gohberg** (Ed.): Time-Variant Systems and Interpolation, 1992, (3-7643-2738-3)
57. **M. Demuth, B. Gramsch, B.W. Schulze** (Eds.): Operator Calculus and Spectral Theory, 1992, (3-7643-2792-8)
58. **I. Gohberg** (Ed.): Continuous and Discrete Fourier Transforms, Extension Problems and Wiener-Hopf Equations, 1992, (3-7643-2809-6)
59. **T. Ando, I. Gohberg** (Eds.): Operator Theory and Complex Analysis, 1992, (3-7643-2824-X)
60. **P.A. Kuchment:** Floquet Theory for Partial Differential Equations, 1993, (3-7643-2901-7)
61. **A. Gheondea, D. Timotin, F.-H. Vasilescu** (Eds.): Operator Extensions, Interpolation of Functions and Related Topics, 1993, (3-7643-2902-5)
62. **T. Furuta, I. Gohberg, T. Nakazi** (Eds.): Contributions to Operator Theory and its Applications. The Tsuyoshi Ando Anniversary Volume, 1993, (3-7643-2928-9)

63. **I. Gohberg, S. Goldberg, M.A. Kaashoek:** Classes of Linear Operators, Volume 2, 1993, (3-7643-2944-0)
64. **I. Gohberg** (Ed.): New Aspects in Interpolation and Completion Theories, 1993, (3-7643-2948-3)
65. **M.M. Djrbashian:** Harmonic Analysis and Boundary Value Problems in the Complex Domain, 1993, (3-7643-2855-X)
66. **V. Khatskevich, D. Shoiykhet:** Differentiable Operators and Nonlinear Equations, 1993, (3-7643-2929-7)
67. **N.V. Govorov †:** Riemann's Boundary Problem with Infinite Index, 1994, (3-7643-2999-8)
68. **A. Halanay, V. Ionescu:** Time-Varying Discrete Linear Systems Input-Output Operators. Riccati Equations. Disturbance Attenuation, 1994, (3-7643-5012-1)
69. **A. Ashyralyev, P.E. Sobolevskii:** Well-Posedness of Parabolic Difference Equations, 1994, (3-7643-5024-5)
70. **M. Demuth, P. Exner, G. Neidhardt, V. Zagrebnov** (Eds): Mathematical Results in Quantum Mechanics. International Conference in Blossin (Germany), May 17-21, 1993, 1994, (3-7643-5025-3)
71. **E.L. Basor, I. Gohberg** (Eds): Toeplitz Operators and Related Topics. The Harold Widom Anniversary Volume. Workshop on Toeplitz and Wiener-Hopf Operators, Santa Cruz, California, September 20–22, 1992, 1994 (3-7643-5068-7)
72. **I. Gohberg, L.A. Sakhnovich** (Eds): Matrix and Operator Valued Functions. The Vladimir Petrovich Potapov Memorial Volume, (3-7643-5091-1)
73. **A. Feintuch, I. Gohberg** (Eds): Nonselfadjoint Operators and Related Topics. Workshop on Operator Theory and Its Applications, Beersheva, February 24–28, 1994, (3-7643-5097-0)
74. **R. Hagen, S. Roch, B. Silbermann:** Spectral Theory of Approximation Methods for Convolution Equations, 1994, (3-7643-5112-8)
75. **C.B. Huijsmans, M.A. Kaashoek, B. de Pagter**: Operator Theory in Function Spaces and Banach Lattices. The A.C. Zaanen Anniversary Volume, 1994 (ISBN 3-7643-5146-2)
76. **A.M. Krasnosellskii**: Asymptotics of Nonlinearities and Operator Equations, 1995, (ISBN 3-7643-5175-6)
77. **J. Lindenstrauss, V.D. Milman** (Eds): Geometric Aspects of Functional Analysis Israel Seminar GAFA 1992-94, 1995, (ISBN 3-7643-5207-8)
78. **M. Demuth, B.-W. Schulze** (Eds): Partial Differential Operators and Mathematical Physics: International Conference in Holzhau (Germany), July 3-9, 1994, 1995, (ISBN 3-7643-5208-6)
79. **I. Gohberg, M.A. Kaashoek, F. van Schagen**: Partially Specified Matrices and Operators: Classification, Completion, Applications, 1995, (ISBN 3-7643-5259-0)